THE NEW SOLAR SYSTEM

ICE WORLDS, MOONS, AND PLANETS REDEFINED

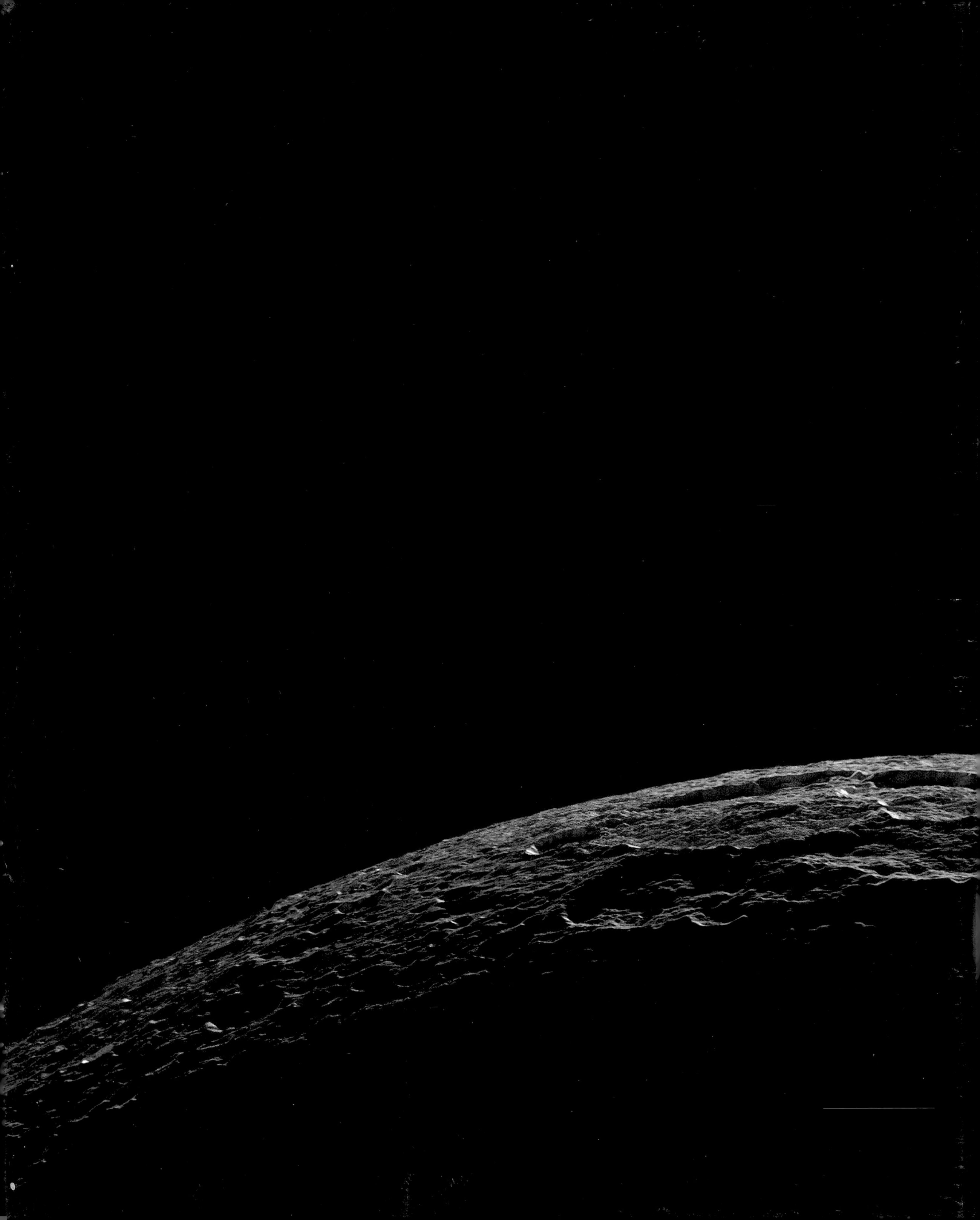

THE NEW SOLAR SYSTEM

ICE WORLDS, MOONS, AND PLANETS REDEFINED

PATRICIA DANIELS

FOREWORD BY ROBERT BURNHAM

NATIONAL GEOGRAPHIC

WASHINGTON, D.C.

CONTENTS

FOREWORD

Our solar system is more than 4.5 billion years old, but paradoxically it undergoes a renewal once or twice a century. Today's solar system—with eight planets, down from nine—is just the latest chapter in a story that began when the earliest humans looked up into the sky and tried to comprehend what they saw.

The book in your hands is a beautiful travel guide to the worlds around us. It presents scientists' latest findings about the planets, asteroids, comets, and other bits of matter that belong to the sun's domain. It is illustrated with gorgeous images from space probes that have indeed boldly gone where humans—at least for now—cannot.

The invention of the telescope in 1608 broke the barrier of unaided eyesight, which had limited our ideas of the solar system since prehistoric times. Slowly but with gathering speed, the first telescopic discoveries painted a new picture of an immensely larger and richer solar system whose movements were as predictable as the newfangled pendulum clocks that tick tocked away in observatories. Throughout the 1600s, telescopic studies of the solar system helped scientists such as Isaac Newton lay the mathematical foundations for all modern physical science.

These two methods of discovery—observation and mathematics—enlarged our solar system in a seesaw fashion. During the century following Newton's death in 1727, the telescope added a new planet to the family—Uranus, in 1781—and starting in 1801, began to fill the gap between the orbits of Mars and Jupiter with small, rocky worlds, the minor planets. Then the mathematicians responded by discovering yet another planet—Neptune, in 1846—that emerged from equations probing minutely observed deviations from the predicted motion of Uranus. The clockwork ran ever more regularly.

Yet as the 20th century opened, the solar system was a backwater area, shunned by every ambitious astronomer. Science's frontier lay in using ever larger telescopes to probe the astrophysics of stars, nebulae, and galaxies far away from the sun's realm. Even the discovery of Pluto in 1930 reflected the lowered status: Amateur astronomer Clyde Tombaugh got the Pluto-hunting job because he was cheaper to hire than a "real" astronomer.

The solar system came back into view quite abruptly in 1957, when the Soviet Union launched a 184-pound satellite into low Earth orbit. This touched off a space race between the Soviet Union and the United States that built upon rocketry developments dating from World War II and earlier. First the moon, and then Venus and Mars, snapped into focus as space probes left Earth and began to reconnoiter the planets from close-up.

By vaulting past the limits of earthbound telescopes, the rocketeers gave humanity an almost wholly new solar system. Debates that festered for hundreds of years over the origins of lunar craters ended with the results of a spaceflight or two.

Martian canals? Gone for good, replaced by a cratered landscape that rivaled Earth's for geological complexity. Venusian swamps and jungles likewise vanished, replaced by tortured lava flows under a crushingly heavy atmosphere hot enough to make lead run like water.

Probes ventured farther outward, surveying gas giants that revealed bewildering details as spacecraft photographed storms bigger than Earth while sailing through mini-solar systems of satellites and rings.

In the past 15 years, as planetary scientists began to inventory the far-flung, icy bodies in the Kuiper belt at the planetary system's edge, the emerging picture no longer resembled what had prevailed since the dawn of the space age. Suddenly a simple question, "What is a planet?" had no simple answer.

Around this time, fittingly, stellar astronomers interested in how stars form and grow began to collaborate with planetary researchers exploring the solar system's origin. Thus a long-standing gap in knowledge finally closed.

And tomorrow's solar system? That's already under way. NASA's New Horizons spacecraft will start imaging Pluto in February 2015, five months before closest approach on July 14. If the past is any guide, a new era—in effect, another new solar system—will begin to take shape right then.

So stay tuned, the journey into infinity has just begun.

A false-color image provided by the Magellan spacecraft highlights the ridged terrain of Fortuna Tessera near Venus's highest mountain, Maxwell Montes.

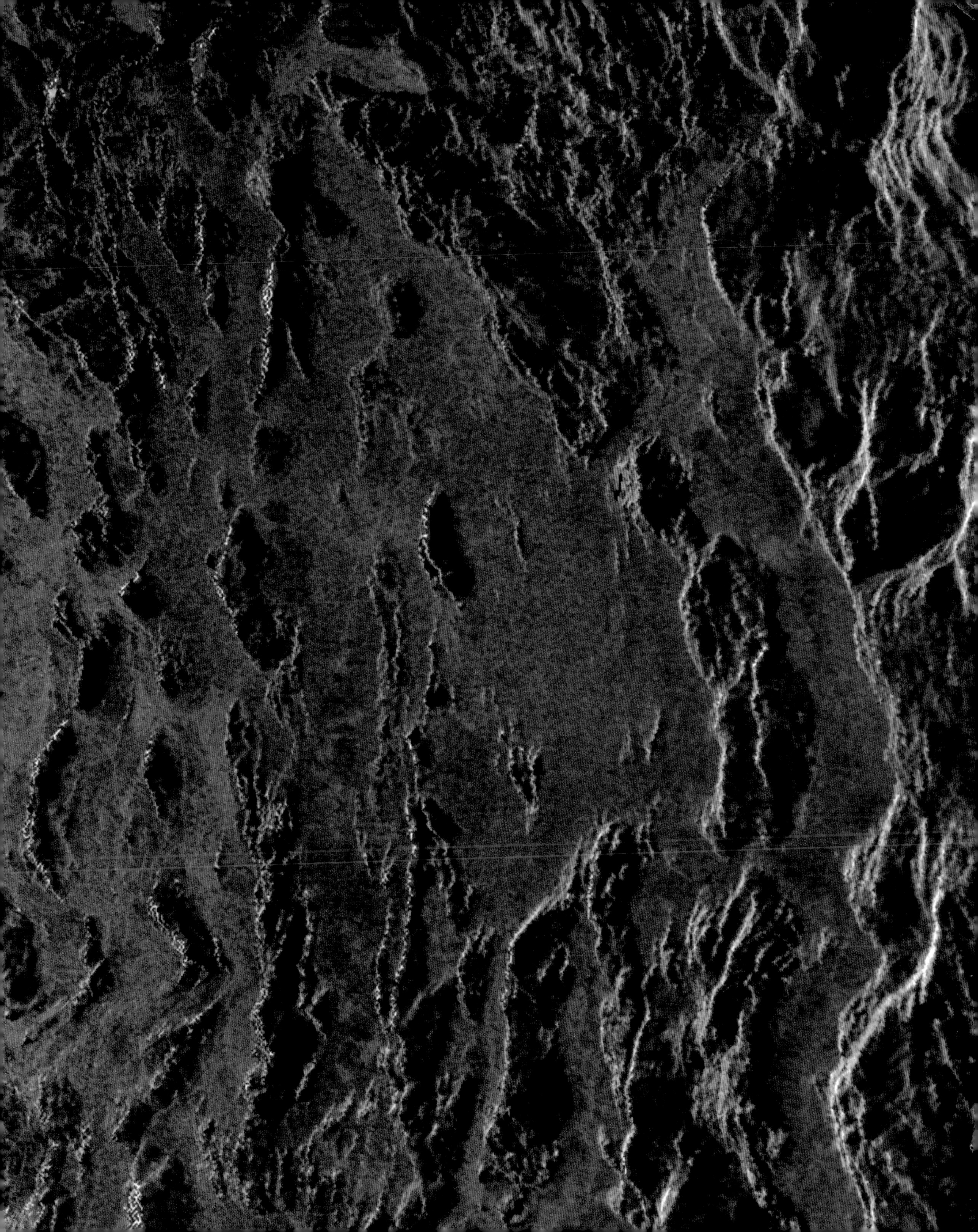

Viewed from their unlit side, Saturn's rings cut across hazy Titan's profile. The Cassini spacecraft, which captured this image, also launched a probe into Titan's atmosphere.

CHAPTER 1

WANDERING STARS

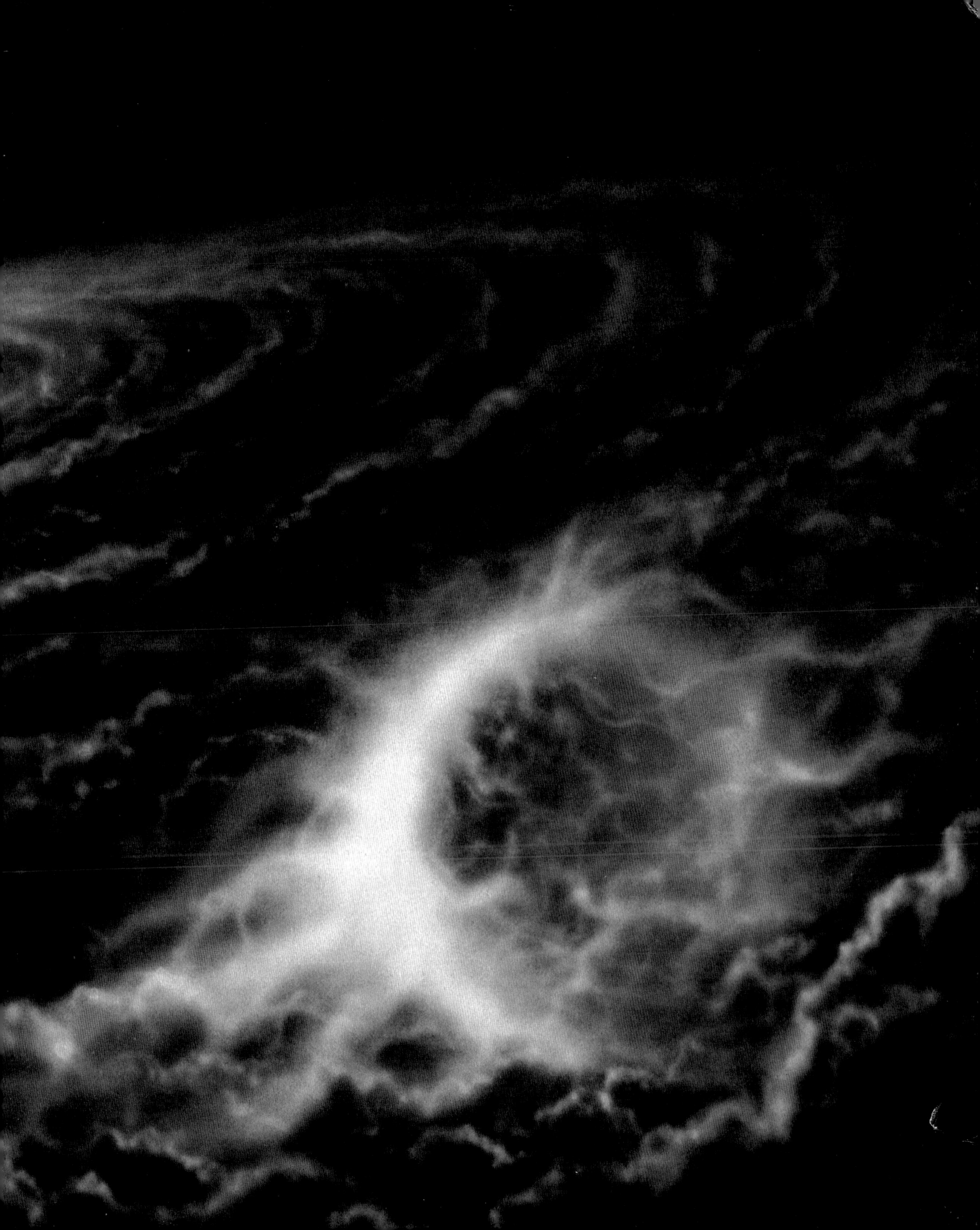

WANDERING STARS

When members of the International Astronomical Union voted in 2006 to demote Pluto from planet to "dwarf planet," the public learned what astronomers already knew: This is not your parents' solar system. In the past 20 years, planetary scientists have discovered new worlds, moons, vast regions of the solar neighborhood, and planets around other stars. The sedate picture of planetary formation has given way to a vision of the early solar system as a pinballing realm of immense impacts and gravitational push and shove. Spacecraft have found water on Mars, rings around every gas giant, and erupting volcanoes on Jupiter's moon Io. For the first time, they have charted the sun's immense magnetic energies and have sipped the icy atmospheres of comets. Earthbound observers have begun, with some alarm, to track traveling asteroids that could wipe out life on the planet. And the search for life is on in a big way, from the surface of Mars to the atmospheres of exoplanets.

Breakthroughs in astronomy don't come from nowhere, of course. Astronomers spend their time studying extraordinarily distant objects, and their discoveries depend upon the tools at their disposal. The history of astronomy can be divided into three overlapping eras: the age of naked-eye observation, the age of telescopes, and the age of spacecraft exploration. We have to assume that naked-eye observation dates back to the first *Homo sapiens* who stared at the sky from the African savanna. Early peoples were blessed with dark skies—no light pollution in Mesopotamia!—and spent their lives outdoors, hunting, farming, and tending animals. They knew the sky intimately. The 3,000 stars visible to the eye, the gauzy swath of the Milky

ca 3000 B.C.
Babylonians observe and make a record of the night sky.

ca 2000 B.C.
The first known recording of a lunar eclipse is made in Mesopotamia.

ca 350 B.C.
Aristotle uses geometry to prove Earth is a sphere.

A.D. 150
Ptolemy writes that Earth is the center of the universe.

650
Maya astronomers produce a calendar that is accurate.

1066
Comet later known as Halley appears, believed to be an ill omen for England.

Glowing trails mark the paths of stars and planets behind one of Mauna Kea's observatories in a long-exposure image. Large, high-altitude telescopes like this one in Hawaii are the product of astronomy's second great era of observation.

Way, and the seven "wandering stars"—sun, moon, and five visible planets—were as familiar to most ancients as our own neighborhoods are to us today. Farmers planned their planting around the seasons and the cycles of the moon. Believing that the planets and stars could guide and predict human events, civilizations employed corps of observers and astrologers to chart the skies. Babylonians, the Maya, and others devoted centuries to detailed, day-by-day charts of planetary movements, lunar cycles, and eclipses.

Into this wealth of observation, the Greeks introduced geometry. No longer content simply to watch the skies, scholars began to inquire about the mechanisms behind the movement of Earth, planets, and stars. Into the days of Nicolaus Copernicus, astronomers devised increasingly intricate arrays of concentric shells and epicycles to account for the way the sun, moon, and planets seemed to swing around the stationary Earth. The Polish cleric's sun-centered model, introduced in the 1500s, simplified the calculations considerably. But it took a century or more before most scientists accepted the heliocentric system as a description of reality. And that change in thinking corresponded, not

1259
Mongol ruler Hulegu builds observatory in Maragheh.

1449
Paolo Toscanelli tracks the path of a comet.

ca 1465
Regiomontanus prints astronomy books using the newly invented craft.

1543
Nicolaus Copernicus publishes *De Revolutionibus.*

1546–1601
Tycho Brahe makes accurate measurements of stars and planets.

1576
Tycho correctly calculates the orbit of Mars.

coincidentally, with the invention of the telescope in the early 17th century.

In the hands of Galileo Galilei and other great astronomers, the telescope revealed the craters of the moon, the disks of the planets, and the moons of Jupiter. Saturn's rings appeared. The solar system was evidently a vast and complex place without any visible gears or spheres. Great thinkers, including Johannes Kepler and Isaac Newton, combined telescopic observations with mathematical insights to consider the forces that moved the planets. Gravitation, mass, and acceleration became part of astronomy's language. Meanwhile, telescopes grew rapidly in size and accuracy, bigger optics bringing distant objects into view for the first time. A singularly fine reflector of modest size allowed English observer William Herschel to detect the first new planet discovered in recorded history, Uranus. And a desert observatory, photography, and Clyde Tombaugh's immense patience brought little Pluto to light in 1930.

Even that most familiar of astronomical objects, the sun, began to reveal its strange and surprisingly stormy nature as observers found ways to study its light without going blind. Sunspots, seen since antiquity, turned out to hold clues to an intense magnetic field. Twentieth-century breakthroughs in atomic physics found their real-life example in the sun's prodigious production of energy, possible only through nuclear fusion.

For a time it seemed as if Pluto, unimaginably distant at 39 times Earth's distance from the sun, marked the edge of the solar system. But there still remained the problem of comets. Some, known as short-period comets, appear in our skies periodically, more or less along the same orbital plane as the planets. Others, long-period comets, pass by just once and can appear from any part of the sky. Where did they all come from?

In 1950, Dutch astronomer Jan Oort proposed an answer to the question of long-period comets. He suggested that the solar system was entirely surrounded by a vast, spherical cloud of comets, icy little worlds left over from the formation of the solar system. Extending part of the way to the nearest stars, this far distant debris shell could not be seen from Earth, but gravitational perturbations might occasionally dislodge a comet that would eventually cut through the solar system. One year later, Dutch-American astronomer Gerard Kuiper theorized that a similar reservoir of icy debris supplied short-period comets,

GIORDANO BRUNO

VISIONARY AND HERETIC

Giordano Bruno (1548–1600), born Filippo Bruno, was an Italian visionary whose radical ideas led to his destruction. Bruno joined the Dominican order at 15 but soon broke with the church to travel Europe teaching about the soul, the universe, and infinity. Though no scientist, he supported the Copernican system and wrote that "innumerable suns exist; innumerable earths revolve around these suns. . . . Living beings inhabit these worlds." These and other heretical words brought him to the attention of the Inquisition; he was burned at the stake in Rome in 1600.

but in this case it consisted of a flat belt beyond Neptune. At the time, neither the Kuiper belt nor the Oort cloud could actually be seen, but they explained the behavior of comets so effectively that most astronomers accepted them as fact. These distant, invisible reaches of the solar system were about to come within reach with the beginning of the space age.

The beeping message of Sputnik I told the world that new opportunities awaited those bold enough to venture into space and examine remote worlds close-up. Spacecraft have been launched toward the sun, every planet, asteroids, comets, and our moon, the only world actually visited by humans. Ingenious robotic craft such as the Voyager missions to Jupiter and beyond and the Mars exploration rovers have enriched our knowledge of the solar system immensely since the 1960s. Among their major findings are frozen water on Mars, with strong evidence for liquid water in the planet's past. The Huygens probe, visiting Saturn's moon Titan, found a startling landscape both like and unlike our Earth: lakes and rivers, but filled with hyper-chilled liquid methane. Meanwhile, Saturn's moon Enceladus and Jupiter's moon Europa seem to possess liquid water beneath their icy surfaces, making them surprising possibilities for the presence of life.

Modern astronomers spend little, if any, time at the eyepiece of a telescope. The world's big observatories are booked months in advance by teams of scientists who spend most of their careers planning their research and then recording and analyzing the results. And the human eye has been surpassed by photosensitive chips and other devices that can read minute variations in a wide range of the electromagnetic spectrum. The vastly improved telescopes and better techniques for analyzing the tiny, distant movements of specks of light have discovered new worlds in the late 20th and early 21st century. Far beyond Pluto, sizable dwarf planets have come into view. And light-years away, wobbling stars have proved to be gravitationally bound to planets of their own. Their presence brings hopes for finding life elsewhere and tells us much about how our own solar system may have formed.

Information comes to astronomers now in streams of data processed by powerful computers. But the questions astronomers ask are still personal ones: Where did our world come from? How do we fit in? Are we alone in the universe?

K. E. TSIOLKOVSKY
ROCKETRY PIONEER

Konstantin Tsiolkovsky (1857–1935), a Russian scientist, is credited with being the father of rocketry. After losing his hearing in childhood, Tsiolkovsky immersed himself in books, including the novels of Jules Verne, and in mathematics. As a young man, he studied aircraft design and built one of the first wind tunnels. In 1896 he began to write the groundbreaking "Exploration of Cosmic Space by Means of Reaction Devices," which outlined the basics of rocket propulsion. Tsiolkovsky's theories laid the groundwork for modern space travel; the satellite Sputnik I was launched on his 100th birthday.

WANDERING STARS
CELESTIAL SPHERES

Most modern humans, awash in a sea of urban light, cannot find or identify a planet in the sky, much less trace its path from night to night. But the stars and planets were familiar and meaningful companions in the lives of early peoples.

Until the invention of the telescope, sky-watchers employed their eyes, careful recordkeeping, and basic mathematical principles to make sense of the heavens. Logic told them that Earth was an unmoving platform at the center of a fixed sphere of stars. Seven *planetes* (from the Greek for "wandering") traveled the skies in their own spheres. The sun, the moon, and the five visible planets—Mercury, Venus, Mars, Jupiter, and Saturn—changed position from hour to hour or night to night. The stars seemed unknowable, but figuring out the planets dominated astronomy until the 19th century. • Although early astronomy was a practical matter, it had its mystical side as well. Cultures around the world worshipped the powerful sun, the mutable moon, or the brilliant beacon of Venus. Even the great Greek/Egyptian observer Ptolemy, cataloger of planetary motion, was moved by the grandeur of the skies. "Mortal as I am," he wrote, "I know that I am born for a day, but when I follow the serried multitude of the stars in their circular course, my feet no longer touch the earth; I ascend to Zeus himself to feast me on ambrosia, the food of the gods."

ARISTOTLE'S SPHERES
Aristotle's universe consisted of concentric spheres holding the planets. From the inside out, they contained:

- **Earth**
- **Moon**
- **Mercury**
- **Venus**
- **Sun**
- **Mars**
- **Jupiter**
- **Saturn**
- **Fixed stars**
- **The Prime Mover**

SKYWATCH

* Planets can be distinguished from stars by the naked eye because they don't twinkle and they are found only along the ecliptic (path followed by the moon and sun).

AMAZING FACT Ancient Chinese people would bang on pots and pans during a solar eclipse to frighten away the dragon eating the sun.

ca 3000 B.C.
Construction begins on the Stonehenge circle in England.

ca 240 B.C.
Eratosthenes measures the circumference of Earth.

46 B.C.
Julius Caesar introduces the Egyptian solar calendar to Rome.

A.D. 1120
Cairo observatory built in Egypt.

1420
Mongol astronomer Ulugh Beg builds observatory in Central Asia.

1608
Dutch spectacle-maker Hans Lippershey invents the telescope.

Maya calendars were elaborate and precise (opposite). England's Stonehenge may have been built, in part, to track the summer solstice (above).

CELESTIAL SPHERES: THE FIRST SKY-WATCHERS

Besides the sun and the moon, the five planets visible to the naked eye have been known since prehistoric times. Unlike the stars, the planets appear in the sky at different times from year to year, sometimes seeming to move in an odd, weeks-long loop-the-loop across the sky. Early astronomers charted these planetary dances against the background stars, as well as the phases of the moon and the changing path of the sun over the seasons, in hopes that they could understand how the behavior of the wandering stars above affected the lives of the humans below.

It's hard to know the true state of astronomy in many early cultures. Ancient civilizations frequently left little in the way of durable records or artifacts. Babylonian tablets and Maya carvings give us a fairly rich picture of their sky-watching traditions. But the people who built Stonehenge or left behind cairns on the American plains may also have been sophisticated astronomers. In the absence of written records, we are left with statistical likelihoods and informed speculation.

Chichén Itzá's El Caracol ("the snail," so named for its spiraling staircase) was probably built as an observatory by Maya astronomers to track the motions of the sun and Venus.

BABYLON

The foundations of Western astronomy were laid in ancient Mesopotamia, the fertile land between the Tigris and Euphrates Rivers. As long ago as 1800 B.C., Babylonian astronomers systematically recorded the motions of the sun, moon, and other heavenly objects on clay tablets. They continued this patient practice for an astonishing 700 years, a vast period of time that allowed them to analyze and predict even rare events such as eclipses. In this they were aided by their system of mathematics, which recognized place value and was based on 60—a method that continues to live on today in our 60-second minute and the division of a circle into 360 degrees.

Astronomy and astrology were one and the same to the Babylonians, as indeed they were to most cultures until the last 300 years. They used their storehouse of knowledge to read omens in the sky and advise rulers on signals, favorable or not, from the gods. Eventually the astrologers compiled a set of 70 such tablets collectively called the *Enuma Anu Enlil,* from the first line of text, which read "when [the gods] Anu and Enlil. . . ." The tablets contained 7,000 heavenly omens from the past that could be used to guide current action. Astrology as we know it today, using the signs of the zodiac, began in Babylonia sometime later when sky-watchers began not just to look to the past, but to predict individuals' futures based on their birth charts.

Babylonians also employed their observations to work out an accurate calendar based on lunar months. Sunset marked the beginning of each new day, and the first appearance of the crescent moon after sunset began the new month. Because the solar year is more than 12 but fewer than 13 lunar months long, the Babylonians added a 13th "intercalary" month at regular intervals: 7 intercalary months every 19 years, now known as the Metonic cycle.

CHINA

China's astronomical tradition goes back at least as far as Babylon's. Modern scholars have found inscribed oracle bones from before 1600 B.C. that depict a nova—a stellar explosion. Like other fields of study in ancient China, astronomy was handled by government bureaucrats who focused either on the calendar or on passing, ominous phenomena such as meteors, comets, sunspots, and eclipses. These professionals were nothing if not organized, and this was reflected in their maps of the sky. They divided the sky into 28 segments, or "lunar lodges," through which the sun, moon, and planets would pass. Official astronomers would record daily occurrences in a way that suggested the emperor exerted control over events above and below. Official astrologers, on the other hand, noted strange happenings in the sky, because they might foretell danger below; similarly, disastrous actions by a Chinese ruler might find their reflections in the heavens.

Like the Babylonians, the Chinese organized their calendar by lunar months, beginning each month with the new moon and adding intercalary months when needed to keep the years on track. Over the centuries they developed tables that predicted the motions of the planets as well as lunar eclipses. So accurate, complete, and extensive are the Chinese records and tables that even today historians use them to accurately date events such as the supernova of 1054 or the appearance of Halley's comet in 240 B.C.

The Chinese, like other early sky-watchers, seem not to have been interested in analyzing the anatomy or mechanics of the sky. China was the center of a flat Earth at the center of a celestial sphere. Stars and planets were affixed to the inside of this rotating sphere, circling about Earth.

THE AMERICAS

Early Mesoamerican peoples—the Olmecs, Inca, Aztec, Maya, and others—were superb observers who believed that sun, moon, planets, and stars had the divine power to influence the events on Earth, and therefore bore careful watching. Perhaps the most remarkable of these sky-watching cultures was the Maya. Their accurate observations of the heavens are captured in carvings on temples, pillars, and the few bark books that survive the unforgivable vandalism of the conquering Spaniards in the 16th century.

The Maya developed several intricate calendars, the most important of which was the sacred calendar of 260 days. The days had 20 names, such as Muluk ("water") or Ix ("jaguar"), and 13 numbers, each day therefore designated by a combination of name and number. The planet Venus was held to be particularly influential in the course of events. The Maya knew that Venus followed a 584-day cycle, divided into periods when it appeared or disappeared in the morning or evening. The time of morning reappearance was particularly dangerous, and war had to be avoided during that period. Jupiter, too, governed activities, and such grim entertainments as ritual sacrifices were timed to the end of Jupiter's retrograde (backward-looking) motion in the sky.

Less is known about early astronomers in North America, but carvings and stone structures left behind by different pre-Columbian peoples suggest that they tracked the yearly movements of the sun and stars. Dozens of rocky wheels dot the Great Plains, of which perhaps the most interesting is the Big Horn Medicine Wheel in Wyoming. Formed from small, flat rocks, the 26.5-meter-wide (87-ft) wheel has 28 spokes radiating from a central cairn. Five more cairns mark certain of the spokes, with one more at the end of a particularly long spoke. This longest line happens to mark the spot where the sun rises at the summer solstice; the other marked spokes are aligned with the rising points of brilliant stars Fomalhaut, Sirius, Rigel, and Aldebaran. The alignments imply—but do not prove—that the Plains Indians monitored the sky.

THE ZODIAC
PATH OF THE PLANETS

As early as 2000 B.C., Babylonian astronomers saw that the sun, moon, and planets followed a narrow path across the sky now known to us as the ecliptic. Astrologers found it significant that this path crosses certain constellations. By about 1000 B.C., Babylonians had identified the 12 ecliptic constellations that Greeks later knew as the zodiac: the Bull, the Lion, the Scorpion, the Water-carrier, the Twins, the Furrow (later the Virgin), the Archer, the Swallow (later the Fish), the Hired Man (later the Ram), the Crab, the Scales, and the Goat-Fish.

More dramatic, but not conclusive either, is the Chaco Canyon sun dagger in northwestern New Mexico, where the Puebloan people thrived between 850 and 1250, constructing impressive buildings and grand monuments. At the summer solstice, rocks propped on a ledge in the canyon direct the sunlight in a long dagger of light across the center of a spiral petroglyph.

NORTHERN EUROPE

Compared with these other civilizations, early Europeans contributed little to astronomy. Where they excelled, perhaps, was in following the sun through its seasons (which were more exaggerated in those climes than in the tropics) and particularly in predicting its solstices (the highest and lowest points during the year) and equinoxes (the days when day and night are of equal length).

Many tombs, monuments, and stone circles throughout Europe appear to be aligned with the sun. The great stone circle of England's Stonehenge is undoubtedly the most famous of these. Begun around 3100 B.C. by Neolithic people, it was constructed in three stages over perhaps 2,000 years. The standing stones that remain form a circle around an inner horseshoe of uprights. A ditch surrounds the entire monument, and a straight path, "the Avenue," approaches it and contains a stone known as the Heel Stone near the entrance.

Many students of the stones, including well-known astronomers, believe that at minimum Stonehenge was designed to mark the summer solstice: The midsummer sun rises directly over the main axis of the momument and its Heel Stone. Other standing stones in France, Ireland, and Britain may have similar orientations. But claims that their builders constructed Stonehenge and other circles as early observatories, allowing them to predict eclipses and the like, are controversial. It's easy to find significant alignments if you are looking for them. It's hard to show that they were intentional.

Even if the makers of Stonehenge had a sophisticated knowledge of the sky, like most early sky-watchers they were probably concerned mainly with concrete observations and predictions. It remained for the Greeks to take the wealth of observations they inherited and to look for the logical system that governed the sun and planets.

CELESTIAL SPHERES: THE GREEKS

Scientific astronomy began with the ancient Greeks. They, in turn, benefited from the data collected by the Babylonians: Seven hundred years of observations are a scientific motherlode. But the Greeks were the first civilization to move beyond the details of calendars and tables of planetary motion in search of the unchanging laws that lay behind them.

As early as the sixth century B.C., Anaximander of Miletus—a Greek colony now part of Turkey—theorized that the universe was organized in concentric wheels circling a floating, cylindrical Earth. Each opaque wheel was filled with fire; the sun, moon, planets, and stars were essentially windows in the rotating wheels through which the flames could be seen. The five planets and the fixed stars were closest to Earth; outside of them was the moon, and in the outermost wheel, the sun. As simplistic as this model seems now, it was a major advance in natural philosophy, and as the first theoretical portrait of the cosmos it formed the basis for Western astronomy until Copernicus.

PYTHAGORAS

The great Greek philosopher and mathematician Pythagoras refined this model soon afterward by showing that Earth must be a sphere. Most likely he noted that, no matter where an observer stands, Earth's shadow on the moon during eclipses is always a circle. Pythagoras and his followers had a deeply held belief that the solar system was symmetrical and harmonious—which was why they, and many subsequent astronomers, were brought up sharp when they had to account for the peculiar motion of the planets. Unlike the stars, the sun, moon, and planets don't swing across the sky in a steady, circular path night after night. Indeed, the planets seem to speed up and slow down; at times, some even dig in their heels and back up in a retrograde motion.

HERACLEIDES AND ARISTARCHUS, AHEAD OF THEIR TIME

The philosopher Heracleides Ponticus, born around 390 B.C. in Heraclea (now in Turkey), was the first on record to propose that Earth itself rotated, causing the apparent motion of the fixed stars. He also theorized that the inner planets moved about the sun. These accurate observations gained no support until the Renaissance, when Tycho Brahe proposed a similar model. Astronomer Aristarchus of Samos, born around 310 B.C., went much further, anticipating Copernicus by some 1,800 years.

According to Archimedes, Aristarchus "brought out a book consisting of certain hypotheses, in which the premises lead to the conclusion that the universe is many times greater than it is presently thought to be. His hypotheses are that the fixed stars and the sun remain motionless, that the earth revolves about the sun in the circumference of a circle, the sun lying in the middle of the orbit, and that the sphere of the fixed stars, situated about the same center as the sun, is so great that the circular orbit of the earth is as small as a point compared with that sphere."

Aristarchus, in other words, contradicted prevailing geocentric models by theorizing that Earth and planets orbited the sun. He also employed valid geometric methods to derive a size for the moon and the distance to the sun. Although his measurements were off, making the moon too large and the sun too close, his technique was correct. However, he suffered the fate of many a contrarian scientist, his conclusions ignored in favor of the later, Ptolemaic system. His legacy is visible today on the moon, where one of the brightest craters on the near side is named in his honor.

Greek philosopher Pythagoras was born on the island of Samos around 580 B.C. His students spread his doctrine of the importance of numerical relationships in nature.

CIRCLES WITHIN CIRCLES

Among the most influential natural philosophers were Plato and his pupil Aristotle, living in the fifth and fourth centuries B.C. Aristotle's writings, surviving into the Middle Ages, guided scientific thinking for more than a thousand years. Like Plato, Aristotle saw the world as a cosmos, an order, a logical, symmetrical, interconnected system. The natural world was made from combinations of the four elements earth, water, air, and fire, and the four qualities hot, cold, wet, and dry. Our world, Earth, was formed primarily of the element earth and was encircled by the seas (water), the atmosphere (air), and a shell of fire extending to the moon. A fifth element, "quintessence," made up the unchanging stars and other heavenly bodies.

The stars seemed clearly to revolve about the Earth in their own fixed sphere. The irregular motion of the planets was tougher to explain. Another of Plato's students, Eudoxus, attempted to solve this problem with a clever and complicated model of nested, concentric, rotating spheres. Some carried the moon, sun, and planets while others spun in opposite directions. Aristotle elaborated on this model in an attempt to show just what propelled all these spheres. In his system, the outermost sphere was the prime mover that put all other spheres into motion. Overall, Aristotle's cosmos had 55 perfect spheres, an astronomical whirligig with Earth at the center.

Even then, the astronomers could not explain why the planets grew brighter and dimmer, or why Earth had seasons, implying that Earth did not maintain a uniform distance to the sun or other heavenly bodies. In the second century B.C., Hipparchus of Nicaea used the ample set of Babylonian observations, as well as their 60-based number system, to redraw the map of the solar system. In his new version, the sun followed an eccentric orbit around Earth, and the moon had its own epicycle—it inscribed little circles as it traveled around a larger circular orbit, the deferent. This dizzying and clever model was appropriated, and elaborated upon, by the greatest of the Greek astronomers, Ptolemy.

PTOLEMY

Little is known of Ptolemy's life. He was born Claudius Ptolemaeus around A.D. 85 in Alexandria, Egypt, a city noted for scholarship. A geographer, mathematician, and astronomer, he is best known for *The Mathematical Collection,* which gained lasting fame as the *Almagest,* a Latin-Arabic title meaning "Greatest." Written around A.D. 150, it portrayed a clockwork cosmos in which the sun, moon, and planets occupied concentric circling shells, orbiting a spherical Earth while simultaneously whirling about in their own epicycles. The *Almagest* also contained a list of more than 1,000 stars, as well as accurate tables predicting planetary motions and other phenomena. Complex and brilliant, the Ptolemaic system held astronomers around the world in thrall for more than 1,000 years.

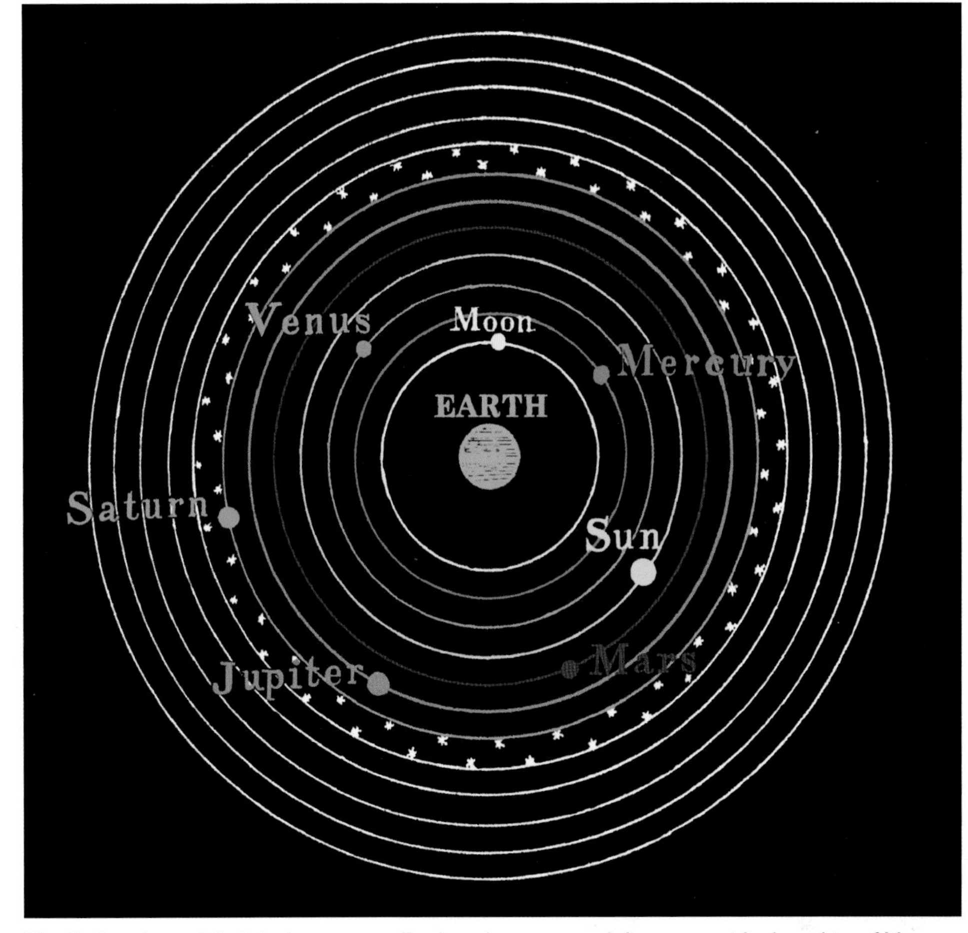

The Ptolemaic model of the heavens put Earth at the center and the sun outside the orbits of Mercury and Venus. The fixed stars circled outside the sphere of Saturn.

SIZING UP EARTH

Greek geometry may have turned the universe into a dizzying mechanical toy, but it also provided an elegant answer to a tough question: How big is Earth? Greek scientist Eratosthenes, born around 276 B.C. in Cyrene, Libya, was an outstanding mathematician, astronomer, and poet. He learned that on the day of the summer solstice, the rays of the sun shone directly down a well in the town of Syene, 500 miles south of Alexandria in Egypt. In other words, the sun was directly overhead that day, and an upright rod would cast no shadow at noon. Eratosthenes was also aware that such was not the case in Alexandria; upright objects would cast a shadow on the solstice. So one midsummer's day, he measured the shadow cast by an upright pointer in Alexandria and found it to be one-fiftieth of a circle, or 7.2 degrees. Knowing as well that the distance between Alexandria and Syene was 5,000 stadia, he concluded that the total circumference of Earth was 50 times that distance, or 250,000 stadia. The accuracy of his answer depends upon just how long a stade was, which is in dispute. But the measurement was probably in the neighborhood of 29,000 miles, fairly close to the modern known distance of 25,000 miles.

NAMING THE PLANETS

Not the least of the Greeks' lasting contributions were the names of the planets themselves. The association of certain planets with certain attributes of gods or goddess goes back to the Sumerians, but the Greek names were directly appropriated and latinized by the Romans, and those (slightly anglicized) are the names we use now. Gaea, Selene, Hermes, Aphrodite, Helios, Ares, Zeus, and Kronos became Terra, Luna, Mercurius, Venus, Sol, Mars, Iuppiter, and Saturnus.

CELESTIAL SPHERES: ISLAM TO COPERNICUS

The 1,400-year interregnum between Ptolemy and Copernicus marked a dark time in European astronomy. With the Roman Empire disintegrating, learning eroded in the West, leaving few scholars able to read Greek and pass on the advances of the ancients. But matters were different farther East. As Islam took hold, intellectually hungry rulers began to acquire manuscripts from Byzantine libraries. In the 9th century, the caliph al-Ma'Mun established the aptly named House of Wisdom in Baghdad, a center for the translation of Greek scholarship. Rendered into Arabic, scientific and mathematical texts traveled throughout the rapidly growing Islamic world, eventually reaching India and China. Islamic scholars also translated scientific and mathematical works from Persian and Sanskrit.

The Copernican system drastically reorganized the known universe by placing the sun at the center of the heavens. The planets then followed in the correct order. This 17-century illustration, from Andreas Cellarius's *Harmonia Macrocosmica,* includes the four large moons of Jupiter.

Arab scientists came to revere Ptolemy's *Almagest* but managed to improve upon his observations using their own, newly designed instruments. Islamic rulers built some of the first observatories, though these tended to rise and fall with the fortunes of their patrons. Arab astronomer Muhammad al-Battani refined Ptolemy's observations of the sun's orbit in a work that Copernicus would later use. In the tenth century, Abd al-Rahman al-Sufi compiled a star catalog superior to Ptolemy's. Today, many of the most familiar stars bear Arab names: Algol, Betelgeuse, Vega, and more.

Despite their contributions, Eastern astronomers did not dispute the Western consensus that the universe orbited the Earth in the clockwork Ptolemaic system. That task fell to a Polish cleric, Mikolaj Kopernik, or Nicolaus Copernicus.

COPERNICUS

Born in 1473 to a wealthy merchant, Copernicus was raised by his uncle, a bishop, and schooled for the church. However, he also studied astronomy and astrology (in those days virtually the same thing) and discovered that the much esteemed Ptolemaic model was based upon an awkward system of equants and deferents; nor could astronomers agree on the proper order of the planets. Copernicus later wrote that previous astronomers "are just like someone taking from different places hands, feet, head, and the other limbs, no doubt depicted very well but not modeled from the same body and not matching one another—so that such parts would produce a monster rather than a man."

We don't know exactly when or how the Polish scholar arrived at his radical solution to these problems—a heliocentric solar system—but by 1514 he was privately circulating a manuscript, the *Commentariolus (Little Commentary),* in which he wrote, "Having become aware of these [defects of Ptolemaic theory], I often considered whether there could perhaps be found a more reasonable arrangement of circles, from which every apparent irregularity would be derived while everything in itself would move uniformly, as is required by the rule of perfect motion."

This "more reasonable arrangement," he wrote, established the sun at the center of the cosmic spheres; the only thing to orbit Earth was the moon, while Earth itself rotated on its axis. He also put the planets into correct order: Mercury, Venus, Earth, Mars, Jupiter, and Saturn. His heliocentric model eliminated one of the Greek's greatest bugbears, the retrograde motion of the planets in the sky. The new model clearly explained how planets orbiting at different speeds would sometimes appear to move ahead of Earth, and sometimes fall behind.

Receiving mostly positive reactions to his essay, Copernicus went on to write his longer, groundbreaking book, *De Revolutionibus Orbium Coelestium Libri VI,* or *Six Books Concerning the Revolutions of the Heavenly Orbs.* In it he restated his heliocentric views, writing, "The stations of the planets, moreover, as well as their retrogradations and [resumptions of] forward motion will be recognized as being, not movements of the planets, but a motion of the earth, which the planets borrow for their own appearances. Lastly, it will be realized that the sun occupies the middle of the universe. All these facts are disclosed to us by the principle governing the order in which the planets follow one another, and by the harmony of the entire universe, if only we look at the matter, as the saying goes, with both eyes."

Although Copernicus had not received major opposition to his earlier statement of these ideas, he was still hesitant to publish his longer work. The church held that Earth was the center of creation, and certainly some church officials would find the new work heretical. After delaying more than 30 years, Copernicus finally sent the manuscript via a colleague to a printer in Nürnberg. There it fell into the hands of Lutheran theologian Andreas Osiander. Without Copernicus's knowledge, Osiander added a timid foreword, ostensibly from the author, which said that the book's findings were hypothetical only and that astronomy could never really find the truth of heavenly phenomena. (Reading the fraudulent foreword later, Johannes Kepler was furious: "He [Copernicus] thought that his hypotheses were true," he wrote. "And he did not merely think so, but he proves that they are true.") The printed copy of the book reached Copernicus only on his deathbed in 1543; it is not known whether he was able to read it with its unwanted preface.

Legend has it that *De Revolutionibus* caused an uproar in church circles. However, the reality is that the church took no official position on it until 1616, when it was censured. Many contemporary astronomers admired it and adopted at least some of its principles to simplify their complicated cosmology. However, the heliocentric model did not truly take hold until it was developed and defended by the independent-minded German astronomer Kepler.

MONGOL ASTRONOMY

Belying their reputation as mindless barbarians, many Mongols took a keen interest in astronomy. In fact, two of the great observatories of the medieval world were built by Mongol leaders.

The Mongol ruler of Persia, Hulegu, built the first in Maragheh in 1259. Under the direction of the Persian astronomer Nasir al-Din al-Tusi, the observatory housed a group of sky-watchers who compiled an impressive set of astronomical tables. Its ruins can still be seen today. The Mongol ruler of Samarkand, Ulugh Beg, built the second observatory in the 1420s (a portion of its giant sextant is seen above). A descendant of the conqueror Tamerlane, Ulugh Beg was an intellectual and poet. Astronomers at his three-story observatory compiled an important star catalog before their sponsor met an all-too-common fate: He was assassinated by agents of his son, and his observatory fell into disuse.

WANDERING STARS
TELESCOPE ERA

Galileo's little spyglass opened up vast new vistas in astronomy. The craters of the moon, the satellites of Jupiter, the rings of Saturn, and more appeared to human eyes for the first time through the telescope, astronomy's most valuable tool.

The invention of the telescope ushered in a new age of astronomy and with it, a new way of understanding the cosmos. It is no coincidence that some of the greatest names in science, Galileo and Newton, pioneered the use of telescopes in their pursuit of knowledge. • After 1610, telescopes quickly became a necessity for even the most amateur of astronomers. Over the decades, moving into the 18th and 19th centuries, they grew larger and more powerful, their light-gathering capacities increasing dramatically. As a result, new planets and moons joined the solar list and the freshly revealed details of planetary surfaces and atmospheres paved the way for future space flight. • In the 20th century, telescopes moved beyond visible light. Radio, infrared light, and other waves within the electromagnetic spectrum carried astronomical information to specialized receivers. With the addition of computer technology, modern telescopes have steadily expanded their view not only to dwarf planets far beyond Pluto, but even to planets around other stars. And telescopes in space are combining the best of two worlds: the capacity to scan large areas of the sky with the ability to fly beyond the blurring confines of Earth's atmosphere.

WORLD'S LARGEST OPTICAL TELESCOPES (by aperture size)

- **Gran Telescopio Canarias,** Canary Islands
- **Keck and Keck II,** Mauna Kea, Hawaii
- **SALT,** South African Astronomical Observatory
- **Hobby-Eberly,** Mount Fowlkes, Texas
- **Large Binocular Telecope,** Mount Graham, Arizona
- **Subaru,** Mauna Kea, Hawaii
- **VLT Interferometer,** Cerro Paranal, Chile
- **Gillett,** Mauna Kea, Hawaii

SKYWATCH

• Hint for beginning sky-watchers: A pair of 7 x 50 binoculars will give you a close-up view of the moon's craters as well as Jupiter's four big moons.

AMAZING FACT The world's highest observatory, the Indian Astronomical Observatory, sits at 4,517 meters (14,800 ft) in the Himalaya.

1609–10
Galileo, using reflecting telescope, studies the moon, Venus, Jupiter, and Saturn.

1656
Christiaan Huygens discovers the rings of Saturn and Saturn's moon Titan.

1668
Isaac Newton builds a successful reflecting telescope.

1781
William Herschel discovers Uranus.

1846
Johann Galle discovers Neptune after existence predicted using mathematics.

1930
Clyde Tombaugh discovers Pluto.

Mars was an early telescope target (opposite). Built in 1897, Wisconsin's Yerkes Observatory was one of the first truly modern observatories (above).

TELESCOPE ERA: REVOLUTION IN THE RENAISSANCE

Despite Copernicus's convincing work, the geocentric model died a slow and prolonged death. It took three of the greatest astronomers in history, living in Europe between 1546 and 1642, to make the case for the Copernican system. And one of them, Tycho Brahe, had no intention of doing so.

TYCHO BRAHE

Tycho Brahe, born in 1546, was an eccentric character even by the standards of the time. A member of a noble Danish family, he was kidnapped as a toddler by his childless uncle and raised as the man's heir. During a nighttime duel with a fellow student at a Christmas party (some say over the issue of who was the better mathematician), he lost the top half of his nose and wore a gold-and-silver nosepiece the rest of his life.

Tycho's eccentricity did not extend to astronomy. He was a meticulous and methodical observer who realized that contemporary astronomical tables were out of whack and resolved to improve upon them. In 1572, his sharp eyes spotted a remarkable sight in the night sky—a brilliant new star (now known to be a nova) in the constellation of Cassiopeia. Traditional astronomy taught that the heavenly spheres were perfect and unchanging, but Tycho's measurements showed that the new object was definitely in the heavens; it was more distant than the moon. The changeable nature of the sky was further confirmed by the visit of a comet in 1577. Aristotle had taught that comets were part of the atmosphere, but in a 200-page thesis, Tycho proved that the fiery object was astronomical, moving just as though the "spheres" did not exist.

Blessed by royal patronage, Tycho built a spectacular observatory, Uraniborg, on the island of Hven and outfitted it with the best assistants and instruments available. For years he and his team pulled together superbly accurate observations of stars and planets, including detailed records of the orbit of Mars. Although he died without ever becoming convinced of the Copernican system—he believed that Earth was the center of the solar system, while the other planets orbited the sun—he bequeathed his invaluable observations to an invaluable assistant: Johannes Kepler.

JOHANNES KEPLER

Kepler's life was tougher than Tycho's, but he surpassed his mentor as a mathematician and theorist. The son of an "immoral, rough, and quarrelsome soldier," in his own words, the intelligent youth was educated as a Protestant theologian. One of his teachers introduced him to the writings of Copernicus, and his first work outlined the geometry of the Copernican system. Kepler's first job, teaching astronomy in Graz, Styria, encompassed astrology as well. Although Kepler criticized the "faulty foundations" of astrology, he was lucky in his predictions and gained a reputation as a prognosticator. Fortunately for the more rigorous sciences, Tycho Brahe, then living in Prague, asked the young scientist to join him. Upon Tycho's death in 1601, Kepler succeeded to his post as Imperial Mathematician.

Using the Dane's records of planetary motion, Kepler tackled the knotty issue of planetary orbits and arrived at a radical conclusion. The only orbital shape that made sense was an ellipse: an oval with the sun at one focus. Far from being an interlocked set of spherical shells, the solar system was held together by a force (possibly magnetism, Kepler thought) exerted by the sun at its center. Between 1604 and 1621, he published several epochal works, including *New Astronomy* (1609) and *Harmony of the World* (1619), which contained what are now known as the three laws of planetary motion:

1. All planets move around the sun in elliptical orbits, with the sun at one focus.
2. An imaginary line joining the planet to the sun sweeps outward in equal areas in equal amounts of time. (Planets move faster when closer to the sun.)
3. The squares of the periods of the planets (the time it takes to complete one orbit) are proportional to the cubes of their mean distances from the sun.

Kepler's last years were difficult. Six of his ten children died in childhood. His mother, a maker

THE FIRST TELESCOPES

The first telescopes were refractors: They used combinations of glass lenses to magnify and focus light. Galileo's spyglasses were long, slender tubes with a convex objective lens (the lens at the far end of the tube) that focused light on a second, concave eyepiece lens about 30 or 40 inches away. The bubble-filled, green-tinted glass of the lenses, designed for spectacles, and the difficulty of grinding the glass to a perfect shape meant that Galileo's scopes yielded blurry images compared with today's instruments. They also had a small field of view, so that the astronomer could see perhaps one-quarter of the moon at a time. Even so, telescopes immediately became every astronomer's must-have tool, and their design improved steadily over the years.

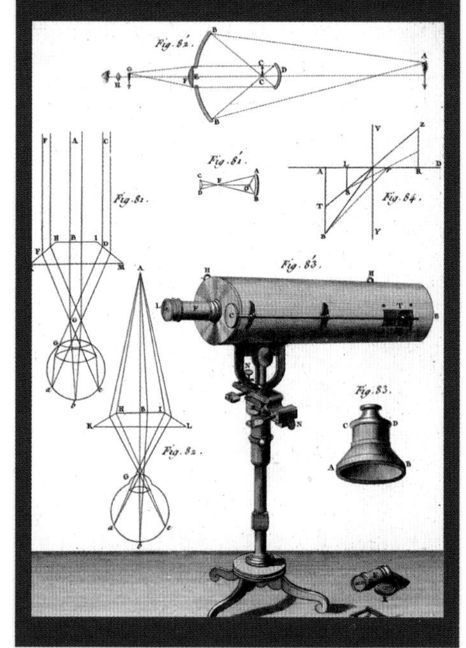

of herbal potions, was accused of witchcraft in 1615. Kepler spent much of the next six years extricating her from the charges. His tolerant religious views got him in trouble with Catholics and Lutherans alike. In 1630, while attempting to collect some money owed him, he died. He had composed his own epitaph, which began: "I used to measure the heavens; now I shall measure the shadows of the Earth."

GALILEO GALILEI

By the 1590s, Kepler had begun to correspond with an up-and-coming Italian mathematician, Galileo Galilei. Originally from Pisa, the aristocratic young scientist had already investigated the mechanics of pendulums and the effects of gravity before becoming a professor of mathematics at the University of Padua. In 1609, the same year that Kepler published his *New Astronomy,* Galileo got word of an intriguing new device being used in Holland: a tube with two glass lenses that could make distant objects seem nearer. The actual creation of the telescope is still in dispute, although it may fall to Dutch optician Hans Lippershey. But it was undoubtedly Galileo who first showed off its true astronomical abilities.

The Paduan soon built his own telescopes, first with an 8x magnification, and then a much improved version with a 20x power. Through it he saw amazing sights never seen before: the mountains of the moon, the phases of Venus, sunspots, and the four largest moons of Jupiter. He demonstrated the powers of his first telescope to fellow scientists and church officials, with mixed results. Some simply refused to believe what they saw through the lens. "Galileo Galilei," said one, "came to us at Bologna, bringing his telescope with which he saw four feigned planets. . . . I tested this instrument of Galileo's in a thousand ways, both on things here below and on those above. Below, it works wonderfully; in the sky it deceives one. . . . Galileo fell speechless, and on the twenty-sixth . . . departed sadly."

This stubborn skepticism was symptomatic of a much larger problem. Each of Galileo's discoveries was a further blow to the old, Greek ideal of the cosmos and the pre-Copernican theology of the Catholic Church. The telescope's solar system was rough, spotty, changeable—and sun centered.

Galileo demonstrates his new telescope and his discoveries about the irregular surface of the moon and the satellites of Jupiter to local clerics.

In 1610, Galileo published his findings about Jupiter in a short book, *The Starry* (or *Sidereal*) *Messenger.* Subsequent letters about the Copernican system found their way into the hands of the Catholic Inquisition, which instructed Galileo not to "hold, teach, or defend" Copernican theory. In his book on the scientific method (*The Assayer*) he famously wrote, "Philosophy is written in this grand book, the universe, which stands continually open to our gaze. But the book cannot be understood unless one first learns to comprehend the language and read the letters in which it is composed. It is written in the language of mathematics, and its characters are triangles, circles, and other geometric figures without which it is humanly impossible to understand a single word of it." The book gained him an unexpected new admirer: the new pope, Urban VIII, who gave him permission to prepare another book, *Dialogue on the Two Great World Systems,* a discussion of traditional versus Copernican theory—as long as he kept his arguments hypothetical.

Alas for Galileo, the text of his new book was not hypothetical enough. Although he had put his defense of Copernican astronomy in the mouth of a fictional character, the pope's own views were espoused by that character's foolish opponent, Simplicio. Galileo was tried before the Catholic Inquisition in 1633. He recanted his views but was nevertheless convicted of heresy and spent the rest of his life under house arrest.

TELESCOPE ERA: FILLING IN THE PICTURE

Seventeenth-century astronomers fell upon the newly invented telescope like starving dogs on a bone. Between Galileo's demonstration in 1609 and the end of the century, European observers turned their newfound lenses on every known body in the solar system.

MERCURY TO THE MOON

Some planets yielded less information than others. Tiny Mercury, hidden in the blaze of the sun, is hard to make out even with modern telescopes. However, in 1639 the Italian astronomer and Jesuit priest Giovanni Zupus used a telescope barely more powerful than Galileo's to determine that Mercury had phases, like the moon and Venus. This meant that it orbited between Earth and the sun, further confirmation of the sun-centered model of the universe. But little else could be seen of the planet until the late 18th century.

Nearby Venus drew more eyes. English amateur astronomers Jeremiah Horrocks and William Crabtree were thrilled to see a transit of Venus across the sun's face in 1639, a rare event that allowed Horrocks to estimate the planet's size and distance. But other observations of the opaque planet were at best wishful thinking. In 1645 and 1646, Italian astronomer Francesco Fontana claimed to see a moon circling Venus and mountains at the planet's terminator, both impossible. The great Italian astronomer Giovanni Cassini tracked bright and dark patches through his telescope and deduced that the planet's rotation took 23 hours and 21 minutes—off by some 242 days. Earth's moon was far more satisfying through the lens. In 1647, a Polish brewer and astronomer, Johannes Hevelius, published a beautiful, detailed atlas of the moon, *Selenographia,* that was the standard for 100 years.

THE MICROMETER

By the 1650s, astronomers were beginning to use the telescope not just to see heavenly bodies, but also to measure them. The device that made this possible was invented by the amateur English astronomer William Gascoigne in the late 1630s. While looking through a Keplerian telescope one day, Gascoigne noticed that a spider had spun one thread of its web directly across the focal point of its lenses. The sharp little line, he saw, could be used to more accurately point his scope at its target.

Gascoigne went on to invent a telescopic sight made of wire crosshairs. He then devised the micrometer, which allowed astronomers to measure objects seen through the lens. Though Gascoigne was killed in the English Civil War, astronomers across Europe picked up the micrometer and used it to measure everything from distances on the moon to the apparent width of planets.

RINGS AND MOONS

Turning the telescope away from the sun toward Mars, Jupiter, and Saturn yielded even more interesting results. Two major figures in the era's astronomy, Cassini and Dutch astronomer Christiaan Huygens, made major discoveries about all three.

The privileged Huygens and his brother, Constantijn, built their own fine telescopes, grinding the lenses themselves. Between the 1650s and the 1670s, Huygens helped create the first map of Mars, detecting the dark region, Syrtis Major, as well as its south polar cap. In 1656, he discovered the first (and largest) known moon of Saturn, Titan. But he is even better known for discovering Saturn's rings. Galileo had seen them, but to his eyes they looked like puzzling, loopy ears or handles on either side of the planet, inexplicable. In *Systema Saturnium* (1659), Huygens explained that the odd structures were two sides of a flat, solid ring around Saturn.

Cassini, working first in Bologna and then in Paris, also added to knowledge of Mars by calculating its rotation at 24 hours, 40 minutes, which was only 3 minutes off the correct time. His observations of Jupiter and its satellites yielded that planet's rotation as well, with his estimate of 9 hours and 56 minutes almost spot-on. Between 1671 and 1684, he spotted four more Saturnian moons: Iapetus, Rhea, Tethys, and Dione. And in 1675, scrutinizing Saturn's rings, he detected a distinct, dark gap between the

CHRISTIAAN HUYGENS
ASTRONOMER AND INVENTOR

Dutch scientist Christiaan Huygens (1629–95) was a multitalented physicist, astronomer, and hands-on inventor of the sort that seemed particularly to flourish in the 17th century. The son of the famous poet and diplomat Constantijn Huygens, Christiaan made the most of a privileged upbringing in a household visited by philosopher René Descartes and painters Peter Paul Rubens and Rembrandt van Rijn. An astronomer who built his own telescopes, he is famous for mapping Mars and discovering Saturn's rings. He invented the first accurate pendulum clock, observing that its swing had to move in a cycloid shape. Huygens also developed the wave theory of light, published in his *Treatise on Light* (1690). The Dutch scientist knew most of the leading scientific figures of his day, including Isaac Newton, whom he admired. Even so, he considered Newton's theory of universal gravitation "absurd." He was, however, an imaginative thinker who believed that other planets, "equally good fitted worlds like ours," might be inhabited. By visiting these worlds, he believed, we would develop a better appreciation of what is worthy on Earth.

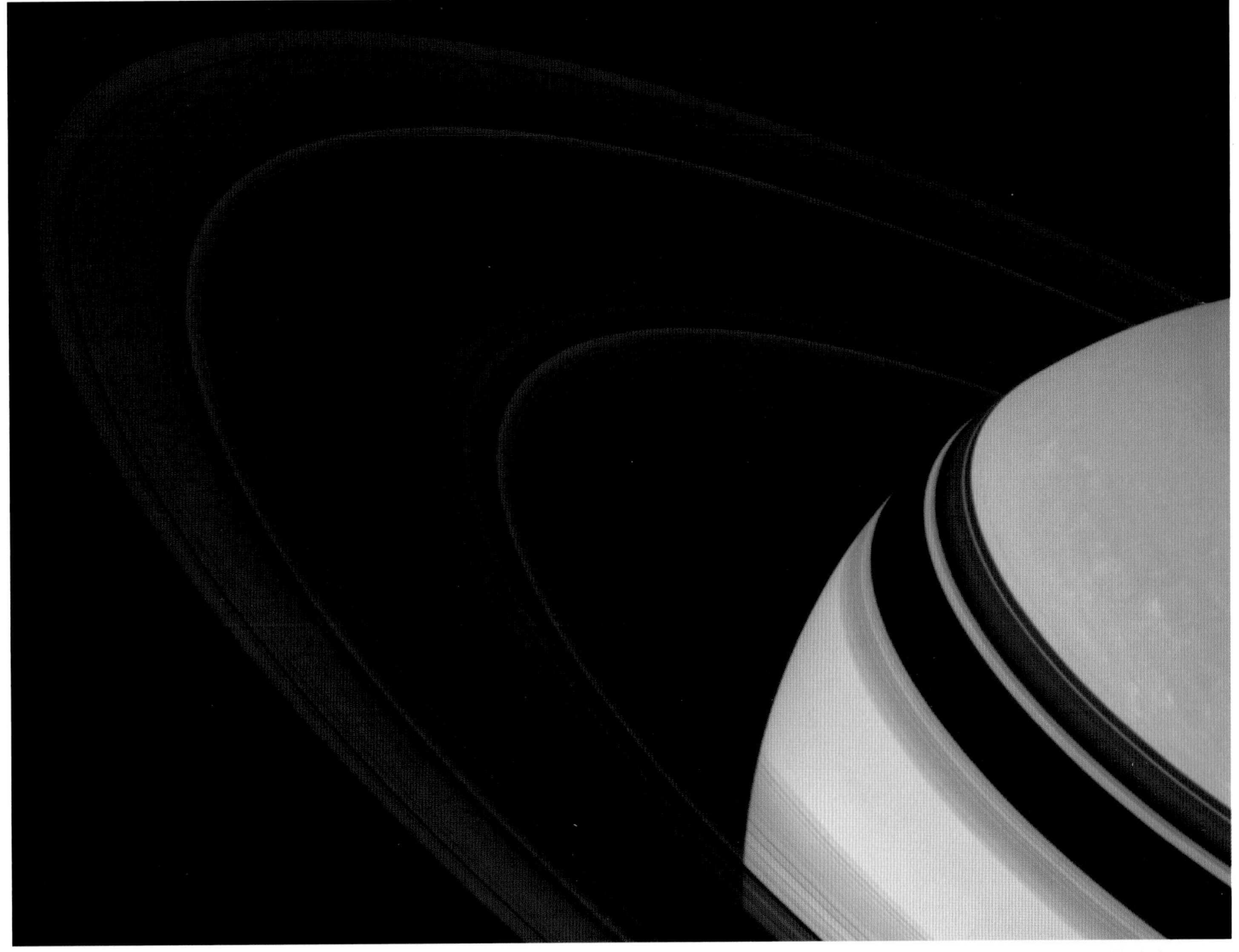

Christiaan Huygens was the first to recognize Saturn's odd appendages as rings (seen here in a modern image from the Cassini spacecraft). A few years later, French-Italian astronomer Giovanni Cassini pointed out that the inner and outer rings were separated by a dark gap.

inner and outer bands, which is now known as the Cassini division. He also realized that the rings were not solid, but swarms of individual little satellites.

LIGHT ITSELF

The rapidly improving telescope of the 17th century also contributed to the first solid calculation of the speed of light. Until that time, the idea that light had a speed at all was controversial. Galileo, ever the experimenter, proposed a test that was later carried out by members of the Florentine Academy: Stand on a hilltop with a shuttered lantern and time how long it takes for the light to reach an assistant on a distant hill, who would unveil his own light as soon as he saw the first flash. However, light traveled too quickly to make that experiment work. Other scientists of the time, including Kepler, believed that light did not travel at all, but appeared instantaneously.

This idea was dashed by Danish astronomer Ole Roemer, working at the Paris Observatory in the 1670s. While watching Jupiter's moons, he noticed a puzzling thing: The revolution of Io around its parent planet could be measured at about 42 hours; yet the orbit seemed to grow shorter when Jupiter was closer to Earth, and longer when it moved away. Roemer concluded that it was not the orbit itself that varied, but the time it took for light from Jupiter to reach Earth. With these observations, he was able to calculate a speed for light at 225,000 kilometers a second (140,000 mps). "'Tis so exceeding swift that 'tis beyond Imagination," grumbled Robert Hooke later. "Why not be as well instantaneous I know no reason." (Light is now measured at 299,792 kilometers a second, or 186,282 mps.)

The great 17th-century passion for observation and experiment, and the rapid accumulation of knowledge about the universe, would soon find an advocate in one of the great minds of modern history, Isaac Newton.

FOR MORE ON OLE ROEMER AND THE SPEED OF LIGHT, GO TO WWW.PBS.ORG/WGBH/NOVA/EINSTEIN/ANCE-C.HTML

TELESCOPE ERA: NEWTON'S UNIVERSE

English genius Isaac Newton's contributions to science include, but are not limited to, explaining the nature of light, inventing a practical reflecting telescope, outlining the universal laws of motion and gravitation, and developing the method of calculus.

By the late 17th century, European astronomers were almost overwhelmed by a wealth of observations: Discoveries about the planets, their distances, new moons, and comets were helping them fill out their picture of the solar system. They were rapidly learning the "what" of the system, but they struggled to understand the "how" and "why." Among the big questions: What laws governed the complex movements of heavenly bodies? What force tied them together?

Kepler had thought a force linked the moon to Earth but a "vortex" guided the planets around the sun. French mathematician and philosopher René Descartes, too, believed that planets orbited within a solar vortex in an interplanetary medium, the ether. In England, scientists connected with Gresham College espoused a magnetic theory of attraction between bodies such as Earth and the moon.

ISAAC NEWTON

Into this ferment came the English phenomenon Isaac Newton, who clarified, organized, and solved some of the most crucial problems in the history of science. Born prematurely to a poor farming family in 1642, he was a neglected child whose father had died before his birth and whose mother remarried a man he hated and then left him with his grandmother. The serious, insecure youth studied math at Cambridge without gaining notice, but had to leave in 1665 when plague shut down the university. At home on the farm, he worked out the mechanics and mathematics of planetary motion and optics and developed the basics of the calculus—in just two years.

And then, for decades, he kept the news to himself. In 1667, Newton took a job at Trinity College, Cambridge, eventually becoming Lucasian professor there. He was not particularly chummy with his peers; in fact, his relationships with other scientists were touchy and defensive. But his collaborations were fruitful. In a correspondence with Robert Hooke, secretary of the Royal Society, he solved many details of his laws of motion. In 1684, British astronomer Edmund Halley visited Newton to ask him what the shape of planetary orbits might be if the force of attraction to the sun was the inverse of the square of their distance from it.

An ellipse, of course, Newton replied, and he had a paper to prove it.

ROBERT HOOKE

PHYSICIST, ARCHITECT, INVENTOR

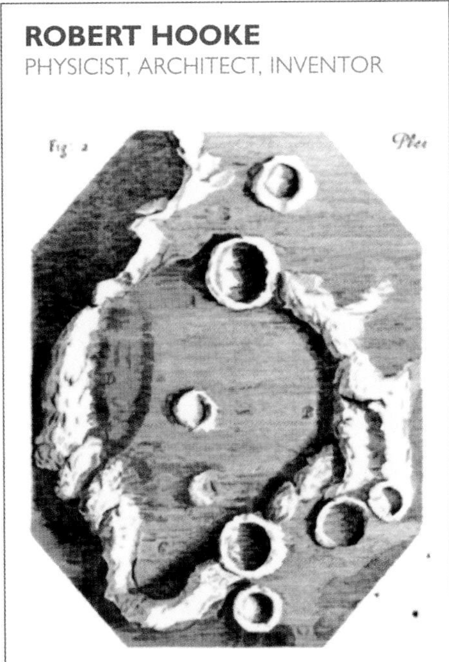

Robert Hooke (1635–1703), brilliant in many fields, may in some ways have been a victim of his own success. The Englishman was an architect who helped Christopher Wren rebuild London after the Great Fire of 1666. He was also an inventor, designing a balance spring for watches, one of the first reflecting telescopes, a compound microscope, and a wheel barometer. Students today may know him best as a biologist, the first to describe cells. And physicists honor him as an intuitive thinker who expounded on the wavelike nature of light and who proposed a universal theory of gravitation and modern laws of motion 13 years before Newton's *Principia* (above, his drawings of craters on the moon). However, Hooke's job with the British Royal Society, which called upon him to devise three or four experiments per week for its members, may have prevented him from taking the time to explore and develop his theories for publication. And so Hooke is remembered for his plant cells—and Newton for everything else.

THE *PRINCIPIA*

That paper was "De Motu" ("On Motion"), and it formed the basis for Newton's *Philosophiae Naturalis Principia Mathematica (Mathematical Principles of Natural Philosophy),* or the *Principia.* Published in 1687 with support from Halley, the *Principia* lays out Newton's laws of motion and universal gravitation.

The three laws of motion, put simply, are:

1. A resting body will remain at rest, while moving bodies will continue to move at a uniform speed in a straight line, unless acted upon by an outside force (law of inertia).
2. A change in motion is proportional to the force applied to it.
3. For every action, there is an equal and opposite reaction.

The law of universal gravitation was also straightforward: Every object in the universe attracts every other object. The force of their attraction is proportional to the product of their masses and inversely proportional to the square of the distance between the two objects.

The *Principia* made Newton famous and formed the foundation for virtually all physical science until the time of Einstein. In astronomy, the mathematics that accompanied his laws explained the forces that shaped elliptical orbits of the planets, allowed for the calculation of their masses, and laid the groundwork for later spaceflight. His perception that gravitation was universal overturned the notion that the forces that governed celestial bodies were fundamentally different from those that guided objects on Earth.

Newton capped his achievements in 1704 by publishing *Opticks: Or, a Treatise of the Reflections, Refractions, Inflections and Colours of Light.* Although the phenomenon of the spectrum was well known, previous thinkers, such as Descartes, believed that light was intrinsically white, and that the colors seen with a prism represented a modification of its basic nature. Newton reversed this, showing that light was made up of many colors that combined to form white light. The individual rays, or particles, of each color would refract at specific angles.

Newton's works brought him international acclaim. But the great man was not gracious in success. Robert Hooke, having seen the material of *Principia* in draft, asked Newton to acknowledge that he, Hooke, had suggested such key points as the inverse square law in their earlier correspondence (which was true). In response, Newton eliminated Hooke's name completely from his work. He also became involved in a long, bitter dispute with Gottfried Leibniz over who invented the calculus (the most likely answer: both of them, independently).

The public adored him despite his ill temper. In 1705 he became the first scientist to gain a knighthood. The poet Alexander Pope wrote: "Nature and nature's laws lay hid in night. / God said, Let Newton be!, and all was light."

Newton died in 1727, still nursing grudges but surrounded by honors, having given science, and certainly astronomy, its most effective tools yet for studying the universe.

NEWTON'S TELESCOPE

The refracting telescope, made famous by Galileo in 1609, had its drawbacks. It was almost impossible to grind its curved lenses accurately enough to avoid blurring, or "spherical aberration." Later in the century, several inventors, including Isaac Newton, designed telescopes that avoided this problem by using mirrors instead of lenses. In 1668, Newton drew upon his understanding of optics to build a small reflecting telescope (below) with an eyepiece on the side rather than the end. Light entering the open end reflected from a curved mirror at the closed end to a second, flat mirror, angled at 45°, which in turn directed the light through the eyepiece. The Newtonian reflector is still favored among many amateur astronomers for the elegant simplicity of its design.

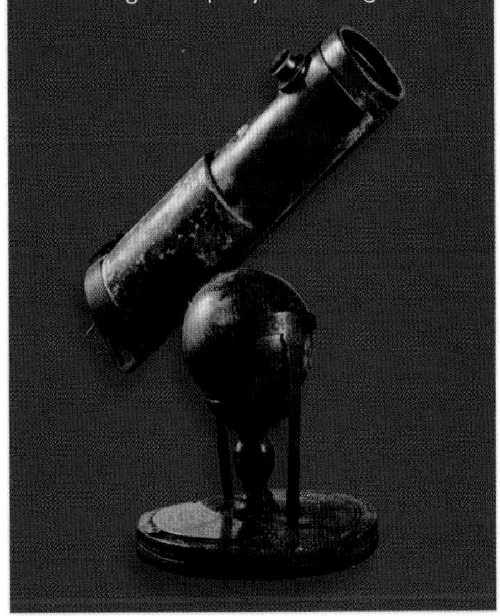

TELESCOPE ERA: THE OUTER LIMITS

With Newton's powerful mathematics in hand, solar system astronomers began to search the planetary neighborhood using numbers as well as telescopes. In the 18th century, several scientists published works that commented on a curious arithmetic rule that seemed to govern the distances of planets from the sun. One was Johann Daniel Titius, a professor of physics at Wittenberg University. In 1766 he noted that planetary distances hewed closely to this pattern: 4 (Mercury), 7 (4+3, Venus), 10 (4+6, Earth), 16 (4+12, Mars), 52 (4+48, Jupiter), and 100 (4+96, Saturn). His announcement drew little attention at the time, but six years later German astronomer Johann Elert Bode popularized the sequence, now known as the Titius-Bode law. Both scientists drew attention to a notable gap in the numbers. Between 16 and 52—that is, between Mars and Jupiter—should come 28, or 4+24. "Can one believe that the Creator of the Universe has left this position empty? Certainly not!" Bode wrote.

And so some astronomers turned their telescopic eyes toward the wide expanse between Mars and Jupiter in hopes of finding the missing planet. But before any such discoveries were made, a serendipitous finding by an English musician gave the Titius-Bode law support.

URANUS

William Herschel, son of a Prussian bandleader, moved to England during the Seven Years' War and became an organist in the spa town of Bath. But his real passion was astronomy. Working with his younger sister Caroline, an accomplished astronomer in her own right, he built his own telescopes, including a beautifully accurate 15.7-centimeter (6.2-in) reflector. On March 13, 1781, as he carefully studied the stars in the constellation Gemini, he spotted "a curious either nebulous Star or perhaps a Comet." Four nights later, it had moved. So not a star—a comet then, he decided, believing with most of his contemporaries that all the solar system's planets had been discovered long ago.

But when other astronomers learned of his discovery and scrutinized the object, its motion and disklike shape through the best lenses confirmed that it was, in fact, a planet, the first discovered in recorded history. Herschel was showered with honors, including the title of Astronomer under Royal Patronage. He wanted to name the new planet the Georgian Star, after his royal patron, King George III, but eventually traditionalists prevailed and it was christened Uranus, after the primordial Greek god. And behold: Its estimated distance matched the next term in the Titius-Bode pattern, 196 (4+192).

ASTEROIDS

Emboldened by the apparent confirmation of the orbital sequence, many observers turned their attentions back to the gap between Mars

Self-taught astronomer William Herschel built more than 400 telescopes of various sizes, including the portable 2.1-meter (7-ft) reflector with a 15.7-centimeter (6.2-inch) mirror, illustrated above.

and Jupiter, where surely a planet would be found. In 1800 a group of 24 European astronomers, the "Celestial Police," even formed an agreement to divide up the sky in a cooperative search for the new body. They intended to enlist the help of Italian astronomer and monk Giuseppi Piazzi, but before he was aware of this he had already made a key discovery in the Mars-Jupiter gap.

Working in Palermo, in January 1801 Piazzi spotted a rather dim new "star" that moved from night to night—clearly not a star, but a member of the solar system. He wrote to Bode and others announcing his discovery. Named Ceres after the patron goddess of Sicily, the new body was briefly considered to be a planet. But Herschel and others soon observed that it was mighty small for a planet. After Heinrich Olbers found another little object (now called Pallas) at about the same distance, poor Piazzi had to give up his hope of being the discoverer of a new planet and settle for being the first to see a new kind of body: an asteroid. By the end of the century, more than 300 others had been seen orbiting between Mars and Jupiter.

NEPTUNE

Mathematics, not telescopes, led to discovery of the next true planet. Ever since the detection of Uranus, astronomers had been puzzling over its orbit. It simply didn't fit the Newtonian equations. Could the gravitational pull of another, massive body be interfering?

Two mathematicians on either side of the English Channel tackled the problem. In 1843, the young Englishman John Couch Adams worked out the approximate orbit of such a new planet and presented his calculations to Britain's foremost astronomer, George Airy, at Greenwich Observatory. And there they languished for two years while Airy was traveling and a discouraged Adams turned toward other things. Meanwhile, French astronomer Urbain-Jean-Joseph Le Verrier also calculated the location of the unseen planet.

But Le Verrier took the important next step of actually enlisting someone to look for it. In 1846 he wrote to the young German astronomer Johann Gottfried Galle, at the Berlin Observatory and asked him to search for the object in its predicted position. That same night, September 23, 1846, Galle and his student studied the sky for less than an hour before they saw it: a new "star" not on any of their star maps. The eighth planet, Neptune, orbited at an astonishing remove of 30 AU, almost twice the distance of the previous record-holder, Uranus. The solar system had just doubled in size.

The thrilling new discovery had an unexpected side effect. Neptune's orbit did not fit the Titius-Bode sequence: The next planet should have appeared at about 39 AU. Astronomers today still aren't sure if the Titius-Bode series has a real physical basis, or whether it is simply a strange coincidence.

As the 19th century rolled toward the 20th, bigger and better telescopes meant astronomers could add finer details to their portrait of the growing solar system. Markings on Mars's surface began to come into focus, and Italian astronomer Giovanni Schiaparelli made the first detailed map of its terrain in 1877. More moons began to join the planetary clan; by 1900, the count was up to 21 (not including Earth's moon). They included two surprising finds practically in Earth's backyard: Phobos and Deimos, the tiny moons of Mars, spotted by American astronomer Asaph Hall in 1877 (see p. 111).

PLUTO

The big observatories did not, however, take part in one of the major astronomical events of the 20th century: the discovery of Pluto. That honor fell to amateur astronomer Clyde Tombaugh at the Lowell Observatory near Flagstaff, Arizona. The diligent young observer, fresh from the wheat fields of Kansas, was hired on at Flagstaff early in 1929. His main task (in addition to giving tours and stoking the furnace) was to search for the semi-mythical Planet X, a body that the late enthusiast Percival Lowell had decided must exist in the far reaches of the solar system. Working steadily night after night, Tombaugh photographed the sky along the ecliptic and a little ways north and south of it. By late winter in 1930, the work had reached as far as the constellation of Gemini, where he hit the jackpot. On two photographic plates, taken six days apart, a tiny speck of light had moved. On March 13, 1930, the anniversary of the discovery of Uranus, Lowell Observatory announced the discovery of the ninth planet, later named Pluto.

When William Herschel discovered Uranus, it looked through his telescope like a "nebulous star." Seen close-up by Voyager, the planet reveals its polar circulation in a false-color image.

LEVIATHAN OF PARSONSTOWN

Bigger is better, as far as telescopes go, and the 19th century saw one of the biggest private telescopes ever built in the Irish Earl of Rosse's huge reflector. Constructed in 1845, the "Leviathan of Parsonstown" had a 183-centimeter-wide (6-ft) mirror in a 17-meter (56-ft) tube. While greatly hampered by its location in cloudy Ireland, it was used to study everything from Jupiter and the moon to distant nebulae.

FOR MORE ABOUT GIUSEPPE PIAZZI AND THE DISCOVERY OF CERES, GO TO www.astropa.unipa.it/Asteroids2001/

WANDERING STARS

THE SPACE AGE

Trapped beneath a turbulent atmosphere, even the best 20th-century observatories were limited in what they could see of the planets. But planetary astronomy soon entered a new golden age when spacecraft began to leave Earth.

Rocketry had developed rapidly during the century, propelled by the ill winds of international dissension and war. Even so, the American public was stunned and not a little frightened when the Soviet Union launched a satellite in 1957: the beeping, beachball-size Sputnik 1. "We knew they were going to do it," exclaimed Werner von Braun, the United States' prized rocketry expert. "For God's sake, turn us loose and let us do something. We can put up a satellite in sixty days." It was more like 16 months before the U.S. managed it successfully, but with the launch of Explorer 1 in 1958 the space race was officially on. • Driven by Cold War competition, early spaceflight was a mixture of hasty failures and a few notable successes. The U.S. Explorer 1 craft hit the jackpot right away when its cosmic-ray detector found belts of radiation around Earth, now known as the Van Allen belts after the detector's designer. The Soviet Union's Luna missions and the U.S. Ranger missions endured a number of launch failures and crashes throughout the 1960s, but Luna 3 took the first pictures of the moon's far side. Following President Kennedy's announcement that the U.S. would put a man on the moon within the decade, the Apollo program resulted in one of the high points in human exploration when it put not one, but two men on the moon in 1969, with ten more to follow before the program ended in 1972. Robotic missions took over exploring the more remote planets, eventually ranging as far as the solar system's outer edge.

TRAVEL TIMES (from Earth orbit via a Hohmann transfer)
- **Mercury:** 105 days
- **Venus:** 146 days
- **Mars:** 259 days
- **Jupiter:** 997 days
- **Saturn:** 2,214 days
- **Uranus:** 5,834 days
- **Neptune:** 11,200 days
- **Pluto:** 17,000 days

SKYWATCH

* More than 9,000 pieces of debris orbit Earth. Most comes from exploded Russian and U.S. satellites. New pieces are continually created as old pieces collide.

AMAZING FACT Helios solar probes reached speeds of 250,000 kilometers an hour (155,000 mph) in their closest approach to the sun.

1957
The Russian spacecraft Sputnik 1 becomes the first artificial satellite.

1965
Russian Venera 3 is the first spacecraft to land on another planet.

1969
American astronauts Neil Armstrong and Buzz Aldrin are first to walk on moon.

1976
Viking landers touch down on Mars and return data for several years.

1990
Hubble Space Telescope is launched into orbit.

2005
Huygens probe lands successfully on the surface of Saturn's moon Titan.

Sputnik 1 was only 58 centimeters (23 in) across (opposite). Twelve years after Sputnik, Apollo 11 astronauts witnessed earthrise over the lunar surface (above).

THE SPACE AGE: MERCURY TO NEPTUNE

NASA's Voyager 1 and Voyager 2 are identical twin spacecraft designed to visit the outer solar system and beyond. Launched in 1977, they were still working well more than 30 years later.

Aside from the moon, most early exploration targeted Earth's nearest neighbors, Venus and Mars. The U.S. Mariner 2 spacecraft flew past Venus in 1962 and returned information about its atmosphere and torrid temperatures. The Venera series, launched by the U.S.S.R., had early failures but then managed to send probes into the sulfurous Venusian atmosphere. In 1970, Venera 7 became the first craft to land successfully on another planet's surface, and its successors lasted just long enough in the intense pressure to send back the first pictures of Venus's dry, fractured surface. En route to Mercury, in 1974 Mariner 10 photographed Venus's atmosphere in ultraviolet light, confirming its circulation.

Spaceflight has always been a risky business. In the first ten years of U.S. spaceflight, 101 out of 164 launches failed. Over the years, missions to Mars in particular have alternated between rousing success and crushing failure, with only 26 out of 43 launched worldwide by 2009 being completely successful. In referring to the Mars failure rate, NASA scientists sometimes jokingly invoke the "Mars curse" or even attacks by the "Great Galactic Ghoul."

Despite the ghoul, early successes for Mars exploration included the Mariner missions of the 1960s and '70s. Mariner 9, in particular, was fruitful, even though it reached Mars in 1971 just as a dust storm veiled the entire planet. When the dust settled after a month, the orbiter was able to map the planet's surface for the first time, revealing its enormous volcanoes and extensive canyons. NASA followed up with the Viking 1 and Viking 2 landers, which touched down on opposite sides of the planet in 1975. The landers conducted several experiments to test for evidence of life in the Martian soil. At first, the results looked positive, to the great excitement of observers. However, subsequent analysis showed that the findings probably came from inorganic chemical reactions (see p. 114).

HEADING OUTWARD

By the 1970s, scientists were looking toward the outer solar system as well. Pioneer 10, launched in 1972, was the first spacecraft to fly through the asteroid belt to the far planets. It reached Jupiter in 1974 and collected close-up images of the planet and its moons, as well as taking measurements of its magnetosphere and atmosphere. Pioneer 11, launched in 1973, picked up momentum in a 1974 flyby of Jupiter and skimmed past Saturn in 1979. It took the first close-up images of the planet and discovered a new ring.

Both Pioneers carried gold-anodized aluminum plaques attached to their antenna support struts. Engraved in the metal are male and female human figures, a drawing of the solar system with distances given in binary code, and symbols of other scientific basics that would presumably be familiar to spacefaring aliens. The spacecraft are now carrying these plaques out of the solar system. Contact with Pioneer 11 was lost in 1995, and Pioneer 10's signals finally faded away in 2003 when it was 7.6 billion miles from Earth and on its way toward the star Aldebaran.

From the late 1970s to the 1990s, increasingly sophisticated spacecraft and telescopes began looking more closely at some of our planetary companions. Mercury had no more visitors, but Venus, concealed in clouds, continued to attract attention.

CLOSE-UPS OF VENUS AND MARS

NASA's Pioneer Venus mission reached the planet in 1978. It consisted of two spacecraft: an orbiter and a multiprobe. The orbiter measured the fierce winds in Venus's upper atmosphere and used radar to pierce the clouds and map much of the planet's surface; the multiprobe launched a tiny armada of five instrument packages that took readings on the different layers of the atmosphere as they descended. In 1983, the U.S.S.R.'s Venera 15 and 16 orbiters also used radar to map the surface from a polar orbit. Not to be outdone, the U.S. returned to Venus in 1990 with Magellan. Over four years, the orbiter made the most detailed maps yet of the planet, surveying 98 percent of its surface and revealing features as small as 300 meters (1,000 ft) in diameter (see p. 92).

After the Viking missions of the 1970s, Mars saw no more spacecraft until the late 1990s. In 1997, NASA's Global Surveyor and Mars Pathfinder both reached the planet successfully. The Global Surveyor, an orbiter, scanned the planet in high resolution, not only creating spectacular maps, but also discovering gullies and deltas that suggested that water once flowed on the surface. Mars Pathfinder, a lander, parachuted and bounced to safety on giant air bags, and charmed the public by deploying a bread box-size mini-rover onto Mars's surface. The rover, Sojourner, trundled around the lander and analyzed soil and rocks that were given nicknames like Yogi and Scooby-Doo.

VOYAGERS

Perhaps the most spectacular missions of the era, and certainly the longest lasting, were NASA's Voyager 1 and Voyager 2 flights. The Pioneer missions had shown scientists that spacecraft could safely navigate the debris-strewn route to the outer planets. The Voyager designers then took advantage of a rare alignment of Jupiter, Saturn, Uranus, and Neptune between 1976 and 1978 to send the two craft past multiple planets. Voyager 1 visited Jupiter and Saturn. Voyager 2 visited both, then went on to Uranus and Neptune as well, in a "Grand Tour." It is still the only craft to have visited these far-distant worlds.

The Voyagers discovered a thin ring around Jupiter and two new moons (Thebe and Metis), and thrilled planetary scientists with close-up views of Jupiter's mottled moon Io shooting volcanic material far into space. Voyager 1's Saturn flyby revealed five new moons and a new ring. Titan was found to have a thick nitrogen atmosphere. Voyager 2 reached Uranus in 1986, discovering ten new moons, details of its thin rings, and a large magnetic field. The intrepid craft then swung on to Neptune, more than four billion kilometers (2.5 billion mi) from Earth, reaching it in 1989. Skimming over its north pole and past its largest moon, Triton, Voyager 2 photographed Neptune's twisted rings and revealed details of its stormy surface, including a Great Dark Spot and a smaller, fast-moving cloud nicknamed Scooter. Voyager 2 also discovered six new Neptunian moons. Both Voyagers then headed out of the solar system on separate trajectories and were still on the move 30 years after their launch.

CHIPS AND TELESCOPES

Planetary discoveries were not limited to traveling spacecraft. By the mid-20th century, astronomers began observing distant stars and galaxies at wavelengths outside the visual range. In 1955, U.S. astronomers Bernard Burke and Kenneth Franklin had turned a large radio antenna toward the Crab Nebula when they began receiving bursts of radio waves from an unknown emitter. Plotting the bursts on a map, they were amazed to see that the source must be the planet Jupiter, sending out signals like a star. Observers on the ground as well as instruments aboard spacecraft began to scrutinize planets in a range of wavelengths, revealing features hidden to visible light.

The late decades of the 20th century saw big improvements in telescope technology and new findings, thanks to computer technology. Charge-coupled devices, or CCDs, began to replace photographic plates in most telescopes (and eventually even in snapshot and home video cameras). Developed in the 1970s, these silicon wafers are divided into pixels—up to several million of them—that register an electric charge when struck by light. Far more efficient than film, CCDs pick up more detail and see more distant objects much faster than older cameras.

CCDs were built into the Wide Field/Planetary Camera of the Hubble Space Telescope when it was launched in 1990. Despite some initial mirror problems that prompted "Hubble Trouble" headlines, with repairs the space telescope became a superb addition to the telescopic arsenal and a public favorite. Most of its observations are directed at distant stars and galaxies, but over the years it has also turned its attention to planets and plutoids from Venus to Eris, capturing storms on Jupiter and the hearts of comets, and discovering two tiny new moons of Pluto.

The Hubble Space Telescope captured the first recorded collision of two solar system bodies. In 1993, a group of astronomers that included Eugene and Carolyn Shoemaker and David Levy were surprised to spot the broken fragments of a comet approaching Jupiter. Comet Shoemaker-Levy 9, as it was soon named, had apparently been torn apart into fragments by Jupiter's gravitational forces during an earlier close approach to the massive planet, and now it was about to take its revenge by smashing into the planet's surface like a series of bombs. For six days beginning on July 16, 1994, the icy chunks bombarded Jupiter repeatedly. At least one struck with the energy of six million megatons of TNT (more than 600 times the explosive power of all of Earth's weaponry); the resulting fireball was more than 3,000 kilometers (1,800 mi) high. In the wake of the impacts, huge, dark clouds spread out through the planet's atmosphere, some of them larger than Earth.

THE VOYAGER GOLDEN RECORD

In the event that Voyagers 1 or 2 should encounter intelligent life in their long journeys, a committee headed by astronomer Carl Sagan prepared a welcome: a 12-inch gold-plated phonograph record, packaged with a cartridge and needle, that contained 115 images and a variety of recorded sounds from Earth. Drawings on the cover demonstrate how to play the record, map our solar system relative to 14 pulsars, and depict a hydrogen atom in its two lowest states, among other images. Sounds include music from different eras, recordings of surf and animal calls, and greetings in 55 languages from Akkadian to Wu. A spot of radioactive uranium was also electroplated onto the record cover; by measuring its decay, an extraterrestrial might be able to deduce the time elapsed since the spacecraft was launched.

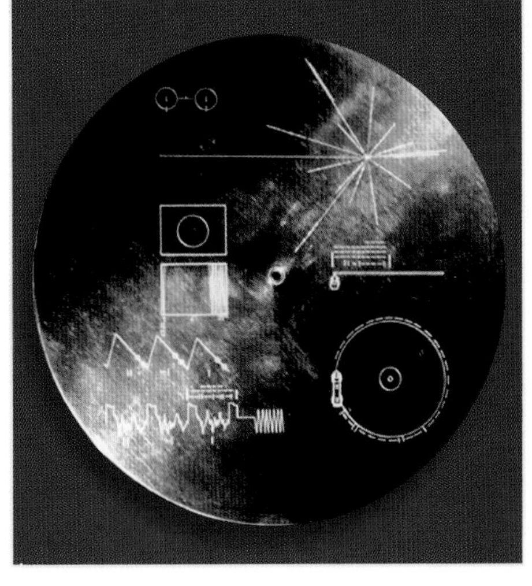

THE SPACE AGE: THE NEW MILLENNIUM

Planetary exploration in recent years has branched out to include ever increasing scrutiny of the sun, comets, and asteroids, as well more familiar destinations such as Mars and Saturn. The level of involvement and cooperation from other countries has also widened: the European Space Agency (ESA) and agencies from countries such as China, Japan, and India have all joined the hunt. Guided by teams of scientists and mission specialists at Internet-linked facilities around the world, spacecraft and rovers today are truly Earth's emissaries, using advanced hardware and software to investigate the solar system.

THE SUN AND SMALL BODIES

Spacecraft have been studying the sun since the early Pioneer missions, but solar ventures picked up in the 1990s. Among those still operating are ESA's Ulysses, which circles from the sun to Jupiter and back, Japan's Yohkoh, an x-ray and gamma-ray observatory, and the ESA/NASA Solar and Heliospheric Observatory (SOHO), which examines, measures, and records data on everything from the solar interior to the solar wind. The solar system's smaller travelers have secured their share of attention as well. Comet Halley's visit in 1986 was met by a host of spacecraft, including the ESA's Giotto. NASA spacecraft NEAR Shoemaker managed to land on asteroid 433 Eros in 2001. Several other missions checked out comets and asteroids from 2002 on, seeking clues to the early history of the solar system as well as information that might help Earth avoid a catastrophic impact in the near or distant future.

MERCURY AND VENUS

In 2008, the most neglected terrestrial planet received its second visitor since 1974's Mariner 10. NASA's MESSENGER mission began the first of three Mercury flybys in January 2008, returning crisp images to earthbound viewers of portions of the planet that Mariner 10 never saw. Information from the flybys confirmed the importance of volcanic eruptions in Mercury's past and also helped to fill in the picture of the planet's extremely thin atmosphere.

Meanwhile, the ESA sent its compact orbiter, Venus Express, to Earth's nearest neighbor in 2005 with the goal of learning more about the planet's thick atmosphere and gaining additional insight into the greenhouse effect. The craft has returned dramatic images of a hurricane-like vortex at Venus's south pole as well as data about the planet's winds and climate.

CROWDED MARS

Mars's close approach to Earth early in the decade unleashed a fleet of spacecraft from various countries. Not all of them made it successfully, proving, perhaps, that the Mars curse is still in operation. In 1999, NASA's Mars Polar Lander reached the planet before communication was lost; presumably it crashed. But when the agency's Mars Climate Orbiter also smashed into the planet in 1999, the cause was not galactic gremlins but a deeply embarrassing, elementary error: The Lockheed Martin team that helped build the craft provided navigation commands in standard English units, while the NASA team used metric. The ESA's

Parched, rocky, and windblown, the surface of Mars looks like a lonely desert in this 360-degree view inside Gusev Crater. Made from hundreds of images captured by the exploration rover Spirit in 2005, the panorama includes the rover's solar panels in the foreground.

first mission to Mars, the economical Mars Express, reached the red planet successfully in 2003 and has been returning valuable information about the atmosphere and surface. However, its Beagle 2 lander, which was intended to sniff out life, disappeared on descent and is presumed to have crashed.

The Mars curse didn't affect NASA's two Mars exploration rovers, which landed in January 2004. The two spacecraft landed spectacularly well, slowing from 19,000 kilometers an hour (12,000 mph) to 19 kilometers an hour (12 mph) in just six minutes to land safely on the rocky surface.

Spirit and Opportunity, each the size of a small dune buggy, contained stereo cameras as well as instruments for digging into and analyzing Martian soil. Like most current Mars missions, their main goal was to find evidence of past liquid water, and therefore, the possibility of life. Among other discoveries, Spirit found a patch of pure silica in the soil, a compound that on Earth is found in hot-springs environments. The tantalizing discovery was complemented by the 2009 announcement that observatories on Mauna Kea had detected a burst of methane gas in Mars's atmosphere, promising either geologic or biological activity.

NASA's Phoenix lander, touching down safely near Mars's north pole in 2008, began making up for the loss of the Polar Lander by uncovering and analyzing water ice, confirming the presence of frozen H_2O just under the surface. Its mission ended in November 2008, as dwindling polar sunlight could no longer keep its solar-powered systems alive.

BACK TO SATURN

Seventeen countries and several agencies, including NASA and the ESA, collaborated in the Cassini-Huygens mission to Saturn and its moons. Launched in 1997, it reached Saturn in 2004. On arrival, the Cassini craft launched its Huygens probe toward Titan; early in 2005, the probe parachuted onto that moon's surface, collecting information about Titan's atmosphere as it dropped (see p. 150). Seen for the first time, Titan's landscape seems to contain rivers and lakes, possibly of liquid hydrocarbons. Meanwhile, the Cassini spacecraft has been repeatedly circling about in the Saturnian system, viewing the planet in various wavelengths of the electromagnetic spectrum and discovering a tiny moon. It also continues to make surprising discoveries about Saturn's satellite, Enceladus, which appears to shoot plumes of water from its south pole. Although the primary mission officially ended in 2008, Cassini remained in good working order and is still transmitting valuable information in an extended mission.

VOYAGERS LEAVE THE PREMISES

The amazing long-running Voyager missions are saying goodbye to the solar system. In 2004, Voyager 1 reached the heliosheath, the boundary where the sun's solar wind meets the thin gas of interstellar space. Voyager 2, moving in a different direction, reached the boundary in 2007. Both explorers were more than twice the distance of Pluto at the time. After traveling through the heliosheath for perhaps seven to ten years, the two craft will officially depart the premises and continue on to the stars.

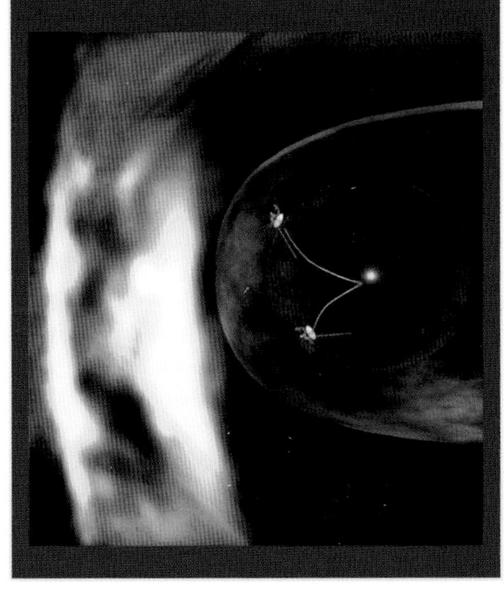

FOR MORE IMAGES FROM THE SURFACE OF MARS GO TO HTTP://MARSROVER.NASA.GOV/HOME/

THE SPACE AGE: NEW WORLDS

Even as spacecraft were getting up close and personal with the sun and planets, astronomers at earthbound telescopes were making discoveries that dramatically revised our notions of what a planet really is.

DWARF PLANETS

In the late 1980s, astronomers David Jewitt and Jane Luu of the University of Hawaii began searching for the theoretical, but never observed, Kuiper belt objects. Using CCDs attached to Hawaii's Mauna Kea telescope, they found the first one in 1992. At an estimated distance between 39 and 44 AU, the orbiting body designated 1992 QB1 was well past the orbit of Pluto, the most distant solar system object ever seen.

Soon astronomers picked up more such objects in their sights, dim and distant as they were. The total mass of the Kuiper belt may be no more than 0.1 Earth mass, but it is literally broken down into millions of chunks of rock and ice. Some of these fragments, knocked out of orbit by gravitational disturbances, will form the nuclei of comets as they speed toward Earth. Others are big enough to own their own moons.

By 2006, astronomers added about 1,000 such bodies to the list. Perhaps the most significant was UB313, now named Eris, which was spotted by astronomer Mike Brown of California Institute of Technology in 2003. With a diameter of roughly 3,000 kilometers (1,850 mi), it was larger than Pluto. It was also extremely remote, its eccentric orbit taking it 97 AU from the sun at the farthest point in its 560-year orbit.

These discoveries revived long-standing and sometimes contentious debates about the definition of "planet" and the status of Pluto in particular. It was becoming clear that Pluto was no different from a number of other rocky, icy bodies circling past Neptune. Amid some controversy and sorrow from dedicated Pluto fans, in 2006 the International Astronomical Union devised a definition for planet that excluded Pluto, since it did not clear its orbital lane of debris (see p. 169). The IAU also added a runner-up category, dwarf planet, that encompassed Pluto, Ceres (formerly the largest asteroid), and Eris.

Since then, the dwarf planet category has been divided into two subsets—dwarf planets and plutoids (dwarf planets that orbit beyond Neptune). By 2008, the dwarf planet category contained one member, Ceres, whereas Pluto, Eris, Makemake, and Haumea were ranked as plutoids. Other sizable trans-Neptunian bodies will undoubtedly be added to the plutoid list as well.

TELESCOPES OF THE FUTURE

Conventional telescopes, even the Hubble Space Telescope, have limits when it comes to distant Kuiper belt objects and exoplanets. Telescope designers constantly seek to increase the light-collecting area and accuracy of their scopes, whether on Earth or in space. Coming soon is NASA's James Webb Space Telescope, an orbiting infrared observatory with a 6.5-meter (21.3-ft) mirror, due to be launched in 2013. Other scientists have suggested orbiting arrays of satellite telescopes, linked by radio—NASA's proposed Terrestrial Planet Finder is one. Linked radio telescopes (below) could also fare well on the moon. And liquid-mirror telescopes have their proponents. With lenses formed from dense, spinning liquids, these telescopes might be suited for the moon, where a steady surface, lack of atmosphere, and shielding from radio interference could give them exceptional vision.

EXOPLANETS

Even while Kuiper belt objects were redrawing the limits of the solar system, far more distant worlds were coming to light. For centuries, astronomers had theorized that planets might orbit other stars. But even the best 20th-century telescopes had no hopes of spotting them visually; not only are the nearest stars light-years away, but any planets in other systems would be relatively tiny, dark, and hidden in their sun's glare.

In the 1990s, astronomers began to get around this problem by observing extrasolar planets indirectly via the parent star's gravitational wobbles. Orbiting planets tug a star this way and that, and scientists analyzing the movement can estimate how many planets circle the star, how massive they are, and their approximate orbits. Surprisingly, the first exoplanets were found circling a pulsar, an extremely dense neutron star. The intense radiation broadcast by such a star makes it unlikely that life could survive on those planets. But since 1992, more than 300 other extrasolar planets have been discovered, some found by measuring the minute dimming of light as the planet crosses, or transits, directly in front of the star. Many are massive, Jupiter-like bodies, probably easier to spot because of their gravitational oomph. But they include multiple-planet systems, planets around giant stars and brown dwarfs, and even one possible free-floating, orphaned body. Some massive planets orbit remarkably closely to their suns, while others circle at a more sedate distance comparable to the orbits of Mars or Jupiter. And at least one system shows evidence of an asteroid belt close to its parent star.

These newest discoveries are immensely helpful in answering some of the oldest questions about our solar system: Just how did it form? How did giant planets and solar system debris interact? In general, both the existence of Kuiper belt bodies and the evidence of exoplanets support the prevailing condensation theory of solar system formation. The new worlds also lend credence to the theory that large gas giant planets such as Jupiter may have migrated inward from their original spots in the solar nebula, disrupting the system and ejecting smaller bodies from their places (see p. 137).

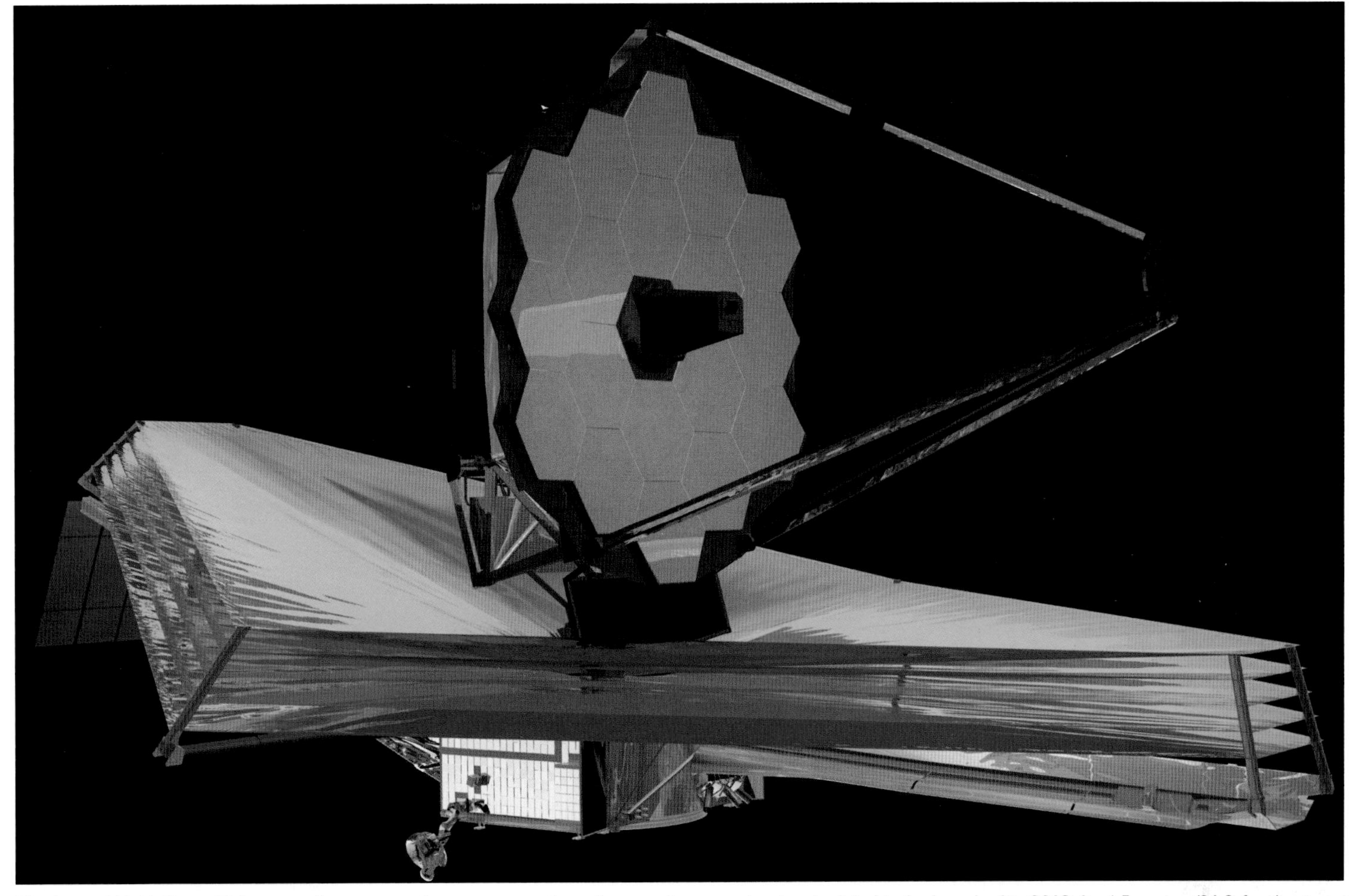

The James Webb Space Telescope, forerunner of a new generation of space observatories, is scheduled to be launched in 2013. Its 6.5-meter (21.3-foot) mirror will pull in light for examination mainly in the infrared range. A sunshield the size of a tennis court will unfold once the telescope is in outer space.

LIFE ITSELF

The discovery of exoplanets has also given new energy to the endless search for astronomy's holy grail: life on other worlds. In our own solar system, the history of liquid water on Mars and the probable existence of liquid water under the surfaces of Jupiter's and Saturn's moons has strengthened the case for finding biological activity even in a wide range of environments. Several missions are under way or in development to look specifically for Earthlike planets around other stars, particularly for Earthlike planets in the habitable zone where liquid water is possible.

Detecting a planet the size of our own Earth, in an orbit relatively close to the bright glare of its parent star, is extraordinarily difficult, however scientists are increasingly optimistic that such a world will be found soon. The best candidate for Earthlike status so far orbits the dim star Gliese 581, 20 light-years away. Five times as massive as Earth, the extrasolar world orbits in the habitable zone.

Proposed telescopes devoted to detecting life on other planets include the European Space Agency's Darwin, planned as a fleet of four or five telescopes floating far beyond the moon at a stable orbital point. The plan is that Darwin would look for exoplanets in the mid-infrared wavelengths, the area where the contrast between the light of the parent sun and the reflected light from the planet is not as bad as in optical wavelengths. Meanwhile, NASA is developing a similar mission (and may eventually partner with ESA). Their Terrestrial Planet Finder might consist of two kinds of space-based observatories: a coronagraph taking in visible light and an array of infrared telescopes all working together.

Even ancient astronomers speculated about the possibility of discovering life on other worlds. In the 21st century, that possibility is very close to becoming reality.

DISTANCE VISION

A million miles here, a million miles there, and pretty soon you're talking about real distances. Astronomers are seeing farther than ever, including to:

- **Pluto's moons,** 39 AU, 4.4 billion km (2.7 million mi)
- **The Kuiper belt,** 30–50 AU, 4.5–7.5 billion km (2.8–4.6 billion mi)
- **Dwarf planet Eris,** 67 AU, 7.6 billion km (4.7 billion mi)
- **Gliese 876 exoplanets,** 945,000 AU, 143 trillion km (89 trillion mi)

THE SUN

THE SUN

By almost any measure, the sun *is* the solar system. As it formed from the gas and grit of the solar nebulae, the enormous sphere sucked in virtually all matter for billions of miles, ending up with more than 98.8 percent of the solar system's mass. The planets and rocky debris that remained in orbit were little more than the dust left behind when the gravitational broom had finished its work. The center of mass for the whole solar system—the barycenter, around which the sun and the planets all revolve—lies within the sun's own atmosphere. As the planets and smaller bodies circle the sun, their ever changing gravitational web tugs the sun back and forth, making it wobble around the solar system's central point.

The sun is a fairly typical star, middling in size, big enough to burn steadily for ten billion years, but not so massive that it will run through its fuel and explode as a supernova. From more than 149 million kilometers (92 million mi) away, it heats and brilliantly illuminates Earth, supplying the energy that makes life possible. Solar radiation is about the closest thing our planet has to a free lunch. Its power drives our winds and waters and its light gives us the basic divisions of day and night that govern our body's rhythms.

It's hardly surprising that almost all of the world's cultures have revered the sun and made it a central part of their mythology. Ancient Egyptians honored it as the chief of the gods, Re or Amun-Re, sailing the skies from east to west during the day and making the dangerous journey through the underworld at night to be born again at dawn. To the Inca, the sun was their ancestor Inti, a benign deity whose consort was the moon, his sister. Early farmers and priests tracked the sun's seasonal shifts across

1375 B.C.
Babylonians record first solar eclipse.

A.D. 968
Sun's corona first observed during solar eclipse.

1543
Copernicus publishes his theory that Earth revolves around the sun.

1715
The sun's corona is first illustrated.

1845
First photograph of the sun is taken.

1870
Jonathan Lane publishes *On the Theoretical Temperature of the Sun.*

the sky. Tombs and temples aligned with the solstices; ancient chronicles listed and interpreted each ominous eclipse.

As astronomy developed, the sun was the key to understanding the geometry of space. When Copernicus made the great non-intuitive leap to putting the sun at the center of the solar system, planetary motions made sense for the first time. Years of dogged observations by astronomers both professional and amateur began to unveil some of the sun's secrets. Sunspots played a surprisingly large role in these discoveries. As soon as the telescope was invented in 1608, a controversy arose over their nature. Some astronomers, such as German observer Christoph Scheiner, argued that they were the dark silhouettes of planets crossing the sun, or else clouds in the sun's atmosphere. Others, including the era's preeminent astronomer, Galileo Galilei, believed that they were part of the sun itself. In 1613, Galileo demolished Scheiner's arguments in his *Letter on Sunspots,* which showed that the spots traveled until they became truncated and then vanished at the sun's edge—proof that the sun itself was rotating and that the spots were part of it. Nineteenth-century observations revealed sunspots followed a regular cycle, possibly tied to Earth's climate, and also showed that they rotated at different speeds at different solar latitudes, indicating that the sun was made of gas. Associated with solar flares and geomagnetic storms, the spots also provided the first clues to the sun's magnetic nature.

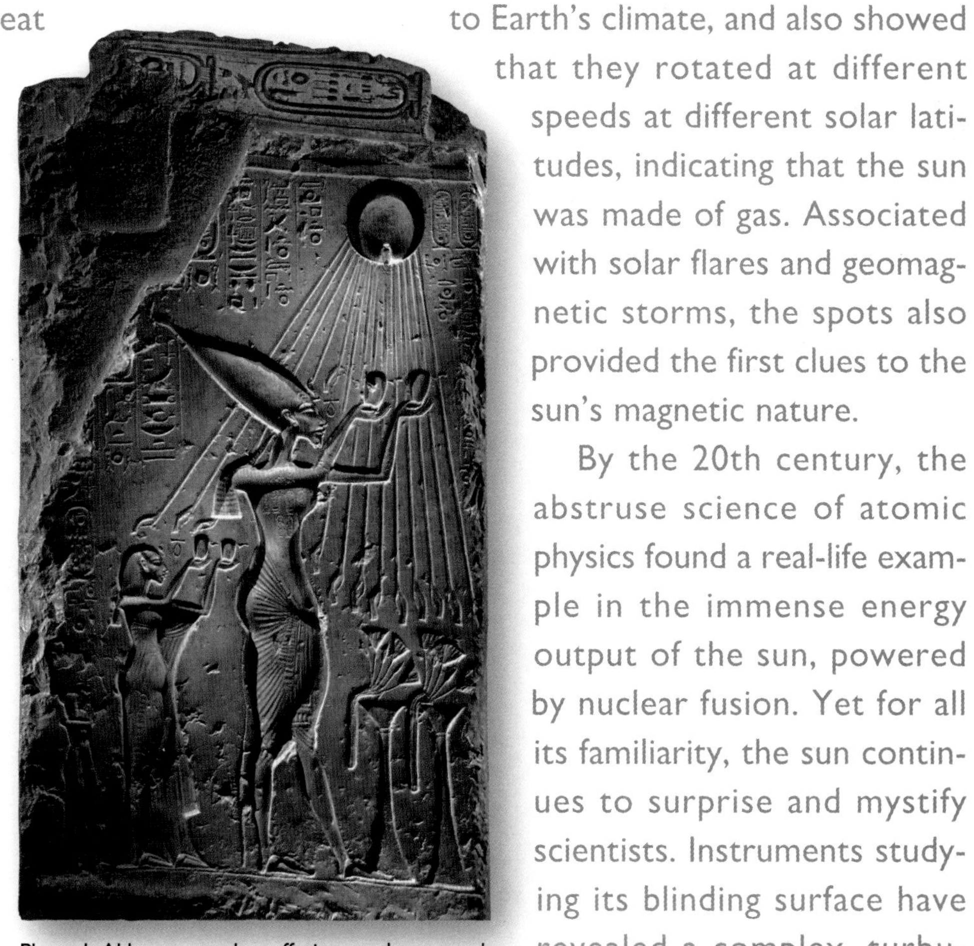
Pharaoh Akhenaten makes offerings to the sun god.

By the 20th century, the abstruse science of atomic physics found a real-life example in the immense energy output of the sun, powered by nuclear fusion. Yet for all its familiarity, the sun continues to surprise and mystify scientists. Instruments studying its blinding surface have revealed a complex, turbulent, and mutable star. Flares explode like planet-size bombs from its surface. Fiery winds tear through its atmosphere, obeying laws we don't yet understand. Far from the star's central heat source, its thin outer atmosphere registers temperatures of millions of

1930
Bernard Lyot invents coronagraph, allowing observations of the sun's corona.

1938
Physicists Bethe, Critchfield determine nuclear reactions make sun shine.

1951
Biermann predicts the existence of solar wind.

1959
Mariner 2 detects solar wind.

1981
Hale invents spectroheliograph, allowing pictures of sun at any wavelength.

1983
SpaceLab provides high-resolution pictures of the surface of the sun.

degrees. A deep-seated magnetic field twists through the sun's enormous body and flows with the solar wind into interstellar space.

In fact, Earth and all of its solar system siblings live within a spectacularly massive solar cocoon, a heliosphere extending far beyond Pluto. The sun's tenuous outer atmosphere, its corona, engulfs the planets through the solar wind. Inside its boundaries, we are bathed in a wide variety of radiation, charged particles, and magnetic storms. Countless trillions of neutrinos, invisible, infinitesimal particles with no electric charge of their own, pour out from the sun's core every second and pass through our bodies and Earth unnoticed.

This intimate connection between the sun and Earth has prompted scientists to turn not only earthbound observatories but also space-based telescopes toward our huge star. The U.S. National Solar Observatory dedicates several high-altitude telescopes to studying the sun, including one on Kitt Peak in Arizona and another on Sacramento Peak in New Mexico. A network of observatories ranging from India to Hawaii to Chile—the Global Oscillation Network Group, or GONG—track the sun's bell-like vibrations for clues to its internal structure. But much of the recent research has been conducted in space, bringing breakthroughs in solar research as well as a portfolio of spectacular images in a range of wavelengths.

Among the most successful of the recent space-based telescopes is SOHO, the Solar and Heliospheric Observatory. The result of a collaboration between the European Space Agency (ESA) and NASA, the spacecraft was launched in 1995 into an orbit about 1.5 million kilometers (almost 1 million mi) sunward from Earth. It orbits the sun in lockstep with our planet, able to record observations during an eternal day. Hundreds of scientists around the world employ its 12 instruments to ferret out answers to some of the sun's big questions: What is its internal structure? Why is the solar corona so hot? What propels the solar wind?

Among the orbiting observatory's groundbreaking discoveries are the structure of the sun's convection zone—the bubbling region just below its surface—and such dramatic and violent phenomena as solar tornadoes, spinning gas twisters with wind speeds up to 500,000 kilometers an hour (310,000 mph). NASA's

ASTRONOMICAL UNITS
A SOLAR SYSTEM STANDARD

Astronomers often refer to distances within the solar system in terms of astronomical units, or AU, the mean distance from Earth to the sun, formally defined as equivalent to 149,597,870 kilometers (92,955,730 mi). In terms of AU, then, the planets orbit at the following average distances from the sun: Mercury 0.39 AU, Venus 0.72 AU, Earth 1 AU, Mars 1.52 AU, Jupiter 5.2 AU, Saturn 9.54 AU, Uranus 19.19 AU, Neptune 30.07 AU. One light-year is equal to 63,240 AU. (If 1 AU equals 1 inch, then 1 light-year equals 1 mile.)

controllers almost lost the spacecraft when, in 1998, errors from ground crew sent the craft spinning out of control. Signals ceased and the spacecraft appeared to be lost. But after a ground-based telescope spotted it tumbling far out in space, controllers were able to signal it to turn its solar panels back toward the sun, eventually reprogramming it to operate without gyroscopes. The dramatic recovery brought the telescope back online and extended its lifetime to the present.

A similar mishap, this one unfortunately unfixable, put an end to the Japanese solar mission Yohkoh. Launched in 1991 and designed to study some of the sun's most explosive artifacts, such as solar flares and coronal mass ejections, the spacecraft also sent home stunning images of the sun taken at x-ray wavelengths. Unlike SOHO, Yohkoh orbited the Earth and thus was knocked out of commission in 2001 by a solar eclipse, which shut down power to its instruments.

Also in Earth orbit, but surviving so far, is NASA's little spacecraft TRACE (Transition Region and Coronal Explorer), which studies the behavior of the sun's looping magnetic fields and its mysteriously superheated corona. Joining it is the aptly named STEREO, actually a pair of observatories ahead of and behind the Earth in its orbit that produces three-dimensional images of solar storms. Japan's Hinode is also scrutinizing the sun's magnetic fields and solar eruptions.

Recent missions have focused on solar magnetism and storms not only because they yield valuable clues to the sun's structure, but because they have major effects on civilization's electronic grid. Satellites, observatories such as the Hubble Space Telescope, and even the International Space Station are also vulnerable to damaging blasts of radiation and particles. Using SOHO and other instruments, the National Oceanic and Atmospheric Administration (NOAA) tracks space weather just as forecasters on Earth track terrestrial storms.

Solar storms aside, even the sun's daily dose of light is not as constant as we might believe. Year by year, the sun is growing brighter—a cheery image until you realize that eventually the sun's increasing heat will boil away our oceans. This will occur long before the sun's expanding body engulfs the Earth, leaving it a charred, wobbling cinder sinking through the solar atmosphere. Humans have a very real, if very long-term, stake in learning about, and understanding our parent star.

EINSTEIN'S LENS

BENDING SPACE AND TIME

One of the fundamental predictions of Albert Einstein's general theory of relativity states that mass creates a curved field of space and time around it. The effect is difficult to measure with small masses, but the colossal sun provides a natural—and relatively nearby—laboratory. According to Einstein's theory, starlight passing close to the sun will be bent around its mass. However, the sun's brilliance overpowers starlight in the daytime sky, so the effect cannot ordinarily be seen.

In 1919, the enormously talented British physicist Arthur Stanley Eddington traveled to Príncipe Island, an island off the west coast of Africa, to photograph a solar eclipse, which took place against the background of the dense Hyades cluster of stars. Later, he photographed the same cluster during the night. Sure enough, the stars had shifted a tiny bit between the two photographs, showing that their light had been bent around the mass of the sun and proving Einstein's theory. Eddington was the first to write about the theory of relativity in English, and one of his many publications, *The Mathematical Theory of Relativity,* published in 1923, was lauded by Einstein.

ORIGINS OF THE SUN
GALAXIES & STARS

On a clear night without light pollution, you can see the Milky Way. The center (hidden behind light-years of dusty gas) lies in the constellation Sagittarius. Telescopes reveal individual stars, but the naked eye is best for viewing the whole expanse.

The sun may rule our solar system, but from a more distant perspective, it is merely one typical star amid an unimaginable number of others. Formed some 13.5 billion years ago (bya) from the primordial gases of the early universe, the first stars collected themselves into galaxies and clusters of galaxies throughout space (just how they did this is currently a matter of scientific debate). Astronomers estimate that the universe now contains roughly 10^{24} stars organized into billions of galaxies. Our own galaxy, the Milky Way, is a barred spiral—a spiral-shaped galaxy with a cylindrical nucleus at its core, from which arms extend in a sweeping circular pattern. The Milky Way holds billions of stars and planets, interstellar dust and gas, and the mysterious substance known as dark matter. The sun lies about two-thirds of the way out one of the Milky Way's spiral bands, near the galaxy's Perseus arm, some 26,000 light-years from the dense galactic center. With the rest of the galaxy, it rotates at 820,000 kilometers an hour (510,000 mph), making one complete circuit every 230 million years. We can't feel this motion, but we can see the Milky Way on a dark, clear night.

THE MILKY WAY
- **Type of galaxy:** barred spiral
- **Diameter:** ~100,000 light-years
- **Disk thickness near core:** 13,000 light-years
- **Disk thickness near edge:** ~1,000 light-years
- **Number of stars:** 200 billion to 400 billion
- **Age of oldest stars:** 13.4 billion years
- **Stellar density:** 1 star per 125 cubic light-years
- **Number of galaxies in local group:** 33
- **Nearest star to our sun:** Proxima Centauri, 4.22 light-years

SKYWATCH

- On a clear night without light pollution, see the irregular band of the Milky Way, its center in the constellation Sagittarius. The naked eye is best to see the whole expanse.

AMAZING FACT If the sun were close to the galaxy's center, stellar radiation would make life impossible.

10^{-6} sec	3 sec	10,000 years	300,000 years	~13.2 bya	4.6 bya
Just after Big Bang Basic forces and subatomic particles of the universe appear	Protons and neutrons form	Era of radiation	Era of matter; neutral atoms form	Milky Way begins to form	Our sun forms

Galaxy NGC 1300 (opposite) is a barred spiral like our own Milky Way. The star cluster NGC 346, in the Small Magellanic Cloud, is a star-formation region (above).

ORIGINS OF THE SUN: BIRTH OF A STAR

Our sun isn't a member of the Milky Way's first generation. The earliest stars formed from condensing hydrogen and helium gases roughly 13.5 billion years ago. Some faded away as white dwarfs, while others, more massive, exploded as supernovae. The tremendous temperatures and pressures involved in these stellar deaths created heavier elements—carbon, nitrogen, oxygen, and others—that were flung into space as each star burst apart in what must have been spectacular light shows. Eventually, the gas and dust of the interstellar medium became seeded with these heavy elements. And although they represented only a small amount (just one percent) of the universe's visible matter, it is a crucial one percent for those of us now living on rocky planets.

Such a supernova may have jump-started solar formation in our own system. Our solar neighborhood began as a solar nebula, a cold cloud of gas, dust, and ice several light-years in width. Around five billion years ago, something triggered the collapse of this cloud. Scientists speculate that the collapse could have been due to an uneven distribution of matter, or maybe a collision with a passing gas cloud, but it is also quite possible that it was caused by the shock wave from a nearby supernova. Matter in the spinning solar nebula began to fall toward the center, forming a dense, spherical core of gas, a protostar, heated by the increasing pressures of the accumulating gas. After about 100,000 years, the sun became what astronomers call a T-Tauri star (named after the prototype in the Taurus molecular cloud and characterized by erratic changes in brightness). It was bigger and brighter than it is today, surrounded by a planetary disk knotted with condensing protoplanets.

Pressure built up slowly within the T-Tauri star's core until the temperature reached ten million kelvins, hot enough to fuse hydrogen into helium. The sun became a true star, broadcasting radiation and charged particles across its circling planets and asteroids. And now, 4.6 billion years after its birth, we are roughly halfway

Artwork depicts five stages in the life and death of a solar system: from top, condensation of a gas and dust cloud; formation of the solar nebula; accretion of the planets; today's middle-aged system; and old age, with the sun as a red giant.

through the sun's life cycle as our star's fusion consumes its huge quantities of hydrogen.

THE STELLAR FAMILY

How does our sun compare with other stars? We might say that it is a typical, but not average, star. Larger and brighter than most stars in our galaxy, it is nevertheless a common type, known technically as a G2V main sequence star.

What does this mean? Stars are classified by size, color, and temperature. Most are divided into seven spectral types, based on a chart known as the Hertzsprung-Russell diagram, which organizes stars according to luminosity (intrinsic brightness), surface temperature, and spectral class (color). The seven spectral classes are labeled O, B, A, F, G, K, and M (remember it, if you will, with the mnemonic "Oh, Be A Fine Girl/Guy, Kiss Me"). O-type stars are the brightest, hottest, and bluest, whereas M types are cool and red. The yellow-white sun fits into the middle. Its subclass, 2, means that it's relatively hot on the scale of 0 (hottest) to 9 (coolest). And the V indicates its luminosity, which is directly related to size—in the sun's case, again, middling. For historical reasons, all nongiant stars are known as dwarfs; our sun is known as a main-sequence dwarf.

The main sequence runs right down the middle of the Hertzsprung-Russell diagram from upper left to lower right. Main-sequence stars like the sun are in the prime of their life, currently powered by hydrogen fusion at their core. The category excludes protostars as well as brown dwarfs, protostars that were too small and cold ever to reach fusion. It also excludes stars at the end of their life, such as red giants and white dwarfs.

The sun's place in the stellar family and its ultimate fate are directly connected to its mass. For a star, mass is destiny. Protostars with masses up to about 75 times that of Jupiter never build up the core temperatures for fusion and remain brown dwarfs, failed stars. More massive stars will begin fusion and stay for a time on the main sequence, but the more massive they are, the faster they burn through their hydrogen and depart the main sequence for the red giant end stage of their existence. Those greater than eight solar masses (eight times the mass of the sun) will eventually compress their core materials into heavy elements, such as iron. When its core reaches a critical density, such a massive star collapses and then explodes as a supernova.

Neutron stars and black holes represent the extreme end stages of supermassive stars. When fusion ceases inside such a giant star, the outflowing radiation dies away and the overlying mass of the star collapses in on itself. Electrons and protons squeeze together to form a crushingly dense neutron core. If the core's mass is less than three solar masses or so, it remains behind as a neutron star while the rest of the star blows into space as a supernova. One teaspoon of its close-packed matter would weigh 100 million tons. If the remnant core is greater than three solar masses, its collapse proceeds beyond the neutron star stage. The stellar core becomes a black hole, an infinitely dense point. Within a certain radius of the black hole, known as the event horizon, no matter or energy can escape.

Astronomers have identified a number of black holes within our own Milky Way galaxy. The most spectacular is at the galaxy's center. The black hole there, known as Sagittarius A* (pronounced A-STAR), contains about four million solar masses.

No such fate is in store for our sun, however. Lodged securely in the stellar middle classes, the sun will burn for another five billion years or so. It will spend its declining years as a cooling, expanding red giant star and then as a white dwarf, a cold cosmic chip (see p. 75).

ELEMENTS IN THE SUN (BY MASS)

Element	Percent
Hydrogen	71%
Helium	27.1%
Oxygen	0.97%
Carbon	0.40%
Silicon	0.099%
Nitrogen	0.096%
Magnesium	0.076%
Neon	0.058%
Sulfur	0.040%
Iron	0.014%
Other	0.147%

SINGLE VERSUS BINARY STARS

The sun is an only child, a single star born from the solar nebula, and in this it is unusual. About two-thirds of all visible stars come in pairs or larger groupings. Multiple stars form from the same interstellar cloud and then orbit a common center of mass. Binaries, true double stars, are the most common, making up about half of all stars. They aren't necessarily identical twins; many binaries consist of a massive star with a smaller partner, such as the brilliant star Sirius and its white dwarf partner, Sirius B. Stars can also come in triples or even quadruples, two pairs of binaries orbiting a common center of mass.

A single star makes a good planetary parent because it has a wider range of possible stable orbits. However, some binary stars have "safe zones" at certain distances in which an Earthlike planet might orbit. Our closest stellar neighbors, the binary stars Alpha Centauri A and Alpha Centauri B, could conceivably maintain planets at about 3 AU from each star. Even so, our own sun, single, middle-aged, and middle-of-the-road, is an ideal host for life. Whether this is a rare coincidence or an unremarkable event is a hot topic among searchers for extraterrestrial life (see pp. 180–195).

STELLAR NURSERIES

The sky abounds in stellar nurseries, dense, gaseous clouds surrounding glowing newborn stars. Many can be found in the Orion Nebula, a region of star birth visible near Orion's sword, as well as in the Eagle Nebula, which contains the open star cluster M16.

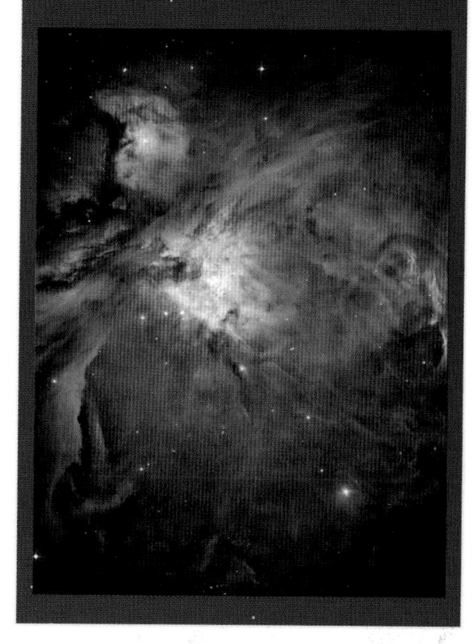

FOR MORE ABOUT STELLAR LIFE CYCLES, GO TO HTTP://ASPIRE.COSMIC-RAY.ORG/LABS/STAR_LIFE/STARLIFE_MAIN.HTML

THE SOLAR POWERHOUSE
THE BIG PICTURE

Early observers saw the sun as a fire, or burning coal, or a glowing planet. Now we know that our star is a bubbling, many-layered sphere of ionized gas whose enormous mass is propped up by the energy radiating from its core.

The sun is an immense sphere of plasma: intensely hot, electrically charged gas, mostly hydrogen and helium. With a diameter of 1,391,000 kilometers (864,000 mi), it dwarfs other members of its planetary family. More than one million Earths could fit into it. So massive is it that a heatproof 45-kilogram (100-lb) human standing on its surface would weigh an intolerable 1,270 kilograms (2,800 lb). • Nuclear fusion in its core powers the sun. The fantastic heat and pressure generated within the sun's center, fusing hydrogen into helium and releasing electromagnetic energy, can take hundreds of thousands of years to move from the core, where the temperatures reach 15.7 million kelvins (around 28 million degrees Fahrenheit), to the sun's visible surface, at a relatively balmy 5800 kelvins (almost 10,000°F). Magnetic fields twisting through its body pull streamers of gas far into space. Occasionally they blast huge plasma storms toward Earth. • The sun dominates the solar system not only through its gravitational influence, which extends to the Oort cloud, up to 200,000 AU away, but also through its solar wind of charged particles, which reaches beyond 100 AU, far past Pluto.

Symbol: ☉
Discovered by: known by the ancients
Distance from Earth: 149,600,000 km (92,960,000 mi)
Rotational Period: 25.38 Earth days
Diameter: 1,391,000 km (864,000 mi)
Circumference: 379,000 km (2,715,000 mi)
Mass: 1,989,000,000,000,000,000,000,000,000,000 kg
Surface temperature: 5777K (5504°C/9939°F)
Core temperature: 15,700,000K (15,555,538°C/28,000,000°F)
Atmosphere composition: Hydrogen, helium, trace elements

SKYWATCH

* Solar astronomy's first rule: Never look directly at the sun, as the light can destroy your retinas. Use specially designed glass or solar viewing telescopes.

AMAZING FACT In space, our eyes would see the sun as bright white, not yellow.

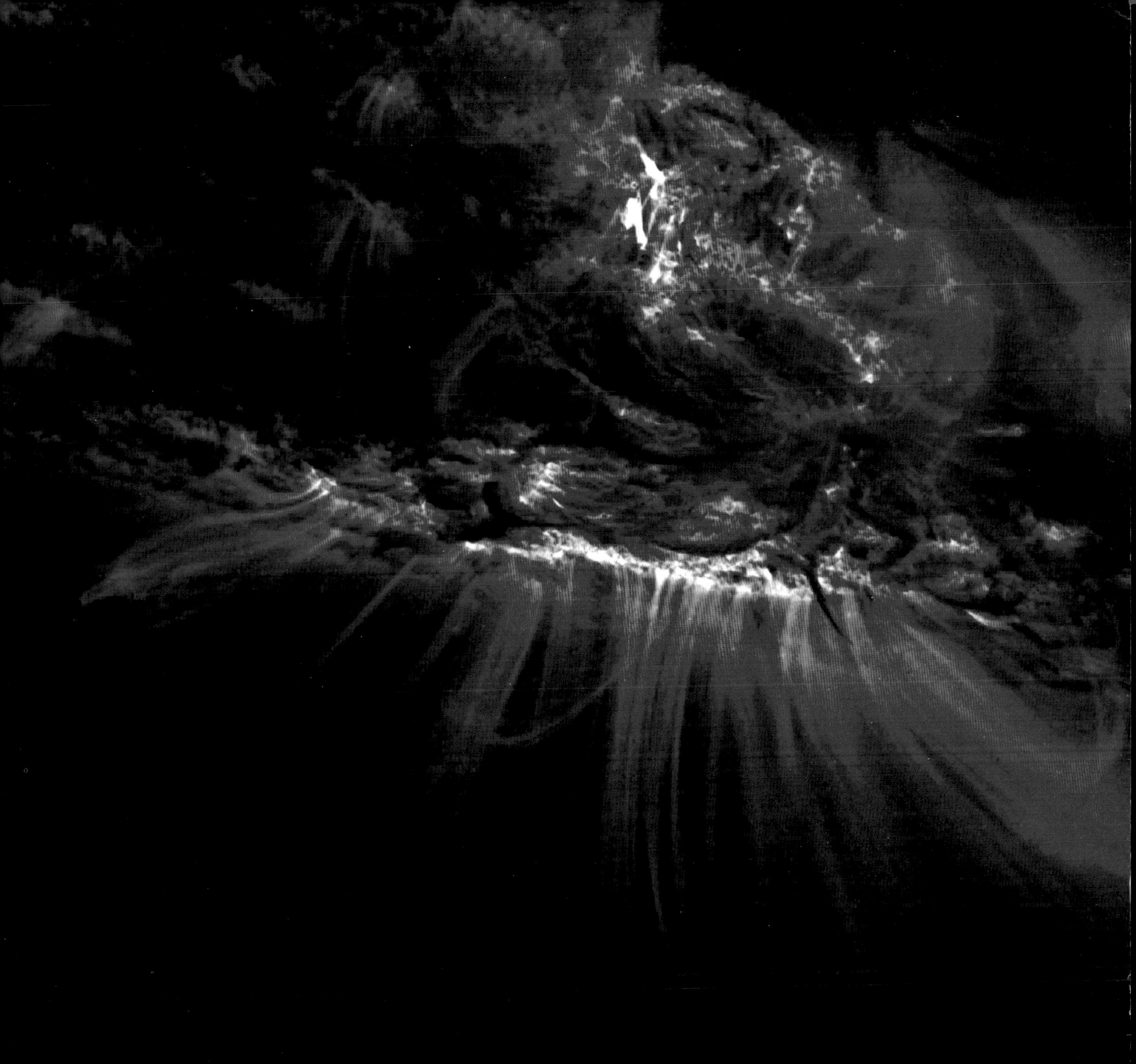

1905
Albert Einstein explains conversion of small masses to massive energy.

1926
Arthur Stanley Eddington theorizes the sun converts mass to energy.

1931
Austrian physicist Wolfgang Pauli proposes the existence of neutrinos.

1932
British physicist James Chadwick discovers the neutron.

1932
American physicist Carl Anderson discovers the positron.

1938
Hans Bethe and Charles Critchfield demonstrate that hydrogen fuels stars.

Hotter and cooler areas on the sun's surface are lighter and darker in this ultraviolet image from the SOHO spacecraft (opposite). Glowing plasma arcs loop out around a sunspot (above).

SOLAR POWERHOUSE: SUN-EARTH CONNECTION

Heat from the sun provides the underlying energy that drives Earth's weather. Thunderstorms, such as the one above, are driven mainly by the heat released when water vapor condenses. A typical storm contains energy equal to a 20-kiloton nuclear explosion.

The sun bathes the rotating Earth in light and charged particles. Its energy propels the planet's atmospheric and ocean currents and feeds its plants. Without it, life would be impossible, Earth's surface a cold, bleak, and desolate wintry wasteland.

Sunlight leaves the sun as electromagnetic radiation and travels virtually unhindered through the near vacuum of space to reach our atmosphere. The inverse-square law dictates that its intensity is inversely proportional to the square of the distance from its source. At 149,600,000 kilometers (92,960,000 mi) away, Earth receives only one-half of one-billionth of the sun's output, but even so, the radiation reaching our planet each year contains more than 20 times the energy of Earth's entire reserves of coal, oil, and natural gas. We wouldn't want more attention from the sun than that. If all the sun's energy were somehow focused on the Earth, within six seconds all the oceans would boil away; in three minutes the planet's crust would melt.

From the perspective of creatures who evolved on Earth, we're in the Goldilocks situation of having just the right amount of solar input. If we floated a flat detector one meter square at the top of the atmosphere, the amount of solar energy reaching its surface would amount to about 1,400 watts—approximately the amount consumed by an electric heater. This figure is called the solar constant, but it is not immutable. Sunspots and solar storms cause small variations in the solar constant over the years, with consequences for our communications and I'd say "probably" our climate. This amount of sunlight warms Earth just enough to allow water to exist on the planet's surface in all three of its states—ice, liquid water, and vapor.

When the energy reaches Earth, some is reflected into space at various atmospheric levels, some is absorbed by the atmosphere, and some is absorbed at the surface, to be reradiated as heat. In the long run, energy coming in balances energy going out, otherwise the Earth would heat up unstoppably. This complex exchange is known as Earth's energy budget. The percentage of energy reflected into space by a surface—its shininess, if you will—is called albedo (see p. 91). The atmosphere reflects about 6 percent of incoming radiation and clouds some 20 percent more. Icy surfaces on the planet also have a high albedo. Clouds, the atmosphere, and particularly oceans and forests also absorb radiation and then reradiate it at longer wavelengths as heat. Some is trapped by water vapor and carbon dioxide in the atmosphere in the greenhouse effect, but overall the amount of radiation leaving the planet matches the amount that first met its atmosphere.

THE SUN AND WEATHER

The unequal heating of Earth's surface is a major factor in the formation and manifestation of its weather patterns. As a sphere, Earth collects radiation unevenly, with sunlight spread thinner across the high latitudes. Tilted at 23.5° to the plane of its orbit, Earth experiences seasons because its hemispheres collect more direct light for more hours per day when leaning toward the sun (summer), and less direct light for fewer hours when leaning away (winter). The Earth's varied terrain reflects or releases heat at different rates. The heat moves through the air and water by convection, making it less dense as it warms, and the differences in pressure drive winds. Sunlight also evaporates water, lifting it into the air, where it condenses in cooler temperatures and then falls back to the ground in the eternal water cycle. Hot, humid air boils up in thunderstorms and hurricanes. The solar energy that propels a hurricane for a single day is equivalent to 200 times the electrical generating capacity of the world. Propelled by the sun's mighty and unstoppable energy, jet streams and ocean currents move heat in generally reliable patterns around the globe, keeping Earth within a comfortable range of temperatures.

INVISIBLE WAVELENGTHS

Because the sun's surface is about 5800K, it emits radiation primarily in three different types of wavelengths: visible, infrared, and ultraviolet. Visible light, obviously, is the range perceived by the human eye, whereas infrared is characterized by longer wavelengths and is registered as heat. Most of the ultraviolet light reaching Earth is absorbed by ozone in the atmosphere, which is fortunate for us, since shorter-wavelength UV radiation can penetrate living cells and cause mutations and cancer. (And unfortunately for us, chlorine that humans have introduced into the atmosphere has broken down some of that ozone layer, particularly over the Poles.) The longest wavelengths are those that tan the skin, but long-term exposure to these can also be harmful. On the other hand, shunning the sun entirely is not the answer, since many animal bodies, including humans', use UV-B light to produce vitamin D. As with many things, moderation is key; some small exposure to sunlight is not only useful and enjoyable, but necessary for good health.

The sun also broadcasts at radio wavelengths, a fact not understood until World War II. In the late 1930s and into the '40s, military specialists were frustrated by the occasional storms of interference that jammed their radio frequencies interrupting communications. British scientist J. S. Hey and American physicist George Southworth theorized that the interfering wavelengths came from the sun, although their findings were classified as secret until after the war. But no such rulings applied to an obscure Illinois tinkerer, Grote Reber. The 25-year-old electrical engineer decided to build a homemade radio telescope, a dish almost 30 feet wide, in his own backyard. The two-story-high, two-ton contraption disturbed the neighbors, but it worked. Reber tracked radio signals from stars all over the Milky Way, and in particular from Earth's own sun. In 1944 he published a map of these galactic signals and became the first to show that the sun was itself a powerful source of radio waves.

Sunlight is not only useful, but it is essential to almost all plants, and by extension to everything in the food chain that begins with plants. Chlorophyll, a pigment found in all photosynthetic organisms, absorbs all visible light except for the green portion of the spectrum (which is why plants typically look green). The energy from light powers a chemical reaction in which water and carbon dioxide are converted to carbohydrates and oxygen. When the first blue-green algae growing in Earth's early oceans began this transformation, it was a worldwide disaster. Oxygen was toxic to the anaerobic life of the time. But in the inexorable process of evolution, green plants prevailed in most ecosystems and new forms of life arose that fed upon oxygen and carbohydrates, releasing carbon dioxide to keep the cycle going.

THE SUN AS POWER SOURCE

Solar energy past and present provides almost all of the energy produced in the world today. It is embodied in the plant and animal material of fossil fuels—coal, petroleum, and gas—laid down in the Carboniferous period. Indirectly, it drives the wind that spins wind turbines. And very directly, it powers solar energy panels when absorbed by photovoltaic cells. Solar panels are already a standard part of most spacecraft. SOHO, or the Solar and Heliospheric Observatory, for instance, pulls its energy from winglike solar panels even as it studies the sun itself.

RAYS FROM SPACE

Earth receives small amounts of radiation from astronomical sources other than the sun—ordinary visible starlight, radio waves from Jupiter, and, occasionally, bursts of high-energy rays from distant active stars. Magnetars, a kind of neutron star, have occasionally blasted Earth with intense gamma-ray and x-ray radiation. Luckily for us, our atmosphere stops the deadly rays before they reach the ground.

SOLAR ECLIPSES

Earth dwellers rarely appreciate the steady presence of the sun so much as when it is blotted out during a solar eclipse. Solar eclipses occur when the moon moves directly between Earth and the sun and casts its shadow on Earth's surface. If the moon is too far from Earth to cover the sun's disk completely, the result is an annular eclipse, when the sun's bright ring, called an annulus, can be seen around the moon. A partial eclipse happens when the viewing site on the ground is in the outer edge of the moon's shadow; the moon seems to take a bite out of the sun. But the truly impressive spectacle is the total solar eclipse. When the moon is at just the right distance to blot out the sun's disk entirely, people living within the narrow band of dark shadow cast by the moon on Earth will see the sun apparently disappear behind the moon. Their surroundings will darken to a deep twilight, the air will cool, and birds return to their branches. For the few minutes of totality, the sun's feathery corona will glow around the moon's edge, reaching far into space.

THE SOLAR POWERHOUSE: A BURNING QUESTION

A turning point in the history of astronomy came when Polish astronomer Nicolaus Copernicus finally proposed that the sun lay at the center of the solar system; with this insight he transformed our understanding of our place in space. But knowing roughly where the sun was did not tell astronomers what it was made of or how it worked. That knowledge arrived much later, in the 20th century.

WHAT IS THE SUN?

Some early astronomers did attempt to reason out the nature of the sun as a natural object, not a god. The Greek mathematician Pythagoras, for instance, held that the Earth revolved around a central fire while shielded by a "counter-Earth," but he also maintained that the sun itself was a different, distant body that circled the Earth. Anaxagoras, the Greek philosopher of the fifth century B.C., theorized that the sun was a "mass of red-hot metal," but Aristotle, in the fourth century B.C., held to the simpler image of the sun as pure fire. And there matters remained, more or less, for centuries.

Even after the invention of the telescope, the nature of the sun remained enigmatic. The sun is difficult to observe without going blind, and almost featureless in small telescopes in any event. The telescope does reveal sunspots, however, and Galileo was able to track their motion and deduce, correctly, that the sun rotates just as do the planets. Moreover, the new perspective on the stars given by the telescope began to convince scientists, among them René Descartes, that our sun was a star like others in the universe. But this told astronomers little about the sun's composition. Even in 1795, no less an observer than the great William Herschel suggested that the sun was an "eminent, large, and lucid planet," surrounded by glowing clouds and inhabited by "beings whose organs are adapted to the peculiar circumstances of that vast globe."

ELEMENTS OF THE SUN

By the 1850s, astronomers were beginning to record such intriguing phenomena as solar flares and the solar corona. But the key to deciphering the nature of the sun was finally found in the laboratory, not through the telescope. The first great breakthrough came with the development of spectroscopy. But not all key discoveries came through the lens. The technique of spectroscopy also added a powerful weapon to the astronomical arsenal. In 1814, German optician Joseph von Fraunhofer had noticed that when sunlight or starlight was split into a spectrum, the bands of color were interrupted by hundreds of dark lines. In the 1850s, German chemists Robert Bunsen (of burner fame) and physicist Gustav Kirchhoff figured out why. Individual elements, heated until they glow, emit characteristic bright emission lines. Elements also absorb radiation from warmer sources, leaving identifiable dark absorption lines in the spectrum.

Kirchhoff learned that three kinds of spectra were produced under three circumstances, summarized as Kirchhoff's Laws:

1. A hot object or hot gas under high pressure gives off a continuous spectrum—one without spectral lines.
2. A hot gas under low pressure gives off emission lines, bright lines against a dark background.
3. When the source of a continuous spectrum shines through a cool gas under pressure, the result is an absorption line spectrum—dark lines across a bright spectrum.

Using spectroscopy, Bunsen, Kirchhoff, and their colleagues were able to identify some of the elements in the sun's atmosphere, including hydrogen. Perhaps more important, Kirchhoff and Bunsen showed that the body of the sun was hot and incandescent within a somewhat cooler atmosphere.

Spectroscopy is now central to astronomical research. Using it, even planetary atmospheres can be studied. Planets are cool but not cold, and so they give off radiation. Astronomers must subtract the lines caused by reflected sunlight and by the interference of solar and terrestrial atmospheres, but what remains tells them which gases belong to the planet.

Not long after spectroscopy was developed, in 1868, French astronomer Pierre Janssen and British astronomer Joseph Norman Lockyer independently discovered an improved technique for observing solar prominences and analyzing their composition. Using this technique, Janssen found a previously unknown element in the sun. Lockyer and a chemist colleague, Edward Frankland, named it helium, after *helios,* the Greek word for sun. The element was not identified on Earth until 1895.

LORD KELVIN

PIONEER OF MODERN PHYSICS

Scottish physicist William Thomson, later Baron Kelvin of Largs (1824–1907), was a multitalented scientist who believed that natural forces would eventually be explained by a single unified theory. His name lives on today in the kelvin temperature scale that he was the first to propose. A measurement of absolute temperature, the scale sets zero (or "absolute zero") as the temperature at which the molecules of a substance have the lowest possible energy: equivalent to -273.15°C or -459.67°F. Each degree has the same magnitude as those on the Celsius scale. Astronomical temperatures, such as the temperatures of stars, are typically measured in kelvins.

THE SOLAR ENGINE

Despite the discovery of some of the sun's constituents, the mystery of its eternal flame remained. What kind of engine could emit such intense heat for so long? In the 19th century,

German physician Julius Mayer calculated that if the sun were made of coal and had unlimited oxygen for combustion, it would burn for only a few thousand years. Perhaps, he suggested, matter falling into the sun supplied it with continuing fuel. Scottish physicist William Thomson, later knighted as Lord Kelvin, debunked this idea. He noted that the combustion would take so much additional mass that the sun's growing gravitational pull would be detectable over time on Earth.

Kelvin and German physicist Hermann von Helmholtz worked out a different answer to solar heat in the 1860s. Gravitational contraction alone could compress the sun's gases, heating them as the pressure rose. Physicists Jonathan Lane and Robert Emden soon took this theory to the next stage, showing that the immense mass of the contracting sun would heat its core to a fantastic 12 million kelvins (based on the absolute scale recently proposed by Lord Kelvin). The Lane-Emden theory of 1907 gave the sun a more believable age of 22 million years, with 17 million more to go before it died.

Alas for this theory, new findings in geology and evolution began to point to an age for Earth on a scale of billions, not millions, of years. Meanwhile, Swiss patent clerk Albert Einstein was overturning the physics apple cart with his theories of relativity and the equivalence of mass and energy. The boom in nuclear physics that followed led to the solution of the solar energy problem. In his 1926 book *The Internal Constitution of the Stars,* British astronomer Arthur Stanley Eddington proposed that Einstein's iconic equation $E=mc^2$ held the answer. Inside the sun, hydrogen atoms were being converted into helium, with an accompanying release of huge quantities of energy. Because the details of the atomic nucleus were still not completely known, he couldn't explain the exact transaction. But Eddington was convinced of his answer. In his 1927 book *Stars and Atoms* he wrote, "To my mind the *existence* of helium is the best evidence we could desire of the possibility of the *formation* of helium. . . . I am aware that many critics consider the conditions in the stars not sufficiently extreme to bring about the transmutation—the stars are not hot enough. The critics lay themselves open to an obvious retort; we tell them to go and find *a hotter place.*"

By the late 1930s, nuclear physics had developed to the point where the mechanism of stellar energy production became clear. In 1939, American physicist Charles Critchfield and German-born physicist Hans Bethe explained how at extremely high temperatures, a chain reaction among atomic nuclei could fuse hydrogen into helium, releasing energy. This proton-proton chain reaction was indeed at the heart of the sun's power—and soon after this discovery, at the heart of the hydrogen bomb as well.

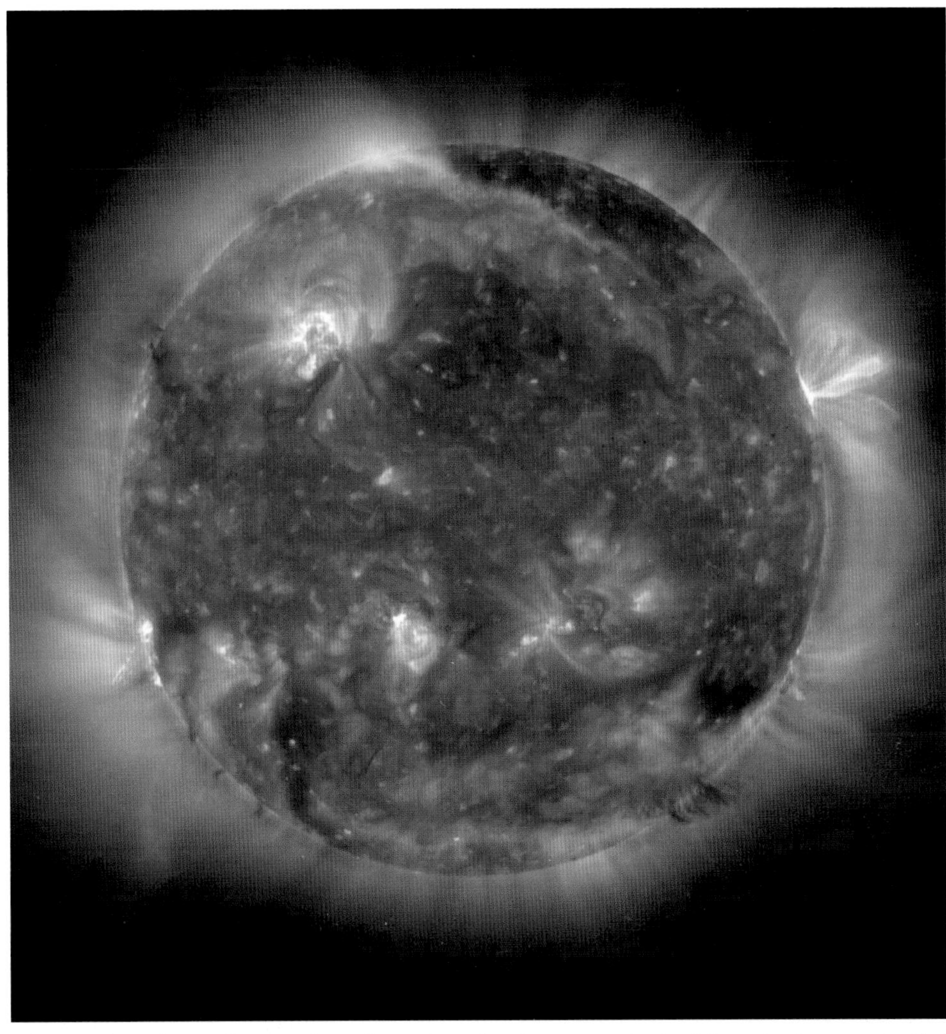

The complex release of energy from the sun's surface is apparent in this false-color image, compiled in three different wavelengths from SOHO's extreme ultraviolet imaging telescope.

HELIUM SHORTAGE ON EARTH

Abundant in stars and interstellar space, helium is rare on Earth. It is typically found trapped with deposits of natural gas, from which it must be distilled, and its supplies are waning. Helium is used to cool magnets in MRI machines, fuel the space shuttle, and even fill party balloons. Demand is growing, but the United States, the largest supplier, is expected to run out of the lighter-than-air gas before the end of the century.

THE SOLAR POWERHOUSE: FUSION

The laws of physics are pushed to extremes in the heart of the sun. Crushed under the star's enormous mass, hydrogen in the sun's core exists at a density 15 times that of lead—and yet it isn't solid, because temperatures in the core are equally high. At 15 million kelvins, the core's heat presses the gases outward, keeping the sun from collapsing.

It is only under conditions this intense that nuclear fusion can occur. Fusion is a mechanism in which light atomic nuclei are fused into heavier ones, releasing energy and tiny particles called neutrinos. Every second, fusion consumes 600 million tons of hydrogen in the sun's core—a mass equivalent to a modest mountain but representing an almost infinitesimal fraction of the sun's total mass. Almost all of that mass becomes helium during fusion, with less than one percent converted to energy. But as we know from Einstein's work, the energy produced by directly converting matter to energy is equal to the mass multiplied by the speed of light squared. Thus, the energy released each second in the sun is equivalent to the detonation of ten billion one-megaton nuclear bombs. And so enormous is the sun that it will continue to radiate steadily in much this fashion for another five billion years.

CHAIN REACTIONS

The sun's energy is thought to be produced mainly by a series of reactions called the proton-proton chain. It proceeds basically as follows:

Step 1. Two hydrogen atoms combine—much harder than it sounds! A hydrogen atom is just a single proton, a positively charged elementary particle. Under normal circumstances, two positively charged protons would repel each other. But in the intensely heated conditions of the solar core, a few are slammed together at high speeds. Once they are within 10 to 15 meters (33 to 49 ft) of one another, they are pulled together by the "strong force," one of the fundamental forces of the universe. The combined protons form an atom called a deuteron, consisting of one proton and one neutron. The reaction releases an antimatter particle called a positron (or antielectron) and a minuscule neutrino.

Step 2. The positron soon encounters an electron, its counterpart in matter, and they

The Sudbury Neutrino Observatory was built 2,070 meters (6,800 ft) underground in the Creighton mine near Sudbury, Ontario, Canada. Its 12-meter-wide (40-ft) acrylic vessel was eventually filled with neutrino-detecting heavy water. The observatory spotted its first neutrino in 1999.

annihilate each other, releasing energy in the form of high-energy photons, or gamma rays. (A photon is a single packet of electromagnetic energy.) Meanwhile, the deuteron combines with another free-floating proton to form an isotope of helium called helium-3: two protons and one neutron. This combination again releases gamma rays.

Step 3. After this reaction has happened twice, two helium-3 atoms collide and combine to make helium-4 (two protons and two neutrons), releasing two more gamma rays.

So for every four hydrogen nuclei (protons) that undergo fusion, the result is one helium-4 atom, two neutrinos, and high-energy gamma rays. This proton-proton chain accounts for more than 98 percent of the sun's energy. Another cycle, more common in more massive stars, creates the rest of the sun's power. This carbon-nitrogen-oxygen cycle uses carbon as a catalyst to start a rotating cycle of unstable carbon, oxygen, and nitrogen nuclei that transform into one another and release energy in the process.

THE NEUTRINO MYSTERY SOLVED

Physicists had worked out the solar fusion process pretty much to their satisfaction in the 1930s, but one piece of the puzzle remained missing. This was the neutrino. Physicists Wolfgang Pauli and Enrico Fermi first proposed the existence of the ghostlike subatomic particle to explain how energy was lost in radioactive decay. In theory, scientists knew, neutrinos should exist, and in huge quantities too. But no one was able to detect one. Could they be somehow massless and thus never interact with the solid world? Or were they *almost* massless, and so elusive that only the finest net could collect them?

To answer that question, in the 1960s scientists constructed a neutrino trap: a tank holding 300 tons of chlorine fluid almost 1.6 kilometers (1 mi) below the surface of Earth in the Homestake gold mine near Lead, South Dakota. Sunk so deep, the tank was shielded from cosmic rays and other interference. Neutrinos, in theory, would pass through Earth easily, but occasionally one should strike a chlorine atom and transform it into radioactive argon in a measurable way. The experiment was barely successful. The Homestake tank detected an average of two neutrinos a week until it was closed in 1993. The existence of neutrinos was confirmed, but the

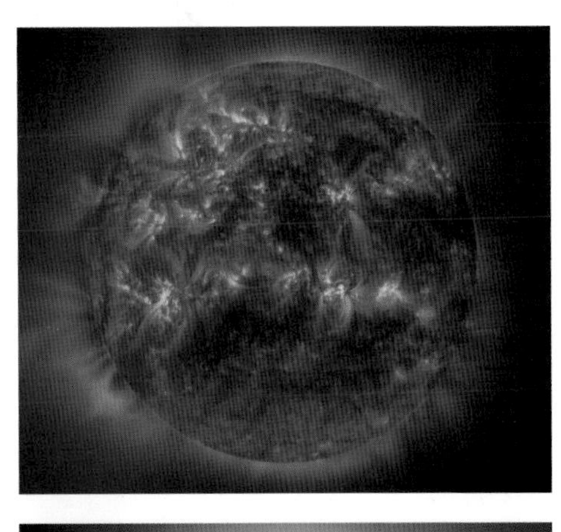

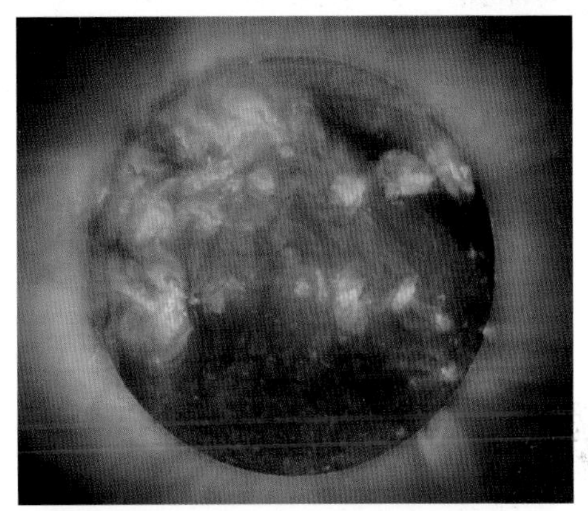

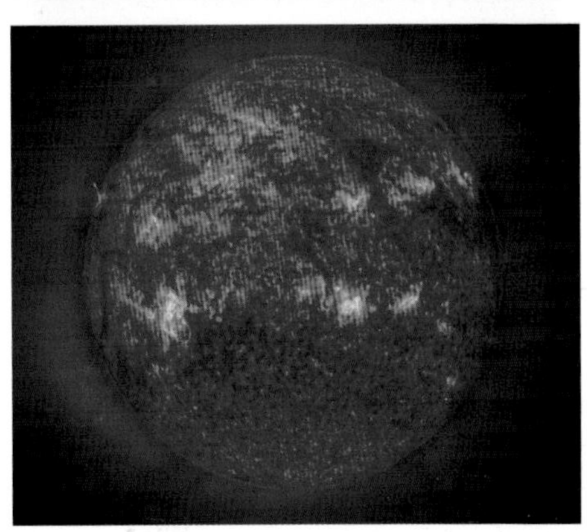

The sun at different wavelengths

rate was only one-third of the predicted value. Two other detectors ran into the same problem. Where were the remaining two-thirds of the solar neutrinos? Or was there a basic problem with solar physics?

The answer arrived with the realization that neutrinos had a "multiple personality disorder," as physicist John Bahcall later described it. Scientists discovered that there were three kinds (or "flavors") of neutrinos: high-energy electron neutrinos, the kind produced in the sun, and also lower-energy muon and tau neutrinos, which can be produced in supernovae or laboratory accelerators. For reasons best left to atomic physicists, the sun's electron neutrinos were oscillating into muon and tau neutrinos on their journey to Earth. In 2001, a new neutrino detector, the Sudbury Neutrino Observatory (SNO), finally detected the predicted amounts of these lower-energy neutrinos as they interacted with 1,000 tons of heavy water in an underground tank in northern Ontario, Canada.

As physicists understand them now, neutrinos are not massless, although their mass is some tiny fraction of the tiny mass of an electron. Traveling at almost the speed of light and having no electric charge, only one out of every ten billion will interact with a particle of matter on Earth. About 65 billion pass through every square centimeter of Earth every second—and almost 65 billion exit the opposite side of Earth without touching a particle of matter, to continue on at nearly the speed of light through space. Astronomers have a keen interest in neutrinos because they travel directly from the solar core, almost unhindered by intervening matter, thus carrying information about the sun and the nature of stars and galaxies.

LIGHT, ETC.

Energy flows from the sun as electromagnetic radiation—energy in the form of electric and magnetic fields that travel in wavelike form through space or through matter. One unit—one quantum—of electromagnetic energy is called a photon. From highest to lowest wavelengths (and highest to lowest energies), the electromagnetic spectrum includes gamma rays, x-rays, ultraviolet light, visible light ranging in color from violet to red, infrared light, microwaves, and radio waves.

THE SOLAR POWERHOUSE: INTERIOR

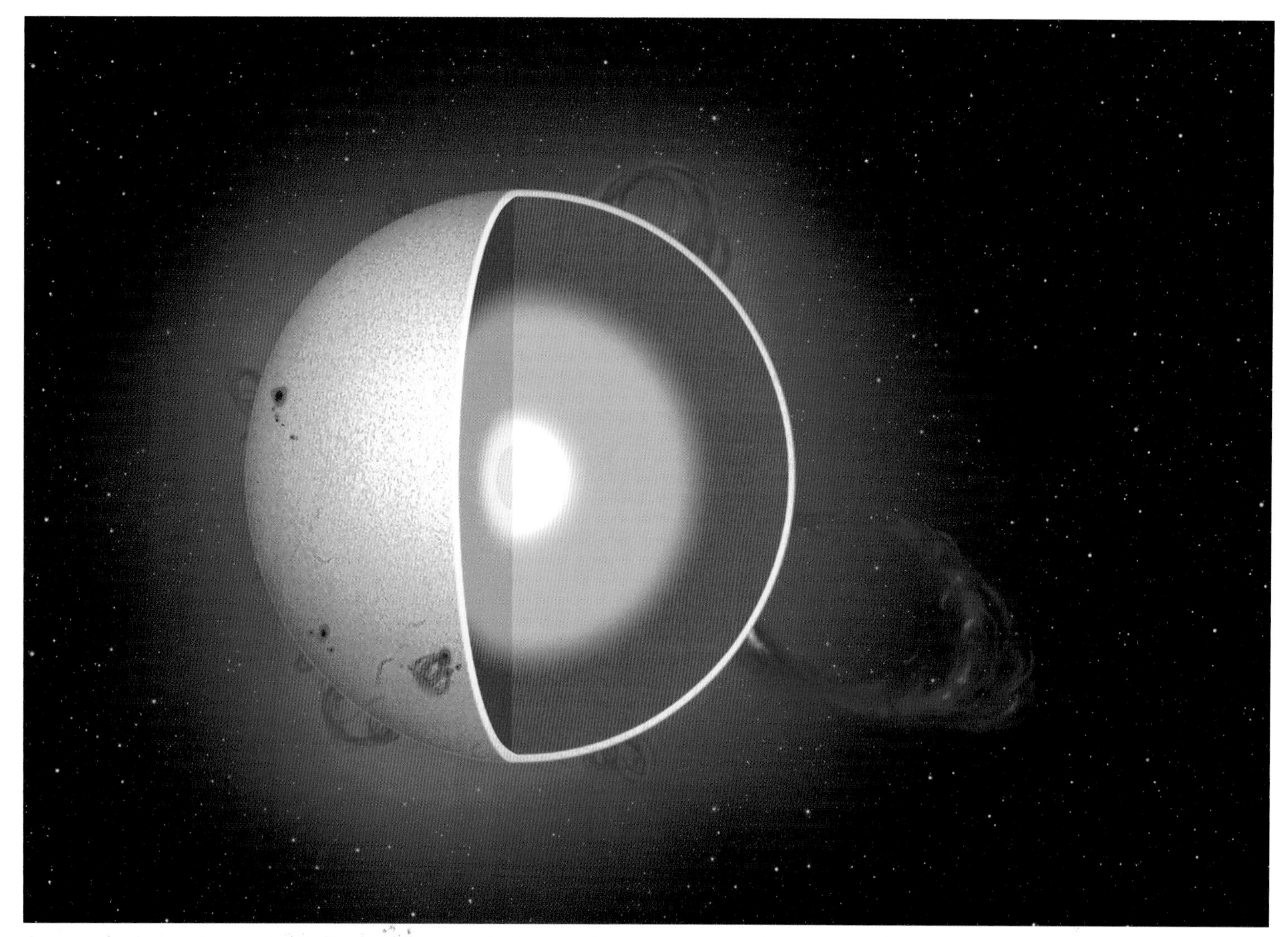

Cutaway art depicts the sun's layers. Energy flows from the fantastically dense, hot core (white) through the radiative zone (yellow) and convective zone (red). Solar flares and arcing prominences leap from the relatively thin outer layer, the chromosphere; around it is the far-reaching corona.

No probes have delved into our nearest star, so how can we be sure what's inside it? We can't. However, we can construct a solar model using what we know of the physics of gases and heat and by studying the way the sun vibrates as waves pass through it. Even so, the sun's complicated, roiling motions leave us with as many questions as answers.

HEAT, PRESSURE, AND WAVES

The sun is fantastically hot, yet its gaseous body does not fly apart. It is immensely massive, and yet it isn't squashed into a dense solar nugget. In fact, the sun maintains a remarkably stable size millennium after millennium. This tells scientists that it is in a state of hydrostatic equilibrium, with the pressure of hot gas pushing outward exactly balancing the weight of gravity pressing in. Researchers plug a number of known factors into the hydrostatic equilibrium model: the sun's mass, luminosity, surface temperature, observed radius, chemical makeup, and more. Then they apply what they know of the physical laws of conservation of energy, mass, and momentum, the ideal gas law, and the transportation of heat. These numbers and more pass through equations that yield an overall description of the sun's composition from core to atmosphere.

Adding to the mathematical model are studies of the sun's oscillating surface. In the 1960s, scientists realized that the sun's outer layers vibrated in regular waves. A new field of study was born and named helioseismology, although the "sunquakes" we see are completely different from earthquakes: The sun has no solid layers and no plate tectonics. By the 1970s, researchers discovered that the sun's quakes were the result of standing sound waves, acoustic vibrations that ring through the solar body like the sound of a bell. Hundreds of thousands of separate vibration patterns, or modes, disturb the sun at a time. As with seismic vibrations on Earth, solar waves of different frequencies can tell us the density and temperature of the material through which they pass.

THE SOLAR CORE

The sun's anatomy can be divided into six regions from core to outer atmosphere. The core, of course, is the sun's center, the powerhouse that produces energy through fusion. With a radius of about 200,000 kilometers (124,000 mi), it is extraordinarily compressed due to the body of the sun around it, containing half the sun's mass in only 2 percent of its volume. A density 200 billion times the pressure of Earth's atmosphere at sea level drives temperatures over 15 million kelvins (28 million degrees Fahrenheit), forcing some of the resident hydrogen atoms together in the fusion process (see pp. 62–63). Fusion releases energy in the form of gamma rays, very high-frequency electromagnetic waves. These rays shoot outward, beginning their long journey to the solar surface. By the time they emerge, they will have slowed down considerably, the gamma ray photons converted to a greater number of lower frequency photons of light and heat.

Although radiation in a vacuum moves at the speed of light, the sun's varying layers impede the photons that speed outward from its core. The time it takes for energy to travel from the center of the sun to its surface varies from 10,000 to 170,000 years, depending upon its random path. (Neutrinos also fly out of the sun's core, but they pass almost unimpeded through the sun and into space in a few seconds.)

RADIATIVE ZONE

In the deep core of the sun, the high-energy photons bounce from proton to proton as they travel to the next shell, the radiative zone. The radiative zone encompasses a large part of the body of the sun, stretching from about 200,000 to 496,000 kilometers (124,000 to 308,000 mi) of the sun's radius. The vast area holds about 48 percent of the sun's mass, ranging from a density that is roughly twice that of lead at its bottom up to to one-tenth that mass at the top of the zone. It's called the radiative zone because the photons there travel by radiation from one atom to another. Swinging from particle to particle in what scientists call the "drunken sailor" or "random walk" motion, they are absorbed and then reradiated over and over again in different directions. Outward progress is colossally slow, taking thousands of years. Temperatures cool as the photons move outward in this zone, dropping from about seven million to about two million kelvins.

CONVECTIVE ZONE

After moving through a relatively thin interface layer, the tachocline, the wandering photons enter the outer layer of the sun's interior, the convective zone, which reaches upward for about 200,000 kilometers (124,000 mi) to just below the solar surface. Cooler than the inner zones as it drops to temperatures of about 6000K, it is also more opaque, because atoms there are better able to hold on to their electrons. Gases in the convective zone are less dense than in lower regions, so although the region is huge in volume, it holds only 2 percent of the sun's mass.

Energy transport changes dramatically in the convective zone. Instead of traveling through radiation, energy here moves in the boiling motion of convection. Heated by radiation from below, huge cells of hot gas rise to the surface, while cooler gas sinks. In the loosely packed plasma, the motion becomes turbulent. As the sun rotates, the gases in the rising cells swirl as well as bubble up and down. There is no actual transfer of mass from one region to another, but heat moves efficiently here, rising through the whole layer in about a week. The cells are stacked in tiers and vary in size. Those at the lowest levels of the convective zone may be as large as 30,000 kilometers (18,600 mi) across, whereas those above them become successively smaller until they are perhaps 1,000 kilometers (620 mi) wide. The tops of these tightly packed smaller cells are visible at the sun's surface, giving it a granulated look.

At this level, the gas is too thin to carry heat through convection or to impede the bright flow of energy. Photons from the sun's core have finally reached the portion of the sun that is visible to our eyes.

LISTENING TO THE SUN

In the 1960s, scientists discovered that the surface of the sun oscillates, or shivers, as though disturbed by multiple "sunquakes," and the science of helioseismology was born. The name is a bit misleading, since the gaseous sun cannot harbor the same kind of seismic action as the solid Earth. Nevertheless, the cause of the sun's oscillations proved to be waves within the sun, just as waves traveling through Earth cause earthquakes. These are acoustic vibrations, enormous sound waves bouncing about within different layers of the sun's interior. The sun is ringing like a singularly complicated bell.

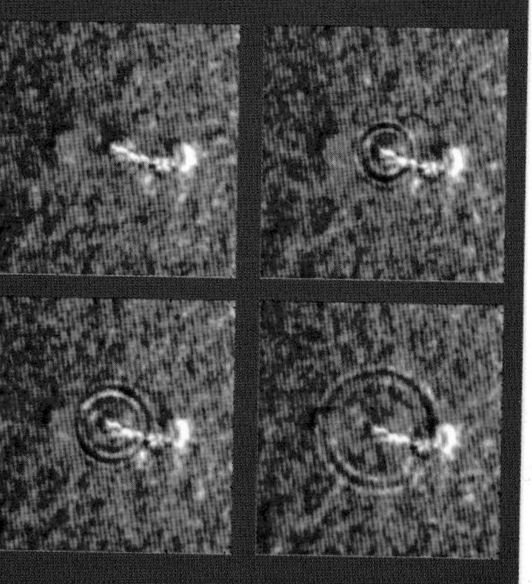

Solar flares precipitate some of these sunquakes. The quake shown above, its ripples moving outward on the sun's surface, began with such a flare releasing energy comparable to a magnitude 11.3 earthquake. Its waves accelerated to 400,000 kilometers an hour (250,000 mph) to a distance of ten Earth diameters before fading away.

The Michelson Doppler Imager, aboard the Solar and Heliospheric Observatory (SOHO) spacecraft, has recorded the bouncing movements of the sun's surface under a barrage of these waves. Traveling from side to side of the sun in about two hours, these huge acoustical waves are too low in frequency for the human ear to detect. However, scientists have sped up the sound file 42,000 times to bring the sun's voice within the audible range. You can hear the buzzing, throbbing song of the sun at the Stanford Solar Center website, *http://solar-center.stanford.edu/singing/*.

THE SOLAR POWERHOUSE: EXTERIOR

The gaseous sun has no real surface, but the outer layer that we see is called the photosphere. About 500 kilometers (310 mi) thick, this is the part of the sun that emits light. In fact, only the hottest, bottom 100 kilometers (60 mi) of the photosphere give off light, which travels through the rest of the cooling layer before entering the sun's atmosphere. The photosphere's brightness, its luminosity, tells us that its average temperature is about 5800K (9980°F).

DIFFERENTIAL ROTATION

Properly equipped telescopes reveal the photosphere as a mottled, spotty, turbulent place. In addition to huge, dark sunspots, which appear and disappear in regular cycles (see pp. 70–71), the photosphere has faculae, unusually bright spots. Bubbling up from the convective zone are granules and supergranules, the tops of convection cells. The sunspots, which have been observed for centuries, travel with the solar surface as the sun spins on its axis and so allow us to measure solar rotation. They tell us that the sun, not being a solid body, is not obliged to rotate as one. Instead it exhibits differential rotation, with different parts spinning at different speeds. The equator rotates about every 25 Earth days, while the poles apparently make one turn every 35 days or so.

The sun can't be said to have its own days—after all, it doesn't rotate relative to its own sun—but each rotation has been numbered since November 9, 1853. This is the date when amateur British astronomer Richard Carrington began observing sunspots. Heir to a brewery fortune, Carrington fell in love with astronomy during his studies at Cambridge and used his wealth to build his own observatory. Over eight years, he recorded sunspot locations every day. His data showed that sunspots were confined primarily to two bands in the north and south solar latitudes and that they gradually drift toward the sun's equator. Differing rates of rotation at various latitudes revealed the sun's differential rotation; the average rotational time was 27.28 days. The sun's Carrington Rotations, named in his honor, were up to number 2,078 as the year 2009 began.

Sharp spicules of plasma, jetting upward at 50,000 kilometers an hour (31,000 mph), rise from a magnetically active area of the sun.

CHROMOSPHERE AND TRANSITION ZONE

Like the Earth, the sun has an atmosphere, a layer of thin gases above its surface. But unlike the Earth's air, the solar atmosphere is thousands of kilometers high, terrifically hot, irregular, and explosive.

The lowest level of the atmosphere, just above the photosphere, is the chromosphere. Extending outward for about 1,500 kilometers (930 mi), it is dim but colorful. Hydrogen in its hot gases gives off a reddish color that has long been visible to astronomers, who can see the vivid chromosphere dancing around the edges when the moon blocks the sun's body during a solar eclipse. Using spectrographs and filters, scientists now can make out a host of other features in the chromosphere as well. Supergranules below the photosphere disturb the chromosphere above them in a weblike pattern. Dark, threadlike filaments streak through the layer, while brilliant plages (from the French word for "beach") look like bright white sand against the darker gases. Spicules jet out of the chromosphere like hot spikes, leaping thousands of kilometers high.

Forming a buffer between the chromosphere and the corona is the transition zone. Thin and uneven, this layer ranges from a few hundred to a few thousand kilometers. It's cooler at the bottom, about 20,000K (19,700°C / 35,000°F), but extremely hot at the boundary with the corona, perhaps 500,000K (499,700°C / 900,000°F). The ionized hydrogen of the transition zone glows only in the ultraviolet range. Spicules poke into it from the chromosphere, while coronal tempests disturb its upper layers.

CORONA

During total solar eclipses, watchers are treated to one of the sun's glories: its corona. Wafting far into space, the thin, opalescent gases of the sun's outer atmosphere surge outward during the sun's more active periods, and cling closer to its body during quiet times. At its farthest edges, the corona becomes the solar wind.

The corona's nature has puzzled astronomers for well over a century. In 1869, American astronomer Charles Young turned a spectroscope toward the corona during a solar eclipse ("the most beautiful and impressive spectacle upon which my eyes have ever rested") and was surprised to see a brief green flash. The spectral line corresponded to no element ever seen before. Subsequent astronomers also saw the mysterious line and decided it must be a new, extraterrestrial element, which they christened "coronium." In 1930, French scientist Bernard Lyot invented the coronagraph, a kind of eclipse in a box, which blocked out the body of the sun and allowed researchers to analyze the corona at their leisure.

In 1941, Swedish astrophysicist Bengt Edlen solved the mystery of coronium with another mystery. He showed that the coronium spectral line was the line emitted by highly ionized iron—iron with half of its electrons stripped away. Other elements in the corona, too, were stripped of their electrons. Only extraordinarily high temperatures greater than one million kelvins can do this. In fact, the corona's temperatures are today estimated to range from one million to three million kelvins. The immense energy contained in its thin gases radiates x-rays and propels the solar wind billions of kilometers into space.

But where does the energy come from? The photosphere emits heat at a relatively modest level of 5800K, so what is lighting a fire under the gases farther into space (where the temperature is just above absolute zero)? We don't yet know. The mystery of coronal heating is a major area of solar research. Most of the tentative explanations currently proposed involve the sun's complex magnetic fields. Rippling waves of acoustic and magnetic energy traveling along the sun's magnetic field lines might speed up and heat the corona's particles. Or magnetic fields that tangle and then release like rubber bands, creating "microflares," might also release energy into the sun's atmosphere. However, these mechanisms don't seem to provide enough oomph to bring the corona to its blistering levels, so the question of coronal heating is still an open one.

MISSIONS TO THE SUN

A small armada of solar spacecraft have enriched our knowledge of the sun in recent years. Among the missions:

- **Ulysses** (NASA, 1990–2008) examined space above and below the sun's poles, measuring interstellar dust and helium
- **Yohkoh** (NASA, Japan, U.K., 1991–2001) studied solar atmosphere.
- **Genesis** (NASA, 2001–08) collected samples of the solar wind and dropped them into the Utah desert, where they were retrieved and are now being examined.
- **SOHO** (ESA & NASA, 1995–present), the Solar and Heliospheric Observatory, has studied the sun's atmosphere and interior, its storms and solar wind, as well as discovering a number of comets.
- **TRACE** (NASA, 1998–present), the Transition Region and Coronal Explorer, examines the sun's magnetic structures and atmosphere.
- **STEREO** (NASA, 2006–present), the Solar Terrestrial Relations Observatory, employs two orbiting observatories to produce 3-D measurements of solar storms.
- **Hinode** (NASA, Japan, UK, 2006–present) uses two telescopes and a spectrometer to observe the sun's magnetic fields.

The SOHO spacecraft is 7.6 meters (25 ft) across with its solar panels extended.

STORMS IN THE CORONA

Spectacular plumes, loops, streamers, and prominences leap through the corona. Most originate in the upper layers of the solar surface, where magnetic field lines twist through superheated gases (see pp. 70–71). They include coronal loops, rings of denser gases that leave and return to the surface near sunspots. Caplike helmet streamers form pointed hats above sunspots, while polar plumes are fingers of gas reaching out from the sun's north and south poles. Immense solar prominences pull flaming gases in huge arches away from the sun. Some contain up to 100 billion tons of solar plasma and are far larger than Earth. But as dramatic as they are, they are cooler than the corona, so they show up as darker bows against the brighter atmosphere. Darker yet are coronal holes, huge, shadowy regions visible on x-ray portraits of the sun. These are probably areas where the coronal gases are particularly thin, allowing the solar wind to escape toward Earth.

The award for most impressive solar display, however, might go to coronal mass ejections (CMEs). These are difficult to spot without specialized equipment, but as early as 1860 a drawing made of the solar corona during an eclipse shows the huge, looping gases characteristic of a CME. Erupting from the gases of the corona itself, these enormous explosions fling billions of tons of matter into space at speeds up to 3,000 kilometers a second (more than 6 million mph). The ejection can last for hours and span an area as wide as a planet. Propelling the solar wind, these solar storms can disrupt electronics on Earth.

THE ACTIVE SUN
SOLAR CYCLES

Our sun's seething, explosive nature is difficult to unravel, but scientists are beginning to understand how regular cycles govern its outbursts and how magnetic field lines, twisting and tangling, pull its plasma into loops and storms.

The sun as magnetic object is a relatively recent notion, but clues to its true nature have always been visible in its dark, changeable sunspots. Sunspots have been a favored subject of solar astronomers since the telescope was invented. Galileo saw and recorded sunspots, as did Johannes Kepler, who at first mistook one for a planet. The first great long-term sunspot observer was Samuel Heinrich Schwabe, a 19th-century German pharmacist and amateur astronomer. Schwabe was actually searching for a planet that he believed might be found between Mercury and the sun. After observing the sun every clear day for 17 years and recording the spots that crossed its surface, he realized that their appearances and disappearances were not random. In an 1843 article, he noted that "it appears that there is a certain periodicity in the appearance of sunspots and this theory seems more and more probable from the results of this year." Schwabe had discovered solar cycles, opening the door to a new and fruitful field of solar research. The most current solar cycle goes from January 24, 2008, to circa 2018. All of the cycles can be seen at *http://solarscience.msfc.nasa.gov/SunspotCycle.shtml.*

RECENT SUNSPOT CYCLES

Cycle 15: August 1913 to August 1923
Cycle 16: August 1923 to September 1933
Cycle 17: September 1933 to February 1944
Cycle 18: February 1944 to April 1954
Cycle 19: April 1954 to October 1964
Cycle 20: October 1964 to June 1976
Cycle 21: June 1976 to September 1986
Cycle 22: September 1986 to May 1996
Cycle 23: May 1996 to March 2008

SKYWATCH

* With proper precautions, you can see sunspots with binoculars. Place no. 14 welder's filters over the objective lenses of the binoculars (*not* the eyepiece).

AMAZING FACT The typical solar flare is the size of Earth.

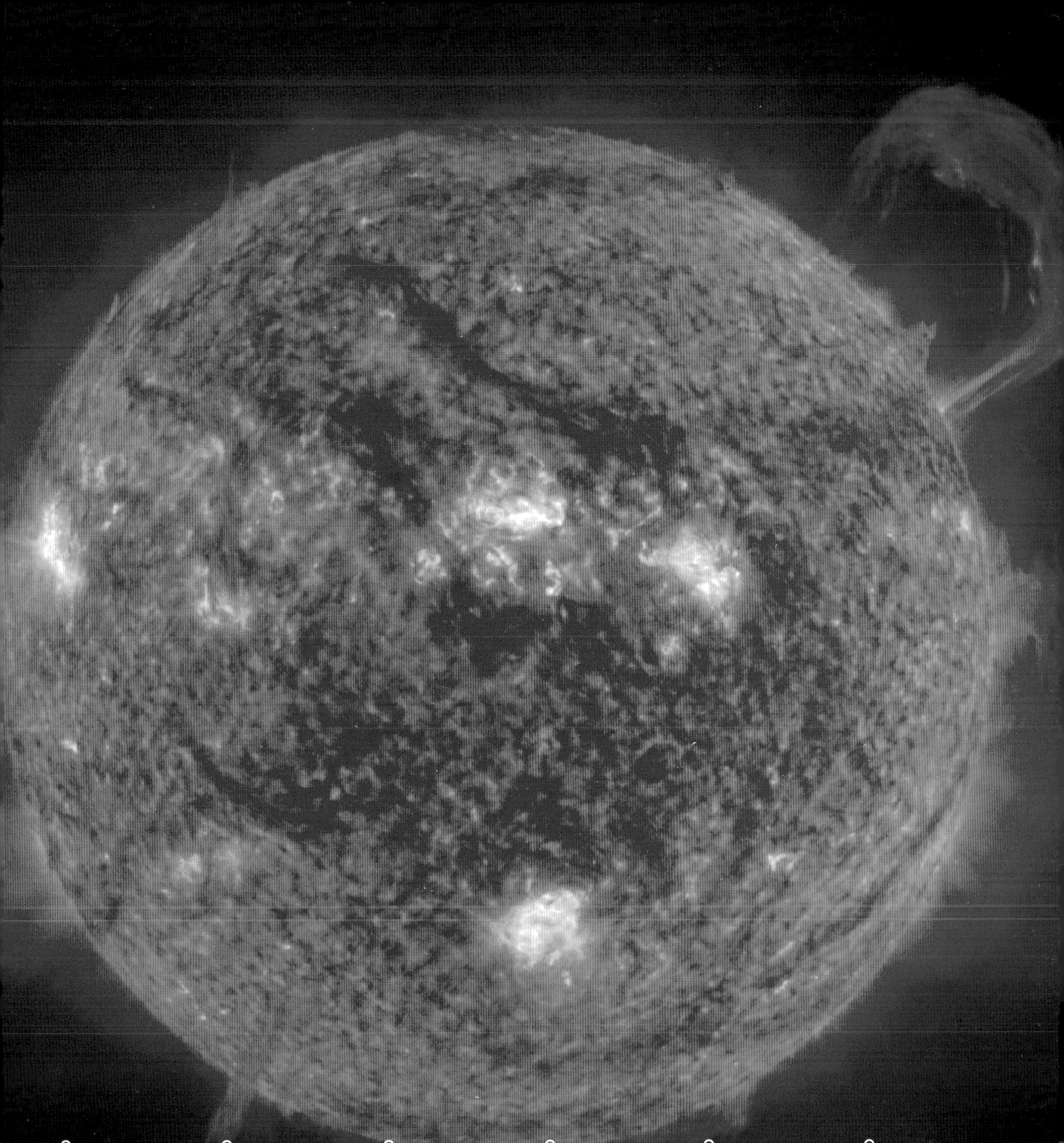

1128
Monk John of Worcester makes first drawing of sunspot.

1610
Galileo observes sunspots through a telescope.

1645–1715
Sunspots vanish during Maunder minimum.

1843
Schwabe discovers sunspot cycle.

1908
Hale proves magnetic nature of sunspots.

1973
Skylab discovers coronal mass ejections.

A major, X-class solar flare shows up in brilliant white in this SOHO false-color image of the sun (opposite). A solar prominence bigger than Earth juts like a handle from the sun in an ultraviolet image (above).

THE ACTIVE SUN: SUNSPOTS & STORMS

Samuel Heinrich Schwabe's observations and those of subsequent astronomers showed that the sunspots waxed and waned in number over a cycle that averaged 11 years. But what, astronomers asked, were these spots? Did their mutable presence on the sun affect Earth? Sunspot students then and now attempted to link them to everything from Earth's climate to human behavior and economic cycles. However, the only obvious connection between sunspots and Earth was, surprisingly, a magnetic one. By 1852, researchers had shown that sunspots seemed to affect Earth's magnetic field. In 1858, scientists Richard Carrington and Richard Hodgson spotted bright solar flares near a large sunspot, followed 36 hours later by a geomagnetic storm.

It took a clever innovation from master telescope pioneer George Ellery Hale to prove the connection. To better analyze the light of the sun, he crafted an improved solar spectrograph, or spectroheliograph, whose selective filter allowed him to study the spectrum of light emitted only by sunspots. He found that the sunspot absorption lines were split down the middle in a pattern known as the Zeeman effect. Named for the Dutch physicist who discovered it in the 1890s, the effect appeared when light was given off by an element in a strong magnetic field. The farther apart the split lines, the stronger the field. Measuring the lines, Hale realized that sunspots have magnetic fields 100 times stronger than the sun's normal field.

Hale also went on to discover that each pair of sunspots is magnetically linked. Observers had seen that sunspots travel in groups. Hale showed that the trailing sunspot in each group had the reverse polarity of the spot in the lead. Furthermore, the polarities were flipped in the

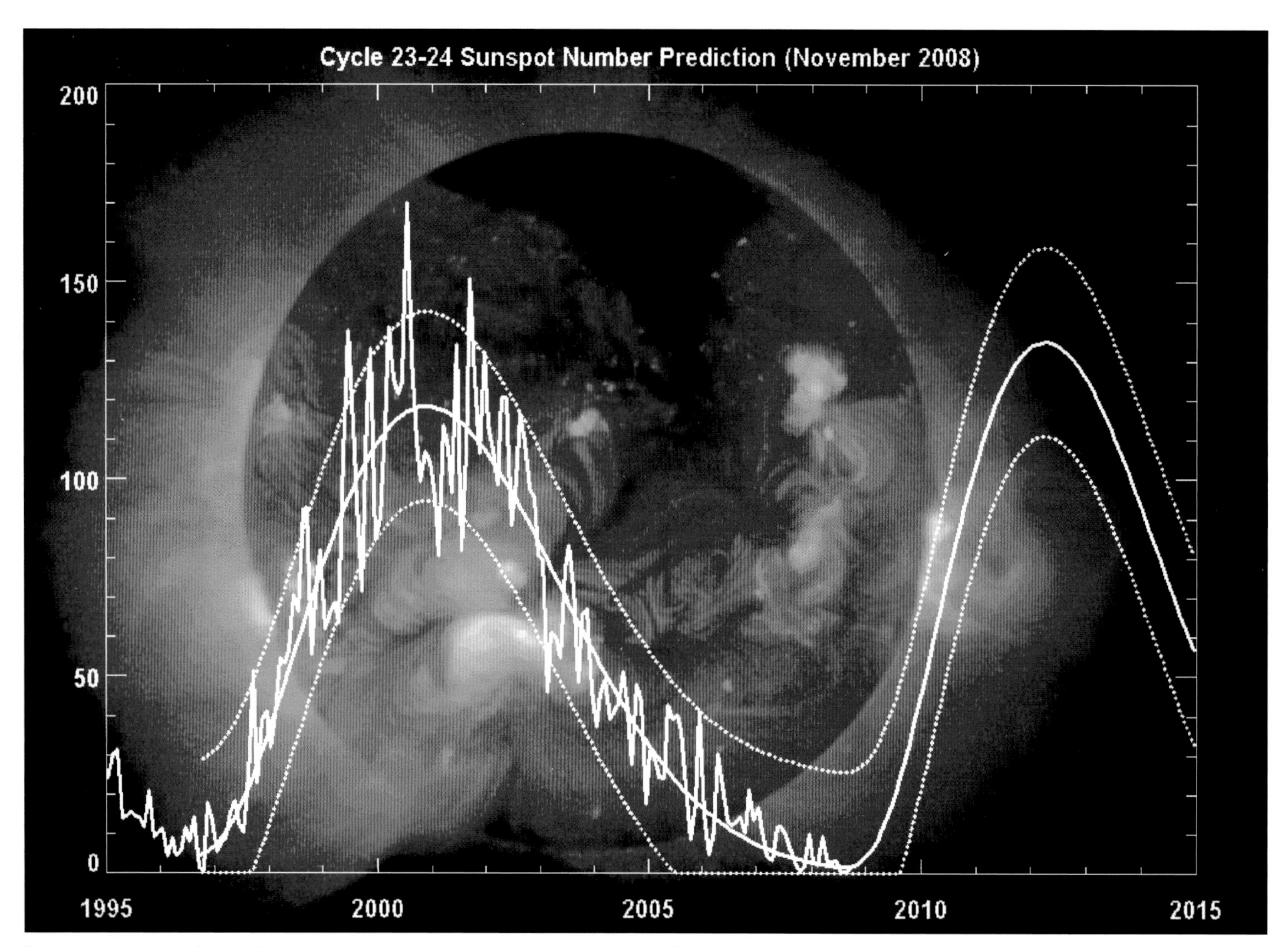

Superimposed on an image of the active sun is a graph of sunspot cycle number 23 and predictions for cycle number 24, which may reach its zenith in the year 2012. Every cycle lasts 11 years from peak to peak.

opposite hemisphere. If the leading sunspot in the sun's northern hemisphere has a "south" orientation, the leader in the southern hemisphere will be "north." But these orientations regularly reverse themselves from cycle to cycle, from north to south over the course of 11 years. So a true solar cycle actually lasts an average of 22 years, as the polarities flip from north to south to north again.

SOLAR MAGNETISM

The study of the sun's magnetism has become a dominant area of solar research in recent years, not least because the sun's magnetically powered storms can strongly affect Earth. Though many mysteries remain, researchers have formed a rough picture of the magnetic sun. Its magnetic fields appear to be generated within the tachocline, the thin layer between the radiative zone and the convective zone. In the radiative zone and below, the sun rotates as a single body, but above it, it exhibits the differential rotation we see on its surface, some parts circling faster than others. The shearing effect of these different speeds in the tachocline produces the magnetic field. Immense magnetic field lines rise from the tachocline, at first running evenly north and south. As the sun's surface rotates, though, the magnetic fields become wrapped around the sun, twisting and tangling. They breach the sun's surface in vast loops. At the base of each end of the loop is a sunspot, with a "north" polarity at one end and a "south" polarity at the other.

Sunspots vary greatly in size, ranging from thousands of kilometers wide to 30 times the width of Earth. Close-up images show a central dark region, the umbra, where the magnetic field line is keeping a lid on the hotter gases below. The umbra is surrounded by a striated brighter region, the penumbra, that radiates from it like the heart of a flower. Sunspot umbras are distinctly cooler than the solar surface, with temperatures averaging 2200°C (4000°F)—so cool that water vapor can actually be found in their gases. Even so, if sunspots were viewed against a dark background, rather than the sun's surface, they would be blindingly bright.

Energy deflected from below the sunspot flows around it to the surface, so the areas around sunspots are a little brighter than average. This means that during a sunspot maximum—when the sun is covered by the highest number of sunspots in the cycle—the surface actually radiates slightly *more* light than at other times.

STORMS AND FLARES

Most of the spectacular solar fireworks that erupt from the sun's surface are linked to magnetic field lines. Solar plasma following the arch of magnetic loops creates solar prominences, for instance (see p. 67). Coronal mass ejections are apparently launched into space when a stable magnetic field is disturbed or detached from the sun. These can be bad news for Earth when they happen to fly in our direction, because the magnetized plasma can cause a magnetic storm in our atmosphere. One such CME on March 13, 1989, knocked out electrical power to more than five million people in Quebec.

Possibly the most destructive of these magnetic explosions is a solar flare. These begin with a magnetic field line arching out of the solar surface through two sunspots. The field lines begin to twist and distort as the sunspots move about. Currents become more intense and the plasma gets hotter, giving off x-rays and gamma rays. At the top of the loop, temperatures can reach 100 million degrees Celsius (212 million degrees Fahrenheit), the hottest temperatures in the solar system. Sometimes the magnetic lines, stressed to the maximum, spring into a simpler shape or "reconnect," in the process releasing huge amounts of energy in a solar flare. Thousands of flares occur each solar cycle, but the biggest are impressive. A flare on November 4, 2003, gave off enough energy to power Earth for 1,000 years. A particularly active sunspot in January 2005 unleased five powerful flares. The fifth released a storm of protons that bombarded Earth in less than 30 minutes, traveling at approximately one-third the speed of light. Apparently the storm occurred when the sunspot was in just the right location to connect to Earth via magnetic field lines.

Flares are spectacular, but their radiation is deadly. Spaceflight planners must take into account the need to protect the human body from flares when they plan for manned missions. If Apollo astronauts had been standing on the moon on August 4, 1972, for instance—in the time between Apollos 16 and 17—they would most likely have been injured or even killed by the radiation from a solar flare.

THE MAUNDER MINIMUM

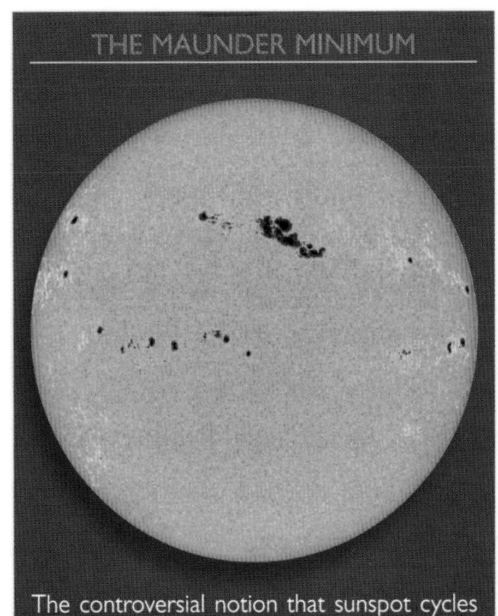

The controversial notion that sunspot cycles could affect the Earth's climate has gained some respectability in recent years. The prime example of such an effect is the period known as the little ice age. In the 16th through the early 18th century, temperatures in northern Europe and North America ran about 1°C to 1.5°C (1.8°F to 2.7°F) below normal: Glaciers advanced in Europe, sea ice surrounded Greenland and Iceland, crops failed, and famine spread. This unusually cold period happens to coincide with a period of very low sunspot activity, first pointed out by astronomers Edward Maunder and Gustav Spörer in 1890 and now called the Maunder minimum. Between 1645 and 1715, virtually no sunspots appeared; over a 30-year interval, only 30 were recorded, as opposed to a more typical 40,000 to 50,000. Few auroras were seen during this era, as well. As we now know, solar luminosity increases slightly during active sunspot periods owing to the energy directed around the spots. When solar activity decreases, on the other hand, decreased UV radiation seems to affect Earth's ozone layer, which in turn may affect the atmosphere and warmth-carrying winds. If this link between sunspots and climate turns out to be valid, it still leaves many questions unanswered. Historic records show other extra-chilly intervals in recent human history, unrelated to geologic ice ages. Do solar cycles explain those cold periods—and if so, can we predict when the next one will occur?

THE ACTIVE SUN: THE SOLAR WIND

Early in the 20th century, a few scientists had speculated that the sun emits more than radiation. British radio pioneer (and psychic investigator) Sir Oliver Lodge theorized in 1900 that the sun gave off "a torrent or flying cloud of charged atoms or ions." In 1932, German geophysicist Julius Bartels, studying magnetic storms on Earth, discovered that they occurred at roughly 27-day intervals, corresponding to one rotation of the sun. Bartels hypothesized that the storms originated from magnetic regions on the sun's surface. But the first solid evidence of the solar wind came from comets—or more specifically, from comet tails.

Astronomers already knew that a comet's tail always points away from the sun; as it moves toward the sun, a comet's tail flows behind it, but after circling the sun, its tail leads the way on the journey outward. Pressure from light itself blows back the dusty appendage. But a comet actually has two tails, the dusty one and a longer, straighter second tail made of ionized gas. The dust tail always points straight away from the sun, but the ion tail can be deflected from that direction by several degrees. In 1951, Ludwig Biermann of the University of Göttingen in Germany showed that cometary ion tails, moving much faster than the dust tails, must be driven by streams of electrically charged particles moving at hundreds of kilometers a second.

Just what were these particle blasts, and what propelled them? In 1957, British geophysicist Sydney Chapman showed that the sun's corona, amazingly hot at one million kelvins or more, must extend beyond Earth itself. And in 1958, University of Chicago physicist Eugene Parker proved that the corona was not just a static shell, but so hot and energetic that its thin gases escaped the sun's gravity and rushed into space, actually increasing their speed as they moved away from the sun's influence. This was, as it came to be called, the solar wind.

The space age brought proof of the solar wind's existence. Early Soviet and U.S. spacecraft, such as Lunik 2 and 3 and Explorer 10, seemed to detect streams of charged particles. In 1962, the U.S. probe Mariner 2, while studying the atmosphere of Venus, provided a more definitive answer. In four months of observations, the craft registered a ceaseless, powerful wind of charged particles emanating from the sun, flowing at thousands of kilometers an hour. The fastest blasts came at 27-day intervals, corresponding to the sun's rotation.

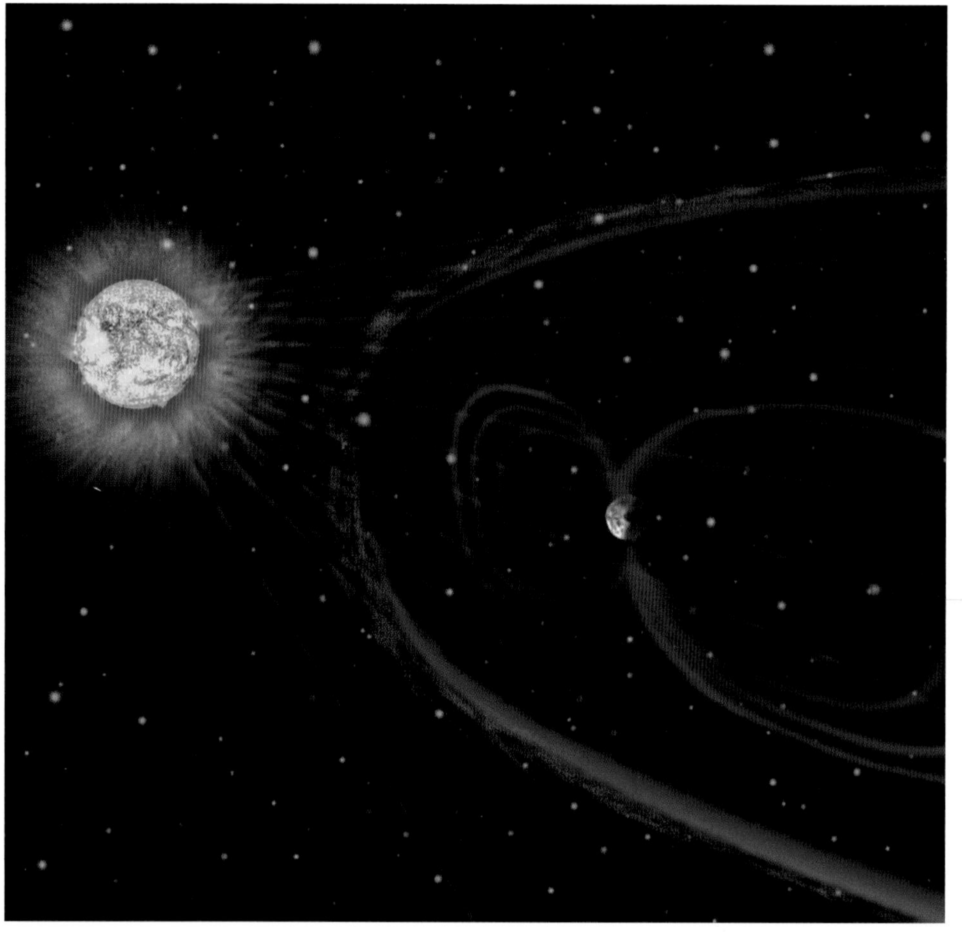

The solar wind plows into Earth's magnetosphere in this illustration. Our planet's magnetic field lines are shown in purple, compressed by wind on the side facing the sun.

AN ENORMOUS PINWHEEL

Today, scientists have put together a rough picture of the solar wind and its effects, although many mysteries remain. Essentially, this wind is the sun's outer atmosphere, its corona, escaping into space. The plasma of charged particles—protons, electrons, and ionized atoms—travels at supersonic speeds between 450 and 600 kilometers per second (280 to 370 mps) as it leaves the sun. It's a thin gas, with only eight protons per cubic centimeter on average, but the sun sheds two million tons of its mass this way every second. Even so, our enormous star has lost less than 0.1 percent of its gas since its formation billions of years ago.

The solar plasma, which functions as an excellent electrical conductor, carries with it the sun's magnetic field, dragging it outward like a comb pulling strands of hair. But because the sun rotates, the magnetic strands swirl around the sun like a pinwheel as they flow outward. Higher speed streams blast out from coronal holes (see p. 67), while lower speed strands flow from coronal streamers. Sometimes the fast-moving streams overtake the slower streams, producing shock waves and accelerating the particles. Occasionally, stormy solar blasts carry with them magnetic clouds, magnetic fields embedded in the erupting plasma.

SPACE WEATHER

The solar wind and its attendant magnetism usually reaches Earth several days after leaving the sun. Because Earth is protected by its own magnetic field, most of the particles are deflected along magnetic field lines, the Van Allen belts, without reachig the surface. The solar wind compresses Earth's magnetic field, condensing it to within ten Earth radii along its leading edge and pulling it out in a long teardrop shape behind. On occasion the wind will let up—for instance, between May 10 and 12, 1999, a relaxation in the solar wind allowed the magnetosphere to expand its volume over 100 times. On the other hand, intense solar storms, such as flares or coronal mass ejections, can drive masses of charged particles in shock waves toward Earth, where they can knock out power or disrupt radio communications, global positioning systems, and more. Currents produced during geomagnetic storms can melt the copper in transformers. Blasts from solar flares can even drive a portion of Earth's atmosphere away into space. However, solar particles, dancing along Earth's magnetic field lines, also create the beautiful high-latitude displays of the aurora borealis and aurora australis (see p. 95 and p. 99).

As charming as the auroras may be, space weather can be highly dangerous to spacecraft, aircraft, and their human cargo. Solar storms heat and expand the upper atmosphere, increasing drag on spacecraft in low orbits. Intense radiation during these same storms can conceivably penetrate aircraft near the poles, bringing on radiation sickness. The same kind of radiation could be fatal to astronauts outside of our protective magnetosphere on missions to the moon or Mars. Like weather forecasters on the surface, government agencies track space weather and issue advisories of upcoming solar storms.

THE HELIOSPHERE

The solar wind hardly stops with Earth. In fact, the intertwined plasma and magnetic fields balloon out from the sun for billions of kilometers, flying at almost 500,000 kilometers an hour (310,700 mph) to about 100 AU from the sun. Along the way, the wind interacts with every planet, asteroid, and comet, deflected around bodies with magnetic fields and bombarding the surface of those without fields. It slows and stops only when it encounters the counteracting pressure of the interstellar medium through which the solar system is moving. The region encompassed by the solar wind and magnetism is called the heliosphere. It is the single largest physical system in the solar system.

Despite its name, the heliosphere is not spherical but comet shaped, compressed on its forward edge where it plows through the interstellar medium, and streaming out behind in a long tail. The region where these two regions interact has several layers. Plowing ahead, the heliosphere creates a bow shock within the interstellar medium. Behind it is the heliopause, the outermost boundary of the heliosphere, where the pressures of the solar wind and the interstellar gases are in balance. The heliosheath, where plasma flows back toward the solar system, streams out behind the heliopause. And behind that, closest to the sun, is the termination shock, where the solar wind slows as it approaches the edge of the heliosphere. Voyagers 1 and 2 have already reached the termination shock on their way out of the solar system (see p. 41).

SOLAR SAILING

The dangerous and erratic solar wind has not yet been harnessed to propel a solar sailboat, but sunlight is another matter. Photons from the sun exert a tiny but perceptible pressure on any object they meet. On Earth, forces such as friction easily counteract the pressure of sunlight. But in space, the impact of innumerable photons can and does move floating bodies. NASA and other agencies have developed prototype spacecraft that would employ football field-size arrays of tissue-thin metallic cloth as solar sails, propelled by sunlight. Pulled by these sails, such an unfueled craft would move slowly at first but pick up speed steadily until traveling at 324,000 kilometers an hour (200,000 mph).

COSMIC RAYS

Any kind of particle bombarding the Earth is known as a cosmic ray (so called because Victor Hess, who discovered them in 1912, originally believed that the radiation he was detecting in the upper atmosphere was electromagnetic). Aside from those from the sun itself, our planet occasionally encounters particles from the Milky Way or from interstellar gases that have interacted with the heliosphere. Like a planet's magnetosphere, the solar system's vast heliosphere shields it from charged particles in interstellar space. Neutral particles, though, aren't affected by the magnetic field and flow through the heliopause at 25 kilometers a second (15 mps). Some get tossed around by the termination shock and become charged and accelerated. Spacecraft sometimes detect these foreign particles, which are known as anomalous cosmic rays and interest scientists for what they tell us about the interstellar environment.

Particles from sources beyond our own galaxy can be intensely energetic. They may originally have been accelerated into space from the outskirts of distant black holes. These rays can hit our atmosphere at nearly the speed of light, colliding with molecules in our air to create "secondary cosmic rays" that shower to the ground.

SOLAR WIND MYSTERIES

The space age has helped us uncover the huge plasma/magnetic bubble that encases our solar system, but many questions remain. We don't know what causes the solar corona's fantastic heat and particle acceleration, the origins of the solar wind. The causes of solar storms and alternating low-energy events are still in doubt. The exact shape and distance of the heliopause are unknown, although Voyagers 1 and 2 may sketch out those dimensions if their radio transmissions hold up. Solar weather forecasting is proving just as complex as its terrestrial counterpart.

THE ACTIVE SUN: THE SUN'S FATE

Happily for the readers of this book, the sun is in its prime. It has spent about five billion years on the main sequence of stellar life, fusing hydrogen to helium in its core, and it has about five billion years to go before that changes. But change it will, with drastic effects for the whole solar system. And long before it leaves the main sequence, gradual increases in its luminosity will probably spell the end of life on Earth.

THE FATE OF EARTH

The sun has been growing in brightness since it entered the main sequence, gaining in luminosity by perhaps one percent every 100 million years. Since its birth, it has become about 40 percent brighter. When life began on Earth, some 3.8 billion years ago, the sun was so dim that the oceans would have been frozen and life impossible, save for the counteracting effects of carbon dioxide in the atmosphere, blanketing the planet with the greenhouse effect.

Multicellular life has appeared relatively recently in the planet's history and has evolved to live in what we consider a pleasant average temperature of around 13°C (55°F). But it will have to continue evolving quite dramatically to keep up with increasing global temperatures as

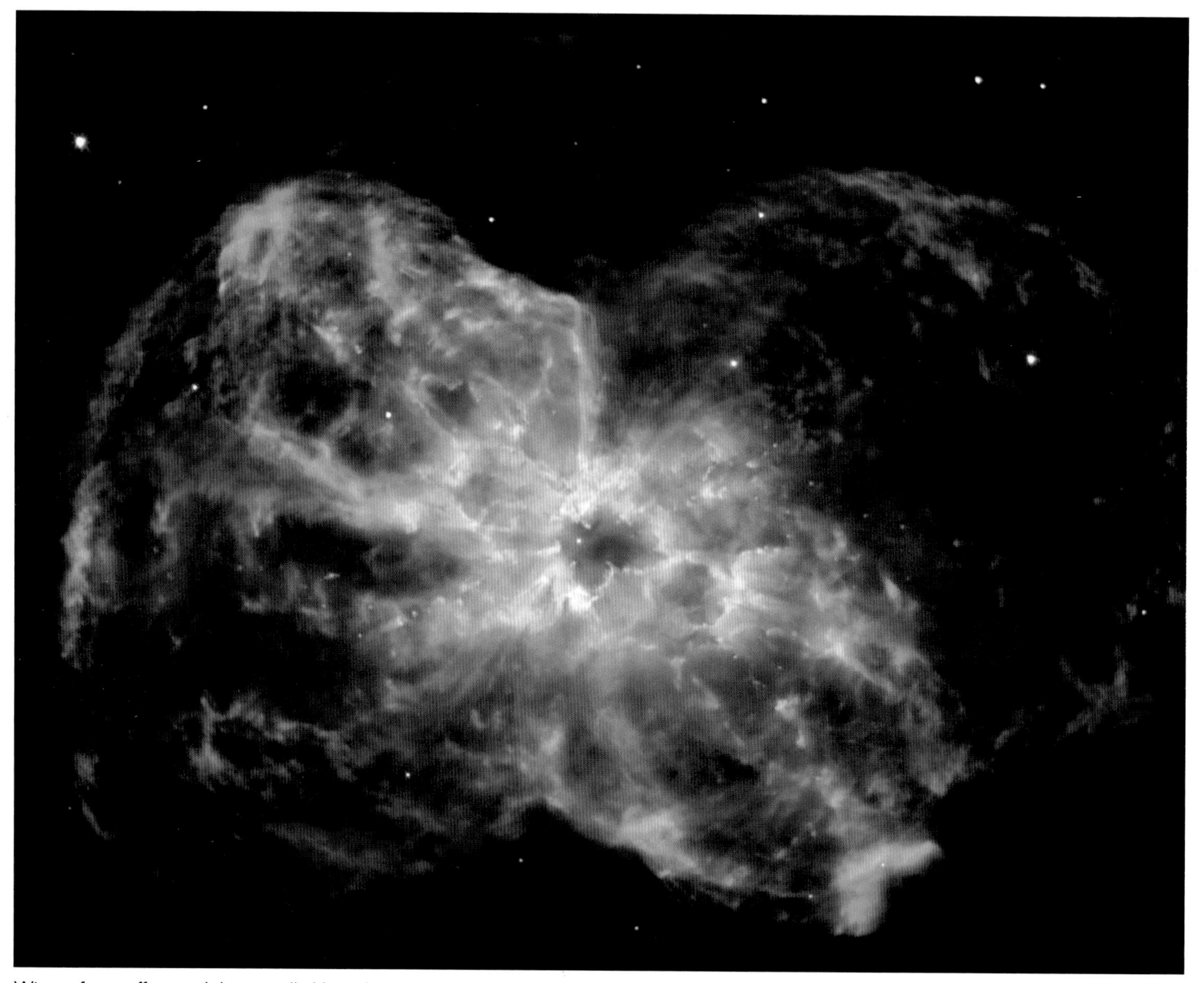

Wings of cast-off gas and dust expelled by a dying star reach far into space, creating planetary nebula NGC 2440. Energized by ultraviolet light from its extremely hot central star, this nebula is more than a light-year wide and about 4,000 light-years away from Earth.

the sun grows brighter and brighter. Even discounting other planetwide events, such as drastic changes in the CO_2 levels, massive flooding or ice ages, in about a half billion years Earth will warm up past an average temperature of 43°C (110°F). On the desert world, carbon dioxide levels will drop and photosynthesis will cease. At one billion years, with the sun 10 percent brighter and hotter, Earth's oceans will boil away.

THE RED GIANT

The indifferent sun will continue on for another four billion years, steadily brightening but remaining on the stellar main sequence. Radiation from fusion in its core will balance the pressure of gases pulled in by gravity. But it does not have an unlimited supply of hydrogen fuel, and when it runs out of hydrogen to fuse in its core, it will undergo a dramatic sequence of events that transform it into a red giant.

Without the outward pressure caused by fusion, the helium core will begin to collapse. As it compresses, it heats again and releases the heat to the layer of hydrogen just outside the core. This gas, heated and compressed, burns in a hydrogen shell. The outer layers of the sun, also known as the solar envelope, expand around this hot shell, ballooning past the orbit of Venus to 100 times its current diameter. Portions of the sun's gassy mass will blow off through the solar system. Its cooler surface will glow a dim red, but because it is so much bigger, the sun will be 2,000 times brighter than it is today.

Meanwhile, the helium core within the fusing hydrogen shell has compressed to a relatively small sphere about twice the size of Earth. But as the fusing hydrogen deposits helium onto the core, it heats up under the increasing mass. When the temperature reaches 100 million kelvins, the helium itself begins to fuse into carbon and oxygen. For about 100 million years the sun stabilizes as helium burns, but when that fuel is used up, it begins to collapse again.

THE WHITE DWARF

The old sun's core becomes denser and denser, reaching a density of one ton per cubic centimeter. Finally it can collapse no more, because the constantly moving electrons in its dense mass maintain an outward pressure. The unstable sun repeatedly ejects its outer layers, forming beautiful widening shells of gas known as a planetary nebula. (This misleading term was coined by astronomer William Herschel because the misty object he observed reminded him of the planet Uranus, which he had recently discovered. Planetary nebulae are stellar and have nothing to do with either planets or galactic nebulae). Gases spreading into space from the dying star are enriched in heavy elements, seeding space with the materials of new solar systems. Inside these gassy shells, the dense core of the old red giant remains as a white dwarf star. No longer burning fuel, it cools and dims over the millennia, eventually becoming a frigid, dark cinder.

And what of our poor old home, planet Earth? Scientific opinions differ on the fate of the planets. Much depends upon the exact gravitational interactions between the ballooning sun and its system. Most likely, under the weaker gravitational pull of a red giant star, the planets would migrate outward in the solar system, with Earth reaching the current orbit of Mars. By that point, however, Mercury and Venus would have been engulfed by the red giant's outer atmosphere, and Earth would long ago have lost its oceans to the brightening main-sequence sun. Even if Earth remained outside of the sun's outer layers, its circling mass would probably create a tidal bulge in the sun's gases. Earth's moon would go first. Its orbit degraded by the sun's nearby pull, it would spiral toward Earth until tidal forces tore it apart. At first, Earth would have a Saturn-like ring of moon debris, but eventually the pieces would cascade from the skies onto Earth's surface.

This would hardly matter, though. By then, tidal forces would be dragging Earth as well into the incinerating body of the sun. By the time the sun shrinks to a white dwarf, its planetary family may consist only of the outer planets and their now watery moons. The distant Oort cloud, released from its tenuous gravitational hold, would drift away into interstellar space. The disruption could be greater, however, if a star happens to pass near the sun at this point in the scenario—that would strip away more material. In any case, the gas giants will move farther out, and for Jupiter, at least, its iceball moons will become waterball moons.

Our sun (seen here in an extreme ultraviolet image from NASA's STEREO satellites) has grown steadily brighter throughout its existence. Long before it becomes a red giant, life as we know it will become impossible on Earth.

RESCUE PLANS

Although Earth's toasty fate does not put a crimp in most people's immediate plans, some scientists have indulged in speculation about ways to save the planet. Most involve shifting Earth to a more distant orbit, which is not as bizarre as it sounds. In the early solar system, gravitational forces routinely shoved the planets about. Even without human interference, a passing star might sufficiently disrupt Earth's orbit at some point that it could be ejected from the solar system. It would, of course, freeze, but bacterial life might remain in hot spots under the sea. Or desperate future humans could create a gravity tugboat, similar to the scenario suggested for averting Earth-impacting asteroids (see pp. 126–29). By changing the orbit of a large asteroid or comet so that it passed near, but did not hit, Earth, Earth's orbit might be slowly altered over the years into a more distant circuit. However, most people agree that the more reasonable course is to find another, younger Earth in another, younger star system, and leave the dying sun behind.

THE INNER PLANETS

THE INNER PLANETS

The inner planets were shaped in the hot embrace of the nearby sun. They ended up rocky, small, and dense—and one, at least, orbited at just the right distance to hold on to watery oceans and host the chemicals of life.

The eight major planets can be divided neatly into two groups of four. Close to the sun are the four inner, terrestrial planets—so called because they are more or less Earthlike. Far from the sun, beyond the asteroid belt, orbit the four gas giants. The inner planets are compact and rocky, with a paltry three moons among them. Their outer siblings are huge and vaporous, possessing rings and more than 160 natural satellites.

The dramatic differences between the two different groups of planets help us to understand the complex manner in which the solar system was formed, a subject of scientific discussion for centuries. Eighteenth-century philosopher Immanuel Kant proposed that the solar system condensed from a disk-shaped cloud of particles (see p. 80). Brilliant French mathematician Pierre-Simon Laplace was the first to tackle the physics of such a disk, or solar nebula. In his 1796 book *Exposition du système du monde (The System of the World)* he explained how the conservation of angular momentum would make a solar nebula spin faster as it contracted, which would force it into a platelike shape. This model of the early solar system is now known as the Kant-Laplace nebular hypothesis. Although it couldn't account for anomalies like the speed of planetary orbits or the retrograde motion of some moons, it was an important step toward understanding the origins of worlds.

The prevailing hypothesis holds that the sun and planets began to take shape about five billion

14–13 bya
Universe begins with big bang.

5 bya
Solar nebula begins to collapse.

4.7 bya
Protosun grows in middle of solar nebula. Disk forms.

4.6–4.4 bya
Particles of metal, rock, ice form in disk's cooler regions. Planetesimals form.

4.6–4.4 bya
Sun begins nuclear fusion. Solar wind blows. Planets reach current size.

years ago (bya), formed from a huge interstellar cloud of gas and dust, the solar nebula. Made almost entirely of hydrogen and helium, the nebula also contained small but important amounts of heavier elements produced in the immense pressures of supernova explosions elsewhere in the galaxy: oxygen, carbon, iron, nitrogen, and others.

One theory posits that matter in the cloud began to condense and clump together simply under its own gravity. Another, more recent theory suggests that a massive shock wave—possibly emanating from a nearby supernova or from the interstellar wind of a supergiant star—jump-started the process by rippling through the nebula and compressing its matter (see Chapter 2). As the cloud collapsed, it rotated.

At first, the nebula was bitterly cold. Ice and dust made of heavy elements floated through the gas, all of it gradually pulled toward the center of the nebula. As the cloud condensed, it rotated more quickly, flattening into a cold disk with a warmer bulge at the center. This warm center, the protosun, was as wide as the current orbit of Mars and glowed red from gravitational contraction. Heat from the protosun began to vaporize ice in the inner solar nebula and push lighter elements, such as hydrogen and helium, toward the outer reaches. Rings of matter begin to circle the infant sun in distinct orbits.

Around the growing protosun, rocky dust and gravel made of heavier elements stuck together in larger and larger clumps, known as planetesimals, ten kilometers (six mi) across or more. Growing rapidly, these collided to create even larger bodies, known as protoplanets, gradually sweeping their orbital lanes clear as their gravity pulled in more material. When the protosun acquired enough mass to begin nuclear fusion, the resulting blast of solar wind and radiation swept through the young solar system like a blowtorch, flinging lighter elements into the farther reaches of the system and heating the surfaces of the inner planets—now fully grown. We know that all of the events of this stage, including the planets' maturation to their current size, happened about 4.5 billion years ago.

How do we know this? Planets don't come with date stamps indicating their time of creation. So how do we know how old the solar system is?

Our answers come from measuring the rate of radioactive decay in rocks

4.6–4.4 bya
Solar wind forces lighter gases outward. Inner planets become rocky.

4.6–4.4 bya
Main asteroid belt forms. Jupiter's field keeps planetesimals from accretion.

4.5–3.8 bya
Frequent collisions occur.

4 bya–3.5 bya
Water on Martian surface vanishes as planet grows dry and cold.

3.56 bya
Cellular life begins on Earth.

from Earth's crust, its moon, and meteorites. Long-lived radioactive isotopes decay at predictable rates, with half-lives—the time that it takes for half the atoms to decay—between 700 million and 100 billion years. Measurements of the relative levels of these elements, such as the uranium isotopes in lead, give us a pretty close approximation of the age of ancient rocks.

The oldest crystals found on Earth date back some 4.3 billion years. However, Earth's surface has been disrupted by the powerful forces of plate tectonics and erosion, so our oldest terrestrial samples don't go all the way back to the beginnings of the solar system. To find that date, geologists study both moon rocks and meteorites. The oldest moon rocks are 4.4 million to 4.5 million years old, whereas most ancient meteorites date back 4.53 billion to 4.58 billion years. Knowing that rocky particles all formed around the same time in the early solar system, we can take the age of meteorites as the age of the solar system in general—or at least, the age at which its solid bits began to condense from the solar nebula.

Incinerated by heat and solar particles, the newly formed inner planets couldn't hold on to ice or gases. They began their adult lives as dense, hot, rocky bodies made of silicates (metallic compounds), iron, and nickel. The torrent of radiation also meant that new planetesimals could not form. However, the solar system was not clear of debris. Massive chunks of ice and rock, some almost the size of planets, were still careering about among the planetary orbits. The young worlds were about to enter the age of collision.

The early solar system was a rough neighborhood. Between 4.5 billion and 3.8 billion years ago, chunks of rock ranging in size from relatively petite boulders to small planets smashed into one another as they hurtled around the sun. The early collisions bulked up the biggest inner planets by accretion. Already blazing hot from their formation process near the primordial sun and from the decay of radioactive isotopes within them, the young planets heated up even further as the planetesimals plowed into them, converting the impact energy into heat. Denser matter, such as iron, gradually sank toward the molten core of the inner planets, while lighter elements remained near their surfaces and cooled off in a process known as differentiation.

IMMANUEL KANT
THEORY OF PLANET FORMATION

German philosopher Immanuel Kant (1724–1804) was the unlikely author of the first nebular theory of planetary formation. Famous now for his writings on knowledge and ethics, Kant was interested in a broad range of scientific issues in his early years and was an avid student of Newtonian physics. In 1755, he suggested that the sun and planets had been formed from a rotating, disklike cloud of scattered particles. Gravity pulled these particles together, he wrote, and chemical reactions bonded them into the heavenly bodies.

The heavy bombardment also left scars, impact craters still visible on the terrestrial planets and their moons. The biggest impacts packed enough punch to permanently deform some planetary surfaces, such as forcing up hills and mountains, or even knock an entire planet askew. Mercury and Earth's moon, in particular, display the battering they took during the era of heavy bombardment. The 1,300-kilometer-wide (800-mi) crater Caloris Basin on Mercury, for instance, is a reminder of a giant blow to the planet in its early days. Directly opposite the crater, on the other side of the planet, is a huge region of ridges and hills pushed up by the impact's shock wave. A similarly massive impact may have sent Venus spinning backward in the retrograde rotation it has today. And it is thought that a titanic collision almost certainly gave birth to Earth's moon (see pp. 102).

Compared with the moon or Mercury, fewer craters are evident on Venus, Mars, and Earth, not because they didn't get walloped, but because their surfaces have continued to change over time. Vast lava flows have smoothed out much of Venus's terrain, while floods, volcanic eruptions, and dust storms have resurfaced Mars. Weathering and the recycling effects of plate tectonics have eliminated most of Earth's old craters, as well.

As the era of heavy bombardment died down after about 3.8 billion years ago, the terrestrial planets cooled and began to evolve into the worlds we see today. Formed from similar materials in the near reaches of the sun, they have a similar composition and structure, at least compared with the outer planets. All have heavy metallic cores composed mostly of iron; hot, medium-density mantles containing iron-rich silicate rocks; and lower density, basaltic and granitic rocky crusts. Decaying radioactive materials continue to heat them from the inside. Their rocky surfaces are fractured and wrinkled into mountains, canyons, and volcanoes. Compared with the gas giants, the terrestrial planets are quite dense, due in part to their iron cores.

The lion's share of rings and moons went to the outer planets. No terrestrial planets have rings—though Earth's moon may break up into rings in the very far future—and only Earth and Mars have natural satellites. Earth's moon is one of the largest and roundest of the solar system's satellites. Deimos and Phobos, Mars's tiny, irregular moons, may be captured asteroids.

AIR AND WATER

All the terrestrial planets have secondary atmospheres (produced after their formation). These range from Mercury's almost nonexistent wisps of hydrogen, helium, and sodium to Venus's crushing smog. Earth's atmosphere is notable for high oxygen content, courtesy of the green plants in its oceans and on its surface. At least three of the inner planets may once have had oceans: Venus, whose seas may have boiled off as part of the greenhouse effect; Mars, whose once liquid oceans might now be frozen under its surface; and of course Earth, the blue planet, orbiting at just the right distance from the sun to maintain liquid water on its surface.

MYSTERY PLANET
MERCURY

Tiny, sunbaked Mercury is not actually very far from Earth as astronomical distances go, but it nevertheless remains one of the most enigmatic members of the vast solar system.

Diminutive Mercury is just 4,878 kilometers (3,031 mi) in diameter, which makes it the smallest major planet in the solar system, slightly larger than Earth's moon. Its orbit is highly eccentric, or off-center, bringing it within 46 million kilometers (28.58 million mi) at perihelion—its closest point to the sun—and as far away as 69.82 million kilometers (43.38 million mi) at aphelion—its farthest point from the sun. Any planet snuggling up to the sun that way is bound to be hot, and Mercury indeed reaches daytime highs of 427°C (801°F), while during its long night the temperature makes an Olympic-size plummet to -173°C (-279°F). Curiously, it's not the hottest planet. Venus beats it out for highest temperature due to the insulating effects of its atmosphere. On Mercury, which is virtually airless, heat is radiated efficiently back into space as its surface turns away from the sunlight, so its temperature extremes are the greatest in the solar system. • The little planet is surprisingly dense. How it got that way, the details of its surface, and many other questions had to await space age exploration before scientists could begin to answer them.

Symbol: ☿
Discovered by: Known to the ancients
Average distance from sun: 57,909,175 km (35,983,095 mi)
Rotation period: 58.65 Earth days
Orbital period: 0.241 Earth years or 88 Earth days

Equatorial diameter: 4,878 km (3,031 mi)
Mass (Earth=1): 0.055
Density: 5.43 g/cm3 (compared with Earth at 5.5)
Surface temperature: -173°C/427°C (-279°F/801°F)
Natural satellites: none

SKYWATCH

* Mercury travels near the sun, so you can see it only in the early evening, low in the west, or early morning, low in the east.

AMAZING FACT The sun in Mercury's sky would seem to grow larger and smaller over the course of a year.

1639
Giovanni Zupus discovers the phases of Mercury.

1889
Giovanni Schiaparelli deduces, wrongly, that Mercury always has a face to the sun.

1965
Astronomers at Arecibo discover true rotation of Mercury, 58.6 Earth days.

1974–75
Mariner 10 makes three flyby visits and maps part of Mercury's surface.

2008–11
MESSENGER spacecraft flies by Mercury three times before settling into orbit.

In 2008, MESSENGER's wide-angle camera recorded portions of Mercury's surface never before seen (opposite). A false-color image (above) highlights gigantic Caloris Basin.

MERCURY: CRATERED WORLD

As the innermost planet, Mercury appears no more than 28 degrees away from the sun in our skies and is typically hidden in the solar glare. Even so, it was familiar to keen-eyed ancient observers. The Babylonians named the fast-moving, sun-hugging body Nabu or Nebu, a record-keeper and messenger of the gods. The Greeks knew it as Apollo in its evening appearance and as Hermes (the messenger) in the morning, an association the Romans retained with Mercurius, the fleet-footed courier.

Centuries passed with little added to our knowledge of the planet. In 1639, astronomer Giovanni Zupus used an early telescope to discover that Mercury has phases, like Venus, further evidence of the Copernican theory. Frustrated astronomers, squinting toward the tiny object, attempted to measure its position and orbit by observing its transit across the face of the sun. In 1607 Johannes Kepler thought he had seen such a transit. He was so thrilled that he ran all the way to the castle of Emperor Rudolf II, his patron, to tell him about it. It turned out that the astronomer had actually seen a sunspot, a phenomenon not known until Galileo documented spots a few years later. "Did I pass off a spot I saw as Mercury?" Kepler asked later. "Then lucky me, the first in this century to observe sunspots."

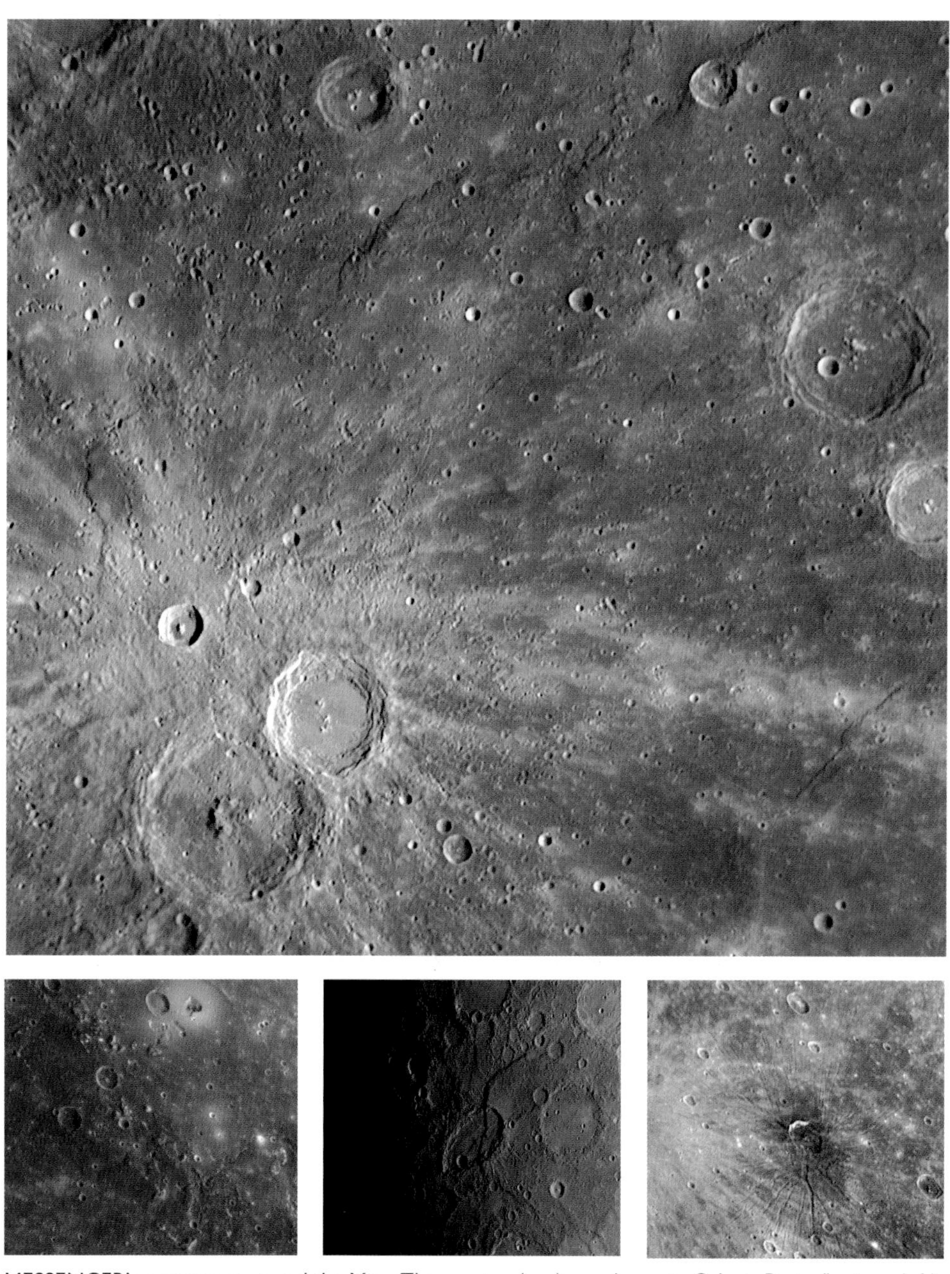

MESSENGER's cameras captured the Xiao Zhao crater (top); a volcano in Caloris Basin (bottom left); Beagle Rupes, a cliff cutting across a crater (bottom middle); and Pantheon Fossae (bottom right).

ONE DAY ON MERCURY

Over time, more powerful and precise telescopes revealed vague streaks and blots on Mercury's surface, features that never seemed to change, no matter when the planet was observed. Some astronomers interpreted this to mean that Mercury's day was the same length as Earth's, so the planet happened to turn the same face toward us at the same time each day. But 19th-century astronomer Giovanni Schiaparelli, now famous for his observations of Mars (see p. 37), thought differently. Like the moon, he said, Mercury was locked into a synchronous rotation, with one side perpetually toward the sun and the other facing outward. In such a rotation, its day would be the same as its year, 88 Earth days. This clever—but incorrect—explanation was the standard model until the 1960s.

In 1965, the American astronomers Gordon Pettengill and Rolf Dyce bounced radar pulses off the planet and used the huge Arecibo radio telescope in Puerto Rico to measure the returning signals. They discovered that Mercury did, as they observed, rotate slowly relative to the sun, spinning once on its axis every 58.6 Earth days, or three times for every two of its years. Because it is moving rapidly around the sun during this slow rotation, its solar day, from sunrise to sunrise for any overheated person on the planet's surface, lasts 176 Earth days. In addition, the three-rotations-in-two-years arrangement forces the same two points on Mercury's equator (at longitudes 0° and 180°) to face the sun at alternating perihelia (the closest points to the sun). Thus these are called the hot poles. The points on the equator at 90° and 270° longitude face the sun at alternating aphelia (the farthest points from the sun); these are termed the cold poles.

Mercury's 3:2 day-to-year ratio is a result of spin-orbit coupling, a phenomenon seen throughout the solar system, including in the 1:1 ratio of the moon's rotation to its orbit around the Earth. The sun's gravity pulls strongly on the nearby planet, actually stretching its solid body a little toward the sun's far greater mass. The pull is stronger on the portion of Mercury nearest the sun, and strongest still when the planet is closest to the sun in its orbit. These tidal forces, a regular pattern of tugs and twists, have settled into a steady resonance that locks Mercury into its day/year pattern.

TRAPPED IN THE GRAVITY WELL

Little Mercury's orbit played a big role in confirming Albert Einstein's theory of relativity. For centuries, astronomers knew that Mercury's elliptical path was not quite right. Mercury's perihelion advances slightly each year, a motion called precession. This is true for all planets, since their orbits are affected by the masses of other bodies as well as the sun. However, Mercury's precession was greater than Newtonian mathematics predicted, a discrepancy of some 43 arc seconds per century. Astronomers scrambled to explain the puzzle, even for a while assuming the presence of a small planet, preemptively named Vulcan, inside Mercury's orbit. Einstein solved the problem with a 1915 paper, "Explanation of the Perihelion Motion of Mercury by Means of the General Theory of Relativity." Einstein's theory of relativity explained that space is distorted in the presence of matter, an effect virtually invisible for most objects but noticeable close to enormously massive bodies, such as the sun. The physicist's equations showed that the curvature of space would advance Mercury's precession by exactly 43 arc seconds.

CRATERS UPON CRATERS

In 1974 and '75, NASA's Mariner 10 spacecraft made three flyby visits to Mercury, during which it was able to map 45 percent of the planet's surface. This mission contributed the only detailed information on the little world until almost 25 years later, when MESSENGER made its first flyby in 2008.

Most of Mercury is pocked by impact craters of various sizes, the largest of which are occupied by smooth plains. Plains also run between craters in the uplands. Long, clifflike "lobate scarps" cut across the surface for up to hundreds of kilometers, while in one region, an area of blocks and troughs makes up a disordered landscape.

Impact craters large and small are a legacy of the solar system's violent and chaotic early history. The period of heavy bombardment, when meteoroids of all sizes smacked into the planets, reached its peak about 3.9 billion years ago and left its marks most obviously on Mercury and Earth's moon. Craters on Mercury range from the typical bowl shape of simple craters to bigger, complex craters with central peaks and terraced rims. Striated markings (ray systems) radiating from larger craters are the scars left by material ejected by impacts. So are secondary craters produced by impact from the resulting debris. Unlike those on the moon, these secondary craters cluster fairly close to the parent crater, testifying to Mercury's higher surface gravity.

The biggest crater seen so far on Mercury is the colossal, multiringed Caloris Basin. Astronomers believe it was created when a huge asteroid slammed into the planet with the force of a trillion hydrogen bombs. This massive asteroid impact most likely occurred during the era of heavy bombardment. So tremendous was this blow that powerful seismic waves echoed through the planet and disrupted the surface on the opposite hemisphere. Today, the hilly, scarred, jumbled region there is known informally as the weird terrain.

In the 1970s Mariner 10 saw only part of the Caloris Basin crater, which lay along the day/night boundary during that mission, but MESSENGER managed to view the entire basin during its first 2008 flyby. It was even larger than previously estimated, with a diameter of 1,550 kilometers (960 mi). Craters scattered within the basin include the strange "spider," now known as Pantheon Fossae, a crater with a spray of troughs radiating from it like an insect with way too many legs. Scientists aren't yet sure how to explain this unusual feature.

MERCURY'S TWO VISITORS

Only two spacecraft have visited Mercury since the space age began: Mariner 10 in the 1970s and MESSENGER, launched in 2004. Both voyages involved gravity-assist maneuvers that swung them past Venus.

Mariner 10, launched in 1973, was the first spacecraft to use a gravity-assist boost to change its flight path. It returned images of almost half of the planet's cratered surface and took readings of Mercury's temperature, atmosphere, and magnetic field.

MESSENGER reached Mercury in 2008—nearly 25 years after Mariner—following two Venus flybys. The lightweight, technologically advanced probe is now proceeding to map almost the entire planet in color and take measurements of its surface, atmosphere, and magnetosphere.

MERCURY: IRON WORLD

Mercury is perhaps most fascinating for what we can't see and what we don't know. Both its surface, marked by an array of puzzling features, as well as its internal structure, seem to be far different from what scientists once expected, leading us to rethink our notions of how the terrestrial planets were born.

Scientists are still debating how to explain the smooth plains that fill areas between Mercury's craters and even out the floors of some larger craters and basins. One theory interprets them as a smooth layer of debris ejected by impacts. Another views them as cooled lava from ancient volcanoes. The volcano theory gained support after MESSENGER's initial flyby, which seemed to show volcanic vents along Caloris's margins.

Lobate scarps, also known as thrust faults, cut through craters and wind across the landscape. Between 20 and 500 kilometers (12 and 300 mi) long and up to 3 kilometers (1.8 mi) high, they may represent wrinkles on Mercury's skin, forced up as the planet cooled and shrank.

HARDBALL

Early clues to Mercury's interior oddity came from the first measurements of its mass in 1835. At the time, the best way to estimate a planet's mass was by its gravitational effects on its moons. Mercury had no satellites, but it did have occasional visitors in the form of comets. German astronomer Johann Encke tracked the swerving orbit of a comet as it approached Mercury and calculated a mass for the planet close to the modern figure of 3.3×10^{23} kilograms—surprisingly massive for such a petite planet. Dividing the little planet's mass by its known volume gives it a density of 5.4 g/cm^3, close to that of Earth.

The most likely explanation for such a high density is a huge iron core. In the case of Mercury, its core would have to make up roughly 75 percent of the planet's diameter (compared with Earth's core, for instance, which extends for about 54 percent of the planet's diameter). Mercury's surrounding mantle and crust, probably made of silicates like Earth's, would be correspondingly thin, perhaps 600 kilometers (370 mi) thick.

Cutaway artwork of Mercury reveals its disproportionately large iron core (yellow) within a silicate crust (orange). The core may have grown when another body smacked into Mercury in its youth.

The long-running MESSENGER mission is delivering reams of scientific data. The spacecraft detected an unexpected magnetic field around the planet, making it the only terrestrial planet other than Earth to possess one. Like Earth's field, Mercury has north and south magnetic poles roughly corresponding to the planet's geographic poles. As on Earth, Mercury's magnetosphere, the area where the magnetic field dominates and deflects the solar wind, is probably compressed on the side near the sun and elongated on the far side, dragged outward by streaming solar particles.

The presence of this magnetic field led scientists to wonder if Mercury, like Earth, had an electrically conducting, partially molten outer core surrounding a solid inner core. Support for this theory comes from recent radar studies of the planet's spin, which wobbles just slightly, as we might expect if it had a liquid layer sloshing inside it. Explaining such a molten core is tough: Earth's might be partially liquid due to high temperatures inside our planet, but Mercury, much smaller, should have cooled all the way through by now. Possibly the outer core is mixed with sulfur, which lowers the melting point of iron. Recent experiments in which molten iron and sulfur were subjected to pressures like those at Mercury's core showed that the condensing iron formed little iron "snowflakes" that drifted downward. If iron snow is falling within Mercury's liquid core, it could create convection currents that would give rise to Mercury's magnetism.

MERCURY'S ORIGINS

Mercury's odd anatomy continues to puzzle planetary geologists, and has led to some rethinking about the formation of all the terrestrial planets. To pick up the amount of iron and sulfur it has now, the growing planet must have pulled in materials from farther out in the solar system, even past the current orbit of Earth. This implies that the early solar system was more of a mixing bowl than previously thought. Even so, the huge iron core and thin mantle need further explanation. One theory holds that conditions in the innermost regions of the solar nebula blew away most of the silicates, but not the iron, as the planet accreted. A second, similar hypothesis says that intense solar radiation and the solar wind knocked the silicates off the forming planet. A third theory, the giant impact theory, imagines the infant Mercury at twice its present size. A head-on collision with a huge planet-size asteroid could have knocked the lighter mantle right off the young planet while merging the impactor's iron core with Mercury's. However the planet originally formed, it seems to have shrunk by at least 20 kilometers (12 mi) in diameter since then as it cooled, wrinkling its thin, rocky surface.

ICE AT THE POLES?

Radar studies of Mercury's north and south poles revealed surprisingly reflective surfaces within some deep craters. Could water ice exist on the planet nearest the sun? It's not as unlikely as it sounds, because craters at Mercury's poles spin in perpetual darkness. Mercury's axis is straight up and down relative to its orbit, so it experiences no seasons and no shift of sunlight across its polar regions. The floors of deep craters would be extremely frigid, below -138°C (-216°F). Water vapor might have outgassed from the young planet and remained frozen within the craters, or alternatively, the ice might have been delivered by impacting comets. On the other hand, the shiny material might be frozen sulfur or even reflective, rocky silicates. Although MESSENGER will not fly over the poles, hydrogen-detecting instruments onboard may help point toward or away from the presence of H_2O.

A BARELY THERE ATMOSPHERE

Surprisingly for such a barren world, blasted by the solar wind, Mercury does have an atmosphere, but it barely earns the name. Vanishingly thin, with a pressure a trillion times less than Earth's, Mercury's atmosphere is more properly called an exosphere. It consists of scattered atoms bouncing about on Mercury's surface—primarily hydrogen, helium, oxygen, sodium, potassium, and calcium. The hydrogen and helium probably arrive on the solar wind. The source of the other elements is unknown. The wispy gases may be released when meteorites strike surface rocks or when particles from the sun sputter atoms from the rocks. Some may leak from the planet's interior. Spectrometers aboard MESSENGER may also clarify the nature and origin of this exosphere.

MESSENGER PLAYS GRAVITATIONAL PINBALL

The long-running MESSENGER mission has accounted for our collective, national, revived interest in the innermost planet. Reaching the planet is a singularly tricky feat in which the spacecraft loops around and around the sun and inner planets in a complex game of gravitational pinball. Launched in 2004, the car-size craft is only the second visitor to the planet in 30 years. It is getting to Mercury using gravity assists from the bodies it passes on the way, picking up little boosts from the planets' own angular momentum to finally put it into position to orbit Mercury in 2011. The entire voyage will cover 7.9 billion kilometers (4.9 billion mi), at times achieving speeds up to 225,300 kilometers an hour (140,000 mph) relative to the sun.

After looping once around the Earth, MESSENGER swung twice past Venus in 2006 and 2007, using Venus's gravity to reshape its path toward Mercury. In the course of three flybys past Mercury in 2008 and 2009, it is collecting images of the planet's surface and exosphere using its payload of seven scientific instruments. Finally, in 2011, its looping path around the sun and planets will slow it enough to settle into orbit around Mercury, about 200 kilometers (124 mi) above the planet's surface. There, protected by a ceramic-fabric sunshade, it will turn to mapping the surface and observing the planet at greater length.

Clearly, Mercury can teach us a lot more than we already know about the early solar system and, by extension, about our own planet as well. The MESSENGER mission, extending into the second decade of the 21st century, should take us toward that goal.

MERCURIAL NAMES

Mercury's surface features, such as craters or cliffs, bear an erudite and diverse set of names. According to the International Astronomical Union's rules, craters, for example, are named after important deceased artists, musicians, or authors: Bartok, Byron, Hawthorne, Zola. Cliffs, or rupes, take the names of ships of discovery or scientific expeditions: Beagle, Fram, Resolution. Fossae (troughs) are named for significant works of architecture, as in Pantheon Fossae. Planitiae, plains, take the names used for Mercury (planet or god) in various cultures, such as Budh, Odin, or Sobkou. Valles, or valleys, honor radio telescopes, as in Arecibo Vallis or Goldstone Vallis.

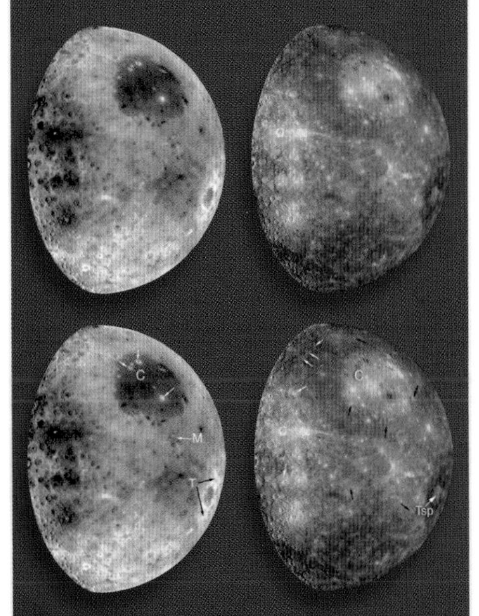

Enormous Caloris Basin is an exception to the rule; it is accurately named "hot," although that description could apply to almost any region of the planet. In the images above, taken though MESSENGER's narrow-band color filters, Caloris is labeled with a C. Below it are Mozart crater (labeled with an M) and Tolstoj basin (Tsp), with bright ejected material radiating from it (T).

EARTH'S EVIL TWIN
VENUS

Our closest planetary neighbor, Venus hides a hellish environment beneath its gleaming clouds. Unlike our own life-nuturing environment, it's the world where nothing went right.

At first glance, Venus could almost be another Earth—or at least, Earth's close sister. The second planet from the sun shares some similar characteristics with our home. Orbiting, on average, 108,208,930 kilometers (67,239,800 mi) from the sun, Venus comes within 38 million kilometers (24 million mi) of Earth at its closest approach every 19 months. In size it's a near match for Earth, only 653 kilometers (405 mi) smaller at 12,103 kilometers (7,521 mi) in diameter. Its mass is only slightly less than that of Earth, and its density and surface gravity are also a close approximation of those on our own planet. Like Earth, it has a substantial, cloudy atmosphere. But space scientists who once hoped to find Venus a welcoming destination were in for a shock as discoveries in the 20th century painted a far different picture of the planet of love. The planet turned out to be a smoggy furnace simmering beneath a crushing acidic atmosphere. The vicious environment rapidly destroyed probes attempting to land on its surface, but in time the spacecraft Magellan, reaching through the clouds with radar, was able to map Venus's rolling terrain.

Symbol: ♀
Discovered by: Known to the ancients
Average distance from sun: 108,208,930 km (67,239,800 mi)
Rotation period: 243 Earth days (retrograde)
Orbital period: 224.7 Earth days

Equatorial diameter: 12,103 km (7,521 mi)
Mass (Earth=1): 0.815
Density: 5.24g/cm³ (compared with Earth at 5.5)
Average surface temperature: 462°C (864°F)
Natural satellites: none

SKYWATCH

* Brighter than any star, Venus is always seen near the sun, up to four hours after sunset or up to four hours before sunrise several months each year.

AMAZING FACT Venus's ancient connection to women: It appears in the night sky for about 260 days a year, about the length of pregnancy.

3000 B.C.
Mesopotamians record appearances of Venus.

A.D. 1610
Galileo observes the phases of Venus: evidence supporting Copernican theory.

1882
Transit of Venus is used to determine the distance from Earth to the sun.

1927–28
Ultraviolet photographs confirm existence of Venus's cloud features.

1966
Venera 3 crashes into Venus, the first spacecraft to impact another planet.

2012
The second and last Venus transit of this century. The next: December 11, 2117.

Data from a variety of Venus missions went into this image of Venus's eastern hemisphere (opposite).
Artwork shows how the solar wind interacts with the planet's thick atmosphere (above), pulling it into space.

VENUS: A GREENHOUSE GONE BAD

Venus is the brightest object in the heavens after the sun and moon, shining brilliantly in both the morning and the evening. Ancient Chinese astronomers called it T'ai-pe, "beautiful white one." Babylonians knew it as Ishtar, mother of the gods, "the bright torch of heaven." The Maya built their calendar around its appearances and disappearances over the course of the year, though for them the bright planet was fearsome rather than alluring. And of course the ancient Romans gave it the name we use now, invoking the goddess of love.

Venus was one of Galileo's first telescopic targets in 1610. His discovery that the planet went through phases, like those of the moon, showed that Venus must actually orbit the sun as predicted by the highly controversial Copernican model of the solar system. Galileo cautiously embedded this revelation in a coded letter to Johannes Kepler. The letters in his message *"Haec immatura a me jam frustra leguntur oy"* ("This was already tried by me in vain too early") could be rearranged to read *"Cynthiae figurae aemulatur mater amorum"* ("The mother of love"—Venus—"imitates the shapes of Cynthia"—the moon).

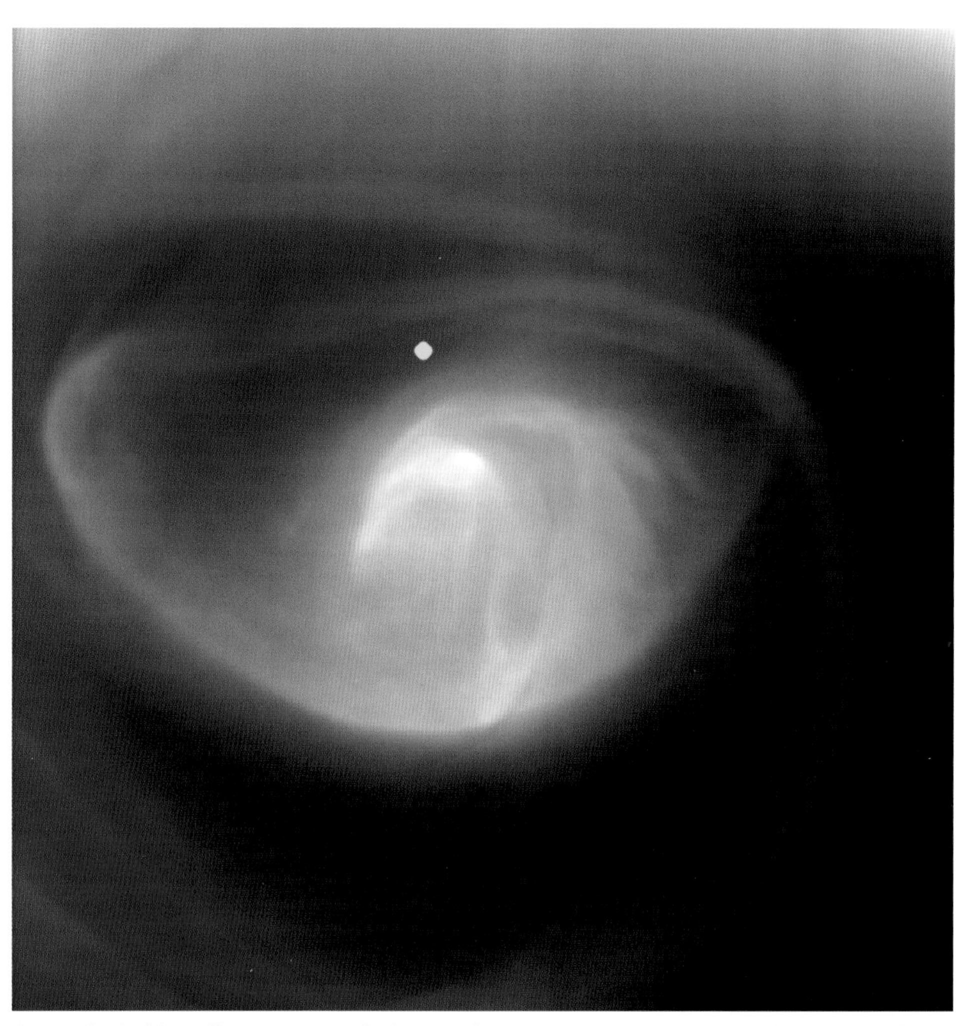

As seen by the Venus Express spacecraft, the eye of an enormous hurricane swirls 60 kilometers (37 mi) above the Venusian south pole (marked by a yellow dot).

SODA WORLD OR INFERNO?

But little else about the planet was obvious. Its impenetrable cloud cover prevented early astronomers from knowing anything about its surface, even its rotation. This lack of information did not deter speculation. Swedish chemist Svante Arrhenius, an early proponent of the greenhouse effect on Earth, vastly underestimated such an effect on Venus. "The average temperature there is calculated to be about 47°C," he wrote (about 112°F). "A very great part of the surface of Venus is no doubt covered with swamps," he noted, supporting "a luxuriant vegetation." This humid hypothesis was knocked down when American astronomers Walter Adams and Theodore Dunham analyzed light from the Venusian atmosphere in 1932 and found evidence for a primarily carbon dioxide atmosphere with almost no water vapor or oxygen. Perhaps, theorized British astronomer Fred Hoyle in the 1950s, Venus was covered with oceans of oil. Other astronomers suggested that the carbon dioxide was mixed into Venus's watery oceans, making them seas of carbonated water.

In the late 1950s, improving technology began to reveal a planet that was stranger, and grimmer, than anyone had imagined. Ultraviolet photographs of Venus's clouds showed that they whipped around the planet from east to west every four days. But radar studies of the surface in 1964 revealed that the planet itself rotated only once every 240 Earth days or so. Its day was longer than its year. Moreover, its rotation was retrograde, or clockwise, with the planet's axis tilted almost perpendicular to the plane of its orbit. Further radio wave and radar research raised estimates of Venus's surface temperature to as high as a blistering 480°C (900°F); its surface pressure, once thought to be close to Earth's, now appeared to be 90 times stronger, a literally crushing level.

A DEATH TRAP FOR SPACECRAFT

The advent of the space age did nothing to rehabilitate Venus's increasingly bad reputation. Both the Soviet Union and the United States made nearby Venus one of the first goals of exploratory spacecraft in the 1960s and

1970s. Ten of the first eleven missions launched by the two countries failed even to reach the planet. None of the early missions that actually entered Venus's atmosphere lasted more than two and a half hours in the destructive heat and pressure.

But as the decades passed, missions to Venus grew more and more successful. In the early 1990s, NASA's Magellan orbiter was able to map more than 98 percent of the planet's surface using radar. And the European Space Agency's Venus Express, which reached Venus in 2006, has contributed valuable information about the planet's turbulent atmosphere. What scientists are learning now is that Venus continues to confound expectations.

Perhaps there was a time in Venus's early history when it resembled the swamp planet of pulp science fiction. But today the greenhouse effect has made Venus a toxic wasteland. The planet's dense atmosphere bears little resemblance to Earth's balmy air. For one thing, Venus simply has more of it. Its clouds form a thick layer rising between 50 and 80 kilometers (30 and 50 mi) above the planet's surface (as opposed to about 12 kilometers/7.5 mi on Earth). These singularly nasty vapors are composed of sulfuric acid droplets, formed by chemical reactions between sulfur dioxide and water vapor in the atmosphere, compounds that may have been vented by volcanoes on the planet's surface. No acid rain ever reaches Venus's surface from these clouds, because any droplets that fall evaporate in Venus's heat, the resulting gas rising again into the clouds.

Below the clouds, the atmosphere is a crushingly dense soup of carbon dioxide (96 percent) mixed with a little nitrogen (4 percent) and traces of water vapor. The pressure at the planet's surface is 90 bars, or 90 times as intense as the air pressure at the surface of Earth. That kind of pressure is the equivalent of what you would encounter if you were 900 meters (2,900 ft) down in the ocean—at almost 1,500 pounds per square inch, it's enough to smash anything as frail as a human body. Before they were destroyed by the hostile environment, Venera landers transmitted pictures from the surface that showed a dim, orange, smoggy light, like Beijing on its worst day. The temperature on the surface averages around 462°C (864°F), hot enough to melt lead. Venus is, in fact, the hottest planet in the solar system, outdoing even Mercury's daytime high. The temperature varies by about 70°C (150°F) between the planet's daytime and nighttime sides, a surprisingly large variation given the thick atmosphere.

THE GREENHOUSE EFFECT

Venus's hellish heat and smog are almost certainly due to a runaway greenhouse effect. The planet may once have resembled the early Earth, with oceans of liquid water. Intense heat from the nearby sun evaporated that water, creating water vapor, a greenhouse gas. Further radiation from the sun broke down the water vapor, while chemical reactions with the surface turned the gases into carbon dioxide, another greenhouse gas. Incoming radiation passed through carbon dioxide to be absorbed by rocks at the surface. Reradiated at a longer wavelength, it couldn't penetrate the clouds, but bounced off the cloud layer back to the surface, heating it further. On Earth, water helps to trap carbon compounds and remove them from the atmosphere, but any liquid water on Venus would have boiled away long ago.

WINDS AND STORMS

Other than that, how's the weather on Venus? It depends on how high up you are. At the clouds' top layers, winds blow at ferocious speeds, up to 370 kilometers an hour (230 mph), traveling in the same east-to-west direction as the planet's rotation. But the planet rotates very slowly, while the winds circle the planet every four days in a "super-rotation." Immense spinning storms, Venusian hurricanes, spiral around each pole, drawing the atmosphere downward like water down a huge drain. Just what causes these storms is still unknown.

Winds above the rest of the planet die down closer to the surface. At the bottom layer of the atmosphere, next to the ground, they move at a poky one meter (three ft) per second. But even these relatively gentle low-level breezes pack some punch because the atmosphere is so dense.

Although an Earth-type thunderstorm is out of the question on a planet without water, instruments have repeatedly picked up visible flashes of light and the bursts of radio waves known as whistlers, typical of lightning. It's not clear just how such lightning would be produced by sulfuric acid clouds.

ALBEDO

Venus is not only the hottest of the eight planets, it's also the shiniest, as determined by its albedo (derived from the Latin *albus*, "white"). Visual albedo measures the fraction of visible incoming light that is reflected directly into space, on a scale of 0.0 to 1.0. Charcoal has a visual albedo of almost 0.0, whereas the whitest snow or the best mirrors come close to 1.0.

Albedo can give us important information about a planet's, moon's, or asteroid's surface. In general, rocky, dusty bodies absorb a lot of light, and icy bodies reflect it. Venus's pearly clouds are highly reflective, giving the planet an albedo of 0.65 and making it visible even at twilight (as in the photo below, seen below the crescent moon). Planets and moons with variegated surfaces will have different albedos across different areas, so their albedos are averages. The albedos of our moon and the solar system's other planets are given below

Mercury:	0.12
Earth:	0.39
Earth's moon:	0.12
Mars:	0.15
Jupiter:	0.52
Saturn:	0.46
Uranus:	0.56
Neptune:	0.51

VENUS: VOLCANIC MYSTERIES

If any observers had lingering hopes for the swampy Venus scenario, these were dashed by the first forbidding pictures from the Russian Venera landers in the 1970s. The Venusian landscape is dry, dry, dry, a parched wasteland of sharp-edged, fractured rocks. These are probably basaltic, igneous rocks formed from cooling lava. Indeed, most evidence to date supports the idea that the planet's surface is shaped largely by volcanic activity.

Perhaps the most pressing issue in geologists' minds when it came to Venus was the question of plate tectonics. Scientists have known for a long time that Earth's geology is dominated by tectonic processes. Our planet's lithosphere—its crust and uppermost mantle—is broken into huge plates that slide over the asthenosphere (see p. 97). Radioactive elements decaying in Earth's core give off heat that rises to the surface, generating convective currents that propel the plates slowly around the globe. The tectonic plates grind into one another in some places, pull apart in others, and occasionally cause havoc for Earth's residents as they slowly but continuously shape the sharply differentiated mountains, trenches, and continents that distinguish our planet.

Mercury, the moon, and Mars had been shown to have no recent plate tectonics. Their surfaces consist of a single, rigid shell. But Venus, of a size similar to Earth's and orbiting at a similar distance from the sun, probably had a hot interior, like our planet. It seemed to be a likely candidate for Earth-type activity.

In the early 1990s, NASA's orbiter Magellan gave us our first detailed map of Venus, using radar to penetrate the planet's dense clouds. It revealed a relatively low, rolling terrain. Two highland regions rise above the plains—Australia-size Ishtar Terra in the north and Aphrodite Terra, about half the size of Africa, near the equator. These cover about 8 percent of the planet's surface.

Aphrodite Terra is marked by cracked, buckled, ridged formations, including long channels that may represent old lava flows. Lakshmi Planum, a vast plateau about 1.500 kilometers (900 mi) across, takes up a good part of Ishtar Terra. Ringing that plateau are mountainous belts that include Venus's tallest peak, Maxwell Montes. At 11 kilometers (7 mi) high, it is higher than Earth's Mount Everest. How these peaks came into existence is unclear, but the mechanism probably involves buckling and wrinkling of the planet's surface as heat was released from the interior.

SEEING WITH RADAR EYES

NASA's Magellan spacecraft, which reached Venus in 1990, created the best map to date of Venus's surface. Circling from pole to pole while Venus rotated slowly beneath it, the orbiter used radar to measure the heights and depths of the planet below. The craft sent out several thousand pulses per second, combining the returning echoes for a finely detailed picture. After each pass it radioed the results back to Earth, where scientists pieced together the map strip by narrow strip, like a quilt, for four years.

The resulting image is a computer-manipulated, three-dimensional, false-color rendering of the planet's topography, down to features no more than 300 meters (1,000 ft) in diameter. Craters, canyons, volcanoes, and coronae (such as the one in the foreground, below) pop out in sharp detail, revealing a dry, eruptive world.

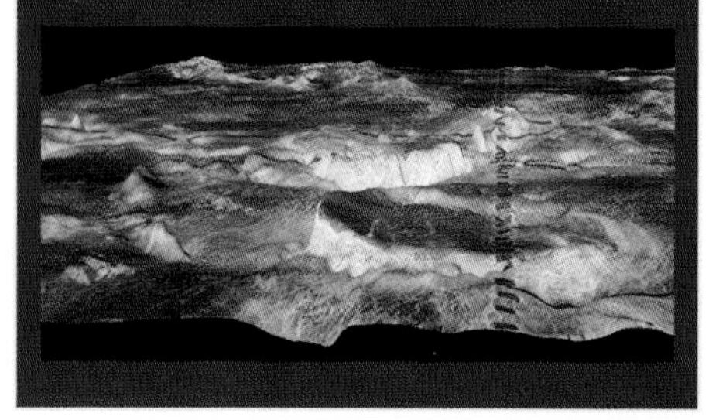

LAVA LAND

Some of these features were suggestive of plate tectonics. But the rest of Venus's surface told another story. When the first radar maps were made, scientists were astonished to see that much of Venus's surface was smoothed out under massive, solidified lava flows, interrupted by volcanic cones and large, sharp impact craters. The craters were surprisingly sparse—only about 1,000 have been counted—and geologically young, none older than 500 million years. The fact that they were scattered fairly evenly about the planet's surface suggested that Venus's surface was all about the same age, or else we would see older areas with many craters and younger areas with few. At least in the past 500 million years, Venus seems to have had an immobile surface, with no plate movement, completely unlike the vigorous Earth. The loss of its water during the runaway greenhouse effect early on may have changed the composition of its lithosphere, stiffening it so that it couldn't break up into plates.

Venus was certainly hit by meteorites billions of years ago, just like the other terrestrial planets, so the youth of the craters implies that the older ones have been erased, probably by fresh lava. Such a resurfacing would have to have been a global cataclysm, a result perhaps of inner instabilities that built up over a long period and then erupted in a worldwide volcanic event. It's possible that Venus undergoes such an event periodically and is currently in a quiet period. Or the planetwide eruption may have been a onetime event, Venus's equivalent of a biblical flood.

VOLCANOES

Its lack of plate movement does not mean that Venus is geologically dead. Quite the reverse—it is the most volcanic planet in the solar system. Taller and wider than most volcanoes of Earth, tens of thousands of volcanic domes dot the planet's surface. These are shield volcanoes like those that make up Earth's Hawaiian Islands, wide-spreading mounds that form over hot spots in planetary crusts where magma punches through to the surface. Calderas at their summits show where lava pooled and then drained. Explosive eruptions are probably rare on the planet, given the intense pressures and heat on the surface.

More than 100 large volcanoes are visible on Venus's surface. The biggest, Maat Mons, is 8.5 kilometers (5.3 mi) high, slightly shorter than Earth's largest volcano, Mauna Kea. But

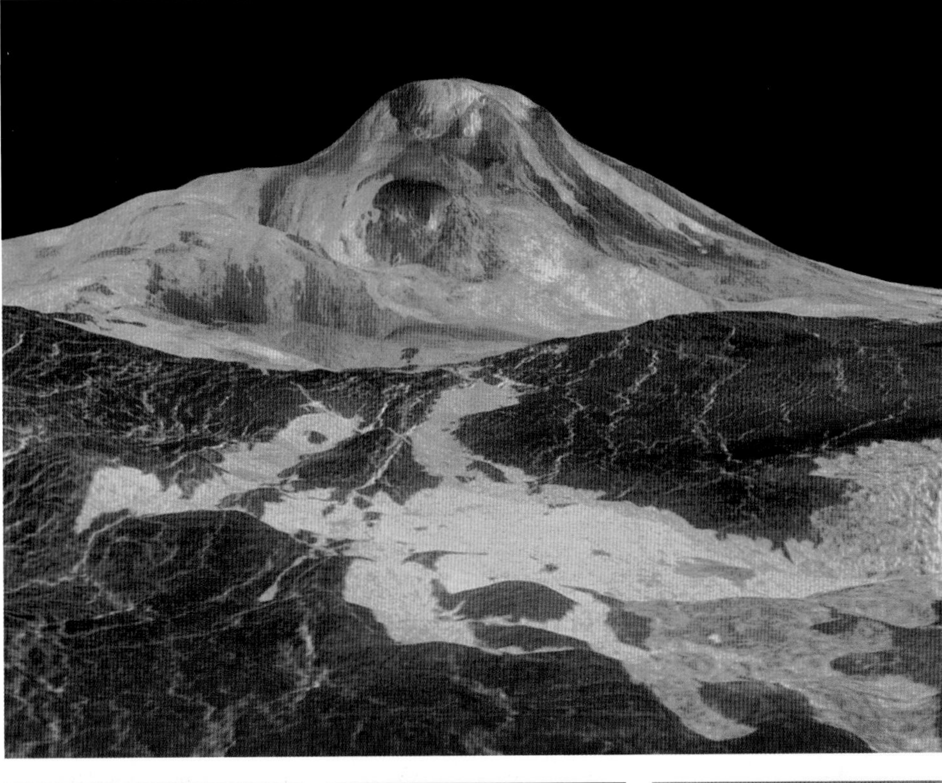

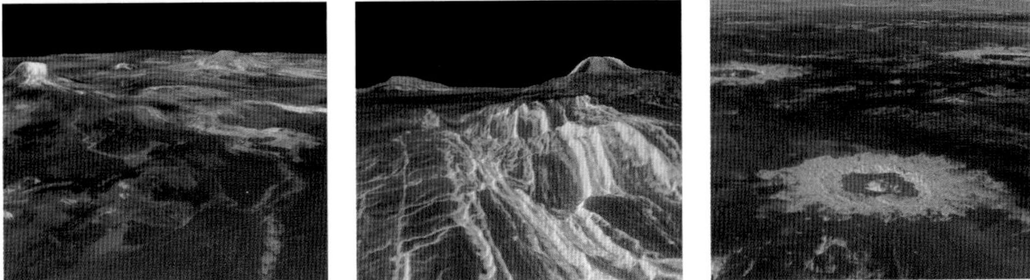

Maat Mons, Venus's largest volcano, rises above lava flows in this 3-D Magellan map (top). Below it are Eistla Regio (left and middle) and craters in Lavinia Planitia (right). Scale in all the images is exaggerated.

it's much wider across than its terrestrial counterpart, 400 kilometers (250 mi) in diameter, as opposed to 100 kilometers (62 mi) for Mauna Kea. Magellan also found many curious, smaller volcanic formations on Venus's surface. Neatly rounded lava domes, 25 kilometers (16 mi) wide, appear in clusters. They are probably the blistered crust left behind after lava welled to the surface and then subsided. Hundreds of huge, circular coronae, surrounded by ridges and cracks, also appear across the planet's surface. Some are slightly raised, some sunken. Like the domes, they probably resulted from plumes of upwelling magma.

Old lava flows are visible everywhere. Smooth sheets of lava cover the plains and flow into the basins of some of the older craters. Sulfur in the atmosphere tells us that Venus has a molten interior that releases the gas, although whether it currently has active volcanoes is still not known.

Unlike Earth, Venus has no magnetic field. Earth's field is probably generated by the reaction of its molten and solid iron core to the rapidly spinning outer planet. Venus, although it seems to have a solid iron core, spins very slowly, and this may account for the lack of magnetism—but like so many other aspects of Venus, we don't know enough yet to say. Without a magnetic field to protect it, Venus is exposed to the charged particles of the solar wind, but its dense atmosphere and electrical currents in its outer ionosphere shield it from the worst of the radiation.

FUTURE MISSIONS

Veiled Venus still holds many mysteries. The ESA's Venus Express has begun to fill in some holes in our knowledge of the planet's atmosphere. Japan's Space Agency, JAXA, will also tackle the climate with its upcoming orbiter, Planet-C. Getting a close look at the planet's surface is tougher, although planetary geologists are eager to learn more—anything, really—about the composition of Venus's crust and rocks. Moreover, a clearer picture of Venus's geologic history would shed light on our own planet's tectonic past. Though much is known about tectonics, there is still much to understand. Getting more data from Venus may help scientists to answer such questions as: Will Earth's plates one day settle into a single rigid plate? Without the constant recycling that plate tectonics brings, would that spell disaster for Earth's climate?

NASA would like to reenter the Venus exploration game with surface stations or mobile surface explorers (like those on Mars), as well as surface sample return missions. But it will take a truly tough machine to overcome the crushing inferno that is Venus.

WHERE WOMEN RULE

With the exception of Alpha and Beta Regio, christened in the 1960s, and Maxwell Montes, named after British physicist James Clerk Maxwell, every feature on Venus bears a female name. Naming conventions, set by the International Astronomical Union, set specific categories for each kind of geological feature. For example, terrae (highlands) are named after goddesses of love, such as Ishtar or Aphrodite (below). Planitiae (low plains) honor mythological heroines, among them Guinevere Planitia, Rusalka Planitia, and Hinemoa Planitia. Paterae (shallow craters) bear the names of famous real women, including Pocahontas Patera, Boadicea Patera, and Garland (as in Judy) Patera. Other features are named after Chinese moon fairies, magical Siberian toads, and singer Billie Holiday.

RESTLESS CRADLE OF LIFE

EARTH

The third planet from the sun is unique in many ways. With liquid oceans and a constantly shifting surface, it is a turbulent home for the only life—that we know of—in the solar system.

Earth has the distinction of being the largest of the terrestrial planets, 12,756 kilometers (7,926 mi) in diameter and 40,075 kilometers (24,901 mi) in circumference at the Equator. It is not completely spherical but bulges slightly around the waist because of its rotation. Denser than the other rocky planets and with a higher surface gravity, it is also the only planet with a liquid-water ocean on its surface. Its surface is varied and dynamic, consisting of crustal plates slowly shifting under a stable, shallow, moist atmosphere. Extending far into space and protecting the planet from radiation is Earth's magnetosphere, a magnetic field thousands of kilometers long. Earth's only moon circles the planet at a distance of 384,400 kilometers (238,855 mi) away. The moon is the largest natural satellite—relative to the size of its parent planet—in the solar system. Earth occupies a unique niche in the solar system: its orbit in a balmy zone that allows for liquid water, its protective magnetosphere, the presence of free oxygen in its atmosphere, and other unique factors have enabled life to flourish on its surface and in its oceans.

Symbol: ⊕
Average distance from sun: 149,597,890 km (92,955,820 mi)
Rotation period: 23.93 hours
Orbital period: 365.24 days
Equatorial diameter: 12,756 km (7,926 mi))

Mass: 5.9737×10^{24} kg
Density: 5.5 g/cm^3
Axial tilt: 23.45 degrees
Surface temperature: -88°C /58°C (-126°F/136°F)
Natural satellites: one

SKYWATCH

* An unaided observer in the Northern Hemisphere can see Earth's moon, up to five planets, 3,000 stars, the Milky Way, and the Andromeda galaxy.

AMAZING FACT Massive planet Jupiter has long served as an asteroid and comet shield for Earth, its gravity deflecting rocks away from us.

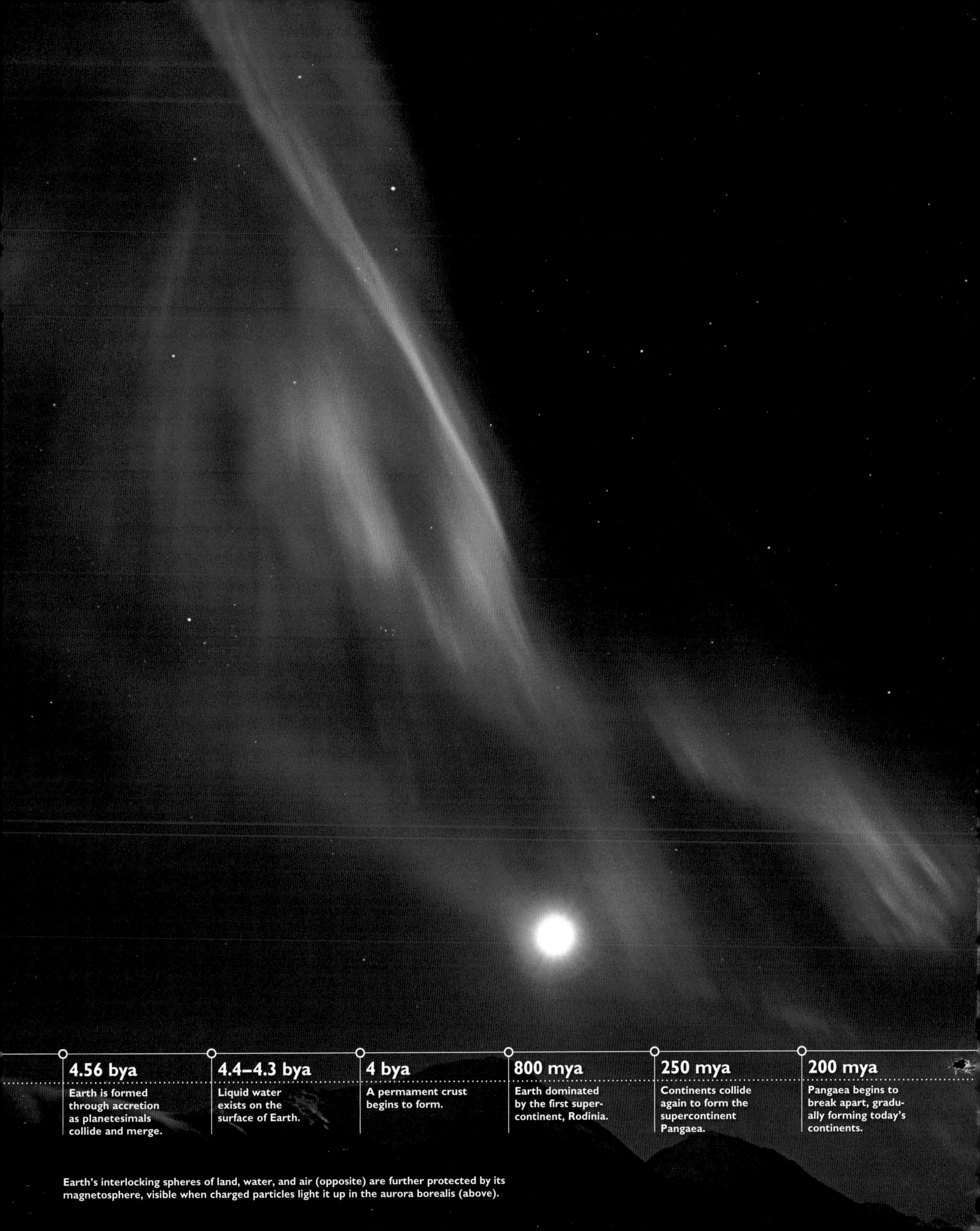

4.56 bya	4.4–4.3 bya	4 bya	800 mya	250 mya	200 mya
Earth is formed through accretion as planetesimals collide and merge.	Liquid water exists on the surface of Earth.	A permament crust begins to form.	Earth dominated by the first supercontinent, Rodinia.	Continents collide again to form the supercontinent Pangaea.	Pangaea begins to break apart, gradually forming today's continents.

Earth's interlocking spheres of land, water, and air (opposite) are further protected by its magnetosphere, visible when charged particles light it up in the aurora borealis (above).

EARTH: IN CONSTANT MOTION

Like its sister planets, the young Earth grew through repeated collisions during the accretion phase of the early solar system, around 4.6 billion to 4.5 billion years ago. At least ten, and possibly many more, massive bodies smashed together to bring Earth to the size it now possesses. Debris created by one of these enormous impacts probably coalesced to become Earth's only satellite, the moon. And one or more such collisions probably knocked the planet askew, giving it its current axial tilt of 23.5° relative to the plane of its orbit. Kinetic energy from these colliding bodies turned into heat in the newborn Earth, keeping it red hot and molten. Radioactive decay of short-lived isotopes inside the planet also contributed to the furnace. As a result, in a process known as differentiation, heavy elements such as iron sank through the magma ocean toward the melting planet's core, while less dense silicates—minerals such as quartz—floated toward the surface to form Earth's mantle and crust.

Asteroids continued to pound the Earth for about 800 million years, the main bombardment tapering off around 3.8 billion years ago. Even afterward, though, asteroids smacked down occasionally, including the object that may have caused a mass extinction 65 million years ago (mya).

THE EARLY ATMOSPHERE AND OCEANS

Geologists divide Earth's history into three eons: the Archaean eon, dating from the planet's origins until 2.5 billion years ago; the Proterozoic eon, when stable continents formed and multicellular life evolved; and the Phanerozoic eon, subdivided into eras and periods that track the evolution of life.

Unlike the outer planets, in the early days of the solar system, inner planets such as Earth were too hot and too close to the torrential solar wind to hold on to gases such as hydrogen or helium as atmospheres. During the Archaean eon, Earth's first atmosphere probably came

Ridges and trenches mark the troubled ground above the San Andreas Fault in western California. More than 1,287 kilometers (800 mi) long and 16 kilometers (10 mi) deep, it marks the boundary between the Pacific plate and the North American plate.

from gases trapped within its molten body and later released to its surface. As the planet cooled, volcanoes continued to vent the gases that made up the earliest air. The steamy mix included water vapor, hydrogen chloride, carbon monoxide, carbon dioxide, and nitrogen, as well as methane and ammonia produced by chemical reactions. Substantial amounts of oxygen appeared only later, about two billion years ago, when the first photosynthetic life spread through the oceans.

Water vapor from Earth's interior cooled and condensed as it reached the surface, falling as the first rain. The original source of this water is not clear. Some was probably primordial, part of the materials of the proto-Earth. Other water may have arrived on comets and water-rich asteroids. Rainwater swelled into rivers that coursed over the planet's surface, picking up the salty minerals that settled into the early oceans, making them saline.

Symmetrical Mount Shishaldin in the Aleutian Islands is one of hundreds of volcanoes along the Ring of Fire, a semicircle of eruptive activity around the Pacific Ocean.

EARLY LANDMASSES

The young planet had no large continents. Instead, it had small tectonic plates and many hot spots of upwelling magma, like the areas that form the Hawaiian Islands today. At first, Earth's surface was scattered with little protocontinents riding on thin crusts. Heat flowing up from Earth's mantle kept the small landmasses moving, preventing them from fusing together. Over billions of years, Earth cooled, the continental crust grew, and protocontinents repeatedly merged into larger landmasses, which then broke up again. Several supercontinents formed and split: Rodinia, which spanned the Southern Hemisphere some 800 million years ago; Gondwana, about 500 million years ago, also near the South Pole; and Pangaea, which drifted north and eventually split into the Americas and Eurasia about 100 million years ago, with the Atlantic Ocean growing between them.

Pieces of Earth's original protocontinents remain embedded in today's landmasses. Called Archaean shields because they date to the Archaean eon, they contain the world's oldest rocks and crystals. The current award winners for oldest minerals on Earth are zircon crystals found in Western Australia, radiometrically dated to approximately 4.1 billion to 4.3 billion years ago. The sedimentary rocks in which the crystals are embedded are younger than the crystals themselves. The oldest actual rocks discovered so far form part of the Acasta Formation in the Canadian Shield, in northern Canada, going back as far as four billion years. Ancient continental crust in Greenland, around the North American Great Lakes, in Swaziland, and in Western Australia also contain rock dating back 3.4 billion to 3.8 billion years.

PLATES ON THE GO

Plate tectonics continue to mold the surface of the planet, a process found nowhere else among the terrestrial planets. Between Earth's crust and the outer core is the planet's mantle. About 3,000 kilometers (1,860 mi) thick, it makes up 80 percent of Earth's bulk. Its iron-magnesium-silicate mix is solid but not rock hard—more like a dense plastic that can be deformed under enough pressure. The upper mantle, a region about 400 kilometers (250 mi) deep, is divided into the asthenosphere, an especially soft layer, topped by the lithosphere, a rocky, brittle layer that includes Earth's thin crust. The crust—about 20 to 50 kilometers (12 to 30 mi) thick under the continents and 8 kilometers (5 mi) thick under the oceans—is the visible part of the solid planet, the thin, wrinkled skin that forms its continents and ocean basins.

Earth's lithosphere is broken into seven large plates and dozens of smaller ones that slowly grind past and dive under and pull apart from each other like a living, combative jigsaw puzzle. Heat from Earth's core rises into the mantle and asthenosphere, where hot convection currents, circling like rollers in the asthenosphere, drive the plates above them by a few centimeters per year.

Plate motions give rise to many of Earth's most prominent surface features. Mountains arise where plates collide: The Himalayan mountains, for instance, have been forced up by the slow but inexorable collision of the Indian plate with the Eurasian plate. Deep-ocean trenches form at subduction zones, where one plate is forced beneath the other. Elsewhere, plates move apart and magma wells up from the mantle, as in the Mid-Atlantic Ridge, 16,000 kilometers (10,000 mi) long. Upwelling magma at plate boundaries builds volcanoes, while the abrupt release of pent-up pressure as plates slide past or under one another ripples through the planet as earthquakes.

Though they have an undeniably profound effect, plate tectonics are not responsible for all of Earth's features. Wind, rain, ice, and other effects of the planet's atmosphere and water also erode and sculpt the surface.

EARTH: IN THE DEPTHS

A lone Adélie penguin strikes a breeding pose amid ice, water, and mist—the three forms of water that coexist on the surface of Earth, making it uniquely suitable for life.

Geologists and astronomers share a common dilemma: Much of what they study is beyond their reach. Without undertaking the legendary journey to the center of Earth, we can't see the inner structure of our planet. The deepest holes ever drilled into Earth bored only 12 kilometers (7.5 mi) into Earth's crust. Students of the deep structure of planet Earth must therefore infer what lies within, using indirect means.

That's where earthquakes come in handy. As destructive as they are, earthquakes are a boon to geologists, because their powerful vibrations send out seismic waves that travel through the planet. Pressure waves (P-waves) arrive at monitoring stations first; they expand and compress the material through which they travel in the same direction as their motion. Shear waves, or S-waves, appear next. These create a motion perpendicular to their travel direction. The speeds of both kinds of waves depend on the density of the matter through which they pass; they can be bent, reflected, or absorbed by different kinds of material along the way. Analyzing just where and when these waves show up at monitoring stations—if they show up at all—allows geologists gradually to build up a picture of what lies along their paths in the invisible depths of the Earth.

THE CORE

At the planet's core is—its core. During Earth's formation, heavy elements sank into the molten planet's heart. Compressed by the immense weight of the overlying material, the metal solidified into a dense mass of iron, possibly with lighter elements such as nickel and sulfur, as the planet cooled. Studies of seismic waves indicate that the solid inner core, about 1,120 kilometers (695 mi) in radius, is wrapped inside a bigger, molten metallic outer core some 2,170 kilometers (1,350 mi) deep. In the 1980s, studies suggested that the inner core is rotating slightly faster than the planet around it—perhaps two-thirds of a second faster per day. The discrepancy between the two spins in these electrically conducting regions may produce Earth's magnetic field. However, the subject of the spinning core is still controversial among geologists. Adding to the fray are recent studies suggesting that contained within the inner core is actually an "innermost inner" core, perhaps half as wide, with a different alignment. The exact composition and relative movement of these hidden regions remains in dispute.

What cannot be disputed, however, is the incredibly intense heat and pressure inside the planet. Temperatures in the inner core would be in the range of 5000K to 7000K (8500°F to 12,000°F), or roughly about as hot as the sun's surface. Only a stupendous pressure four million times that at Earth's surface keeps iron solid in those conditions.

THE MAGNETOSPHERE

Like some other planets, Earth produces a strong magnetic field, generated by currents in its metallic core. It's as if a gigantic bar magnet is buried within the planet, running north to south but slightly offset from Earth's axis. The north and south poles of the magnet drift over time; currently the north magnetic pole is located in northern Canada, while the south magnetic pole is off the coast of Antarctica. Magnetic field lines, invisible to the eye, sprout from the south magnetic pole, curve out around the sphere of Earth, and curve back in at the north magnetic pole.

Without this magnetic field, life as we know it might never have arisen, because the field shields us from the destructive effects of the solar wind. Charged particles streaming out from the sun (see pp. 72–73) get caught up in the magnetic field lines in two doughnut-shaped zones known as the Van Allen belts—one centered around 3,000 kilometers (1,860 mi) and the other about 20,000 kilometers (12,400 mi) above Earth's surface. The entire region where the solar wind interacts with the magnetic field is called the magnetosphere. On the sun side of Earth, the force of the solar wind compresses the magnetosphere to within ten Earth radii at an outer boundary called the magnetopause. On the night side of Earth, the streaming particles drag the field out into an extended magnetotail, possibly several AU (astronomical units) long.

AURORAS

The northern lights (aurora borealis) and southern lights (aurora australis) are splendid side effects of Earth's magnetic field. During particularly active bouts of the solar wind, charged particles spiraling down magnetic field lines near the North and South Poles run into the upper atmosphere. The gases glow when the particles hit them: green or red for oxygen, and blue for nitrogen, billowing and folding in the night sky. In 2008, satellite observations of auroras provided some clues about the reason behind the lights' wavering dance. Magnetic field lines, stretched out by the solar wind, suddenly snap back like rubber bands and reconnect, flinging charged particles back toward Earth. The resulting display flashes and wavers in a shimmering dance.

THE WORLD OCEAN

Earth's blue waters are its most striking feature from space, a testament to the planet's balmy temperatures and protective atmosphere. The vast planetwide ocean covers most of Earth, occupying 71 percent of its surface and containing 97 percent of its water. It serves as a giant radiator and climate regulator for the planet, absorbing heat in hot months and releasing it in cold ones, as well as circulating heat and cold around the world via great rotating currents such as the Gulf Stream. By absorbing about half of the carbon dioxide emitted into the atmosphere, it also stands as a bulwark against a runaway greenhouse effect. And the ocean is the liquid cradle for half of all species on Earth and almost all of its living matter by mass. Phytoplankton, one-celled plants at the ocean's surface, supply half of Earth's oxygen.

Despite the ocean's immense value to human life, we know less about its terrain than we do about the moon's. About 95 percent remains unexplored, buried in the dark under intense pressures. The average depth of the ocean is 3.7 kilometers (2.3 mi), but it varies from the ankle-high shallows of shoreline waters to 11 kilometers (7 mi) deep in the Marianas Trench. Rising from its floor is the world's longest mountain range, the Mid-Ocean Ridge, running for more than 56,000 kilometers (35,000 mi) between the continents, as well as Earth's tallest mountain, measured from its base: Mauna Kea, at about 9,800 meters (32,000 ft). In recent decades, stunning discoveries of new kinds of life in the deep ocean have shown that the ocean can teach us much about the origins of life on Earth and the possibilities of life elsewhere.

ICE AGES

Earth's large-scale climate slowly shifts back and forth between warmer periods, when the planet has little ice, and colder periods known as ice ages, when glaciers (such as Glacier Bay's Reid Glacier, bottom) cover much of the planet. These ice ages may be caused by small but regular and significant shifts in the shape of Earth's orbit and the tilt of its axis. During each long ice age there are glacial and interglacial periods when glaciers advance and retreat, leaving behind features such as New York's Finger Lakes (below). Right now, we are most likely in an interglacial period of an ice age that began in the Pleistocene and reached a recent cold peak some 20,000 years ago.

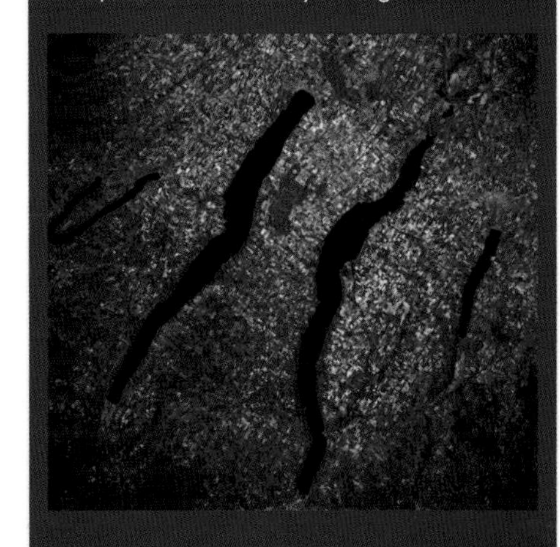

Major ice ages include:

Huronian: 2.7–1.8 billion years ago
Gnejso: 940–880 million years ago
Sturtian: 820–770 million years ago
Varangian: 615–570 million years ago
Ordovician: 440–430 million years ago
Permo-Carboniferous: 330–250 million years ago
Pleistocene: 1.6 million years ago–present

EARTH: LIVING PLANET

Nowhere is Earth's unique nature more evident than in the interdependent spheres of its ocean, atmosphere, and life. Oceans support life and absorb atmospheric gases; water cycles through the atmosphere; and the atmosphere is supplied by the life that depends on water and air.

Earth's current atmosphere is in its second generation. Plant life growing in Earth's early oceans transformed its methane-carbon monoxide mix with a heavy injection of oxygen. Today the atmosphere consists mainly of nitrogen (78 percent) and oxygen (21 percent), with small amounts of argon, carbon dioxide, and water vapor. The gases recycle through Earth's biosphere; nitrogen is a major part of the proteins that govern living cells, while plants take in carbon dioxide and produce oxygen. Animals, of course, do the reverse, using oxygen to fuel their metabolism and exhaling carbon dioxide as waste.

Almost all of the atmosphere's mass is found within 30 kilometers (19 mi) of Earth's surface, with half lying within 5 kilometers (3 mi). (If all 5,000 trillion metric tons of it froze into oxygen and nitrogen snow, it would lie 100 meters/330 ft deep on the planet's surface.) Like the atmospheres of other planets, Earth's has distinct layers. They range from the troposphere near the surface, where nearly all of our weather occurs, to the stratosphere, about 25 kilometers (16 mi) up, where a layer of ozone (O^3) helps to shield Earth's living creatures from ultraviolet radiation, and on through the mesosphere into the ionosphere, which extends into space as increasingly scarce traces of molecules above 80 kilometers (50 mi).

Earth's winds are a consequence of convection currents circulating the air from warmer regions to cooler ones. Hot air rises and expands from ground heated by the sun, then drops and grows denser in cooler upper altitudes. This vertical mixing helps to keep temperatures between the ground and air fairly stable, although uneven heating of different areas on Earth and the clash of higher and lower pressure regions of air also result in winds and weather ranging from a gentle drizzle to a tornado.

THE GREENHOUSE EFFECT

Compared with the size of Earth, the atmosphere is a thin coverlet, but vital for keeping the planet warm. Without it, the planet's average temperature would be -18°C (-4°F), keeping water permanently frozen and life untenable. Much of its warmth is due to the greenhouse effect. Incoming solar radiation is partly reflected by clouds, but the rest is absorbed by Earth's surface and reradiated at different wavelengths. In that form, a good bit of the heat is absorbed by the small amounts of water vapor and carbon dioxide in the atmosphere. The atmosphere, in turn, radiates heat both toward and away from the Earth, keeping the planet a steady 40°C (72°F) warmer than it would be otherwise.

The greenhouse effect is a normal and beneficial feature of Earth's atmosphere. What is neither normal nor beneficial is the recent trend toward man-made global warming. Since the beginning of the industrial revolution in the early 19th century, burning fossil fuels have added increasing amounts of the greenhouse gas carbon dioxide to the atmosphere. The result is a steady worldwide rise in temperatures: about 0.5°C (1°F) in the 20th century but predicted to increase at a higher rate in the 21st. This seemingly modest change could have drastic effects upon world climates, including melting of glaciers and polar ice (already occurring), extreme weather, crop failures, and more.

BACTERIA WORLD

Bacteria (below), Earth's oldest life-forms, exist in almost every environment, from deep ocean vents to Antarctic ice to the tops of mountains to the human gut. They can live off substances ranging from sugar to iron and sulfur and can survive lethal doses of radiation. They are vital to life on Earth, recycling carbon dioxide through decomposition and nurturing plants through nitrogen fixation. It has been said that if all matter on Earth except bacteria were to vanish, the ghostly outlines of the planet and its living things could still be seen, formed from microorganisms.

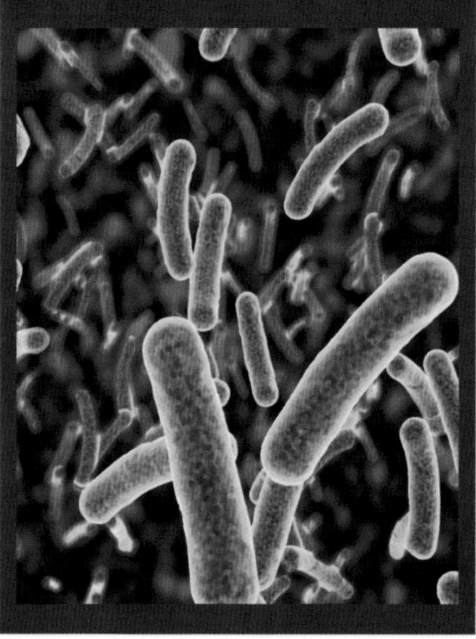

LIFE

Though other planets or moons in our solar neighborhood may turn out to possess life—or to have possessed it once upon a time—none other exhibits the widespread and obvious occupation of living things.

Our definition of life and its requirements may change drastically if we come across it on other worlds. For now, though, we must go by what we see here on Earth. As far as we know, life can arise only on planets, like this one, where the temperature allows for liquid water: a range between -15°C and 100°C (5°F and 212°F). Why is liquid water so important? Because the chemical reactions that keep life going must take place in water, which is the best solvent on the planet. Water also plays a key role in shaping the enzymes that propel chemical reactions, as well as serving as the transportation system for all sorts of substances from one place to another (see pp. 182–83).

Earth, orbiting in the balmy habitable zone around the sun, certainly has plenty of water, and has had for roughly 4.2 billion years. Before that time, massive asteroid impacts would have vaporized any water as well as essentially sterilizing the world in intense heat. So sometime between about 4.2 billion and 3.5 billion years ago, when photosynthetic chemicals appear in ancient rocks, life on Earth began.

No one knows just how that happened. Experiments have shown that amino acids, the building blocks of proteins (which are in turn the building blocks of life), will form from the chemicals present in primordial oceans when prompted by lightning. However it happened, at some point the basic molecules of life, including sugars, proteins, and nucleic acids, were present and interacting in the first oceans. The nucleic acid RNA, once formed, has the ability to replicate itself, and so the earliest life-forms may have been based on RNA, evolving external membranes to protect the chemicals inside. Perhaps the first cells arose in the warm water around deep-ocean vents, where even today extremely primitive life-forms can be found.

FROM BACTERIA TO BABIES

The first life-forms, going back at least 3.5 billion years, were prokaryotes, simple organisms without cell nuclei, chromosomes, or other organelles. (Bacteria are prokaryotes.) Between 1 billion and 500 million years ago, more complex cells containing nuclei, chromosomes, and distinct internal structures arose to join them. The first plant cells appeared and, through photosynthesis, began to produce oxygen. Reproducing and spreading through the oceans over millions of years, these early algae transformed Earth's atmosphere into the oxygen-rich mixture we know today. Toward the end of this transformation, perhaps 600 million years ago, life began an explosive diversification into a wide variety of multicellular forms, including flowering plants and mammals, evolving and spreading over an increasingly green world. Only in the most recent years of life's history did humans arise. Perhaps 4.4 million years ago the first hominins left the trees to walk upright. Modern humans, *Homo sapiens,* appeared a mere 200,000 years ago.

MASS EXTINCTIONS

The failure of individual species is a necessary part of evolution, but at times life on Earth has undergone die-offs on a larger scale, mass extinctions of huge numbers of species. We don't always know the reasons for these deaths, but scientists examining the fossil and geologic record have theories for what lies behind most of them. The biggest mass extinctions that we know of include the Permian-Triassic extinction 251 million years ago, which eliminated 95 percent of all species. An asteroid impact, volcanic disaster, or a combination of the two may have brought it about. The Cretaceous-Tertiary extinction 65 million years ago was probably due to an asteroid impact. Forty-seven percent of marine genera and 18 percent of land vertebrates, including dinosaurs, died out.

Hydrothermal vents along the deep-ocean Mid-Atlantic Ridge spew black "smoke"—actually hot, mineral-rich water that nourishes ancient forms of life.

THE BIOSPHERE

The result of all the energetic reproduction in Earth's youth was the addition of a biosphere to Earth's other three interlocking "spheres": the atmosphere, the lithosphere (land), and the hydrosphere (water). Drawing its energy from the sun, the biosphere is a worldwide ecosystem that is dependent upon and feeds into the other terrestrial systems. As a physical zone, the biosphere extends from the ocean floor, 11 kilometers (7 mi) deep, to the upper regions of the troposphere, some 10 kilometers (6 mi) above Earth's surface. Most life is concentrated into a narrower region, from about 200 meters (650 ft) below the ocean's surface to six kilometers (4 mi) above sea level.

Human beings make up a small part of the world's biomass—vastly outnumbered, for instance, by bacteria—but their influence on the planet is out of proportion to their numbers. Human consumption of world resources, depletion of forests, and burning of fossil fuels have changed Earth's climate rapidly in recent centuries. Human populations are booming overall, but unevenly, growing rapidly in much of the less developed world while declining in other regions. Yet while struggling with problems on their home planet, humans continue to look outward for signs of life in their planetary neighborhood and for future homes for humans themselves.

THE LITTLE MUSEUM NEXT DOOR
EARTH'S MOON

Because Earth's moon—the planet's only natural satellite—lacks erosion and an active geology, it preserves a record of the solar system's early days. It's the closest astronomical body to do so, making it a beacon for space travel.

The moon isn't the largest satellite in the solar system, but it is the biggest relative to the size of its parent planet, measuring fully one-quarter the size of Earth. Orbiting 384,400 kilometers (238,900 mi) away, on average, it exerts a strong gravitation pull on our planet. This is evident not only in Earth's tides but also in Earth's fairly stable orientation in space. Without the steady tug of the nearby moon, Earth's axis, now tilted at 23.5°, might wobble between 0° and 85°, with catastrophic effects on seasons and climate. • The Earth-moon gravitational relationship also explains one of the moon's most obvious characteristics: The same side always faces Earth. Tidal forces between Earth and the moon have tied them together in their solar system dance so that the duration of the moon's rotation is exactly the same length of time as its orbit. Because of this, until the space age, we had never been able to see the far side of the moon. (And note that the far side is not the "dark" side, Pink Floyd notwithstanding. The sun shines on the moon's far side during every orbit—we just can't see the hemisphere.)

Symbol: ☾
Discovered by: Known to the ancients
Average distance from Earth: 384,400 km (238,855 mi)
Rotation period: 27.32 Earth days
Orbital period: 27.32 Earth days

Equatorial diameter: 3,474 km (2,159 mi)
Mass (Earth=1): 0.0123
Density: 3.34 g/cm^3 (compared with Earth at 5.5)
Surface temperature: -233/123 °C (-387/253 °F)
Magnitude: 0.21

SKYWATCH

* The best times to observe the moon are its first or last quarters; pay particular attention to the terminator, the line between its bright and dark portions.

AMAZING FACT The last year there was a month without a full moon was in 1999; the next will be 2018.

Dark basins and bright ejecta rays illustrate the moon's battered history (opposite). Apollo 11 astronauts looked toward their sunlit home as they skimmed over the moon's Smyth Sea in July 1969 (above).

EARTH'S MOON: A HISTORY OF VIOLENCE

Just where did Earth acquire its massive companion? Several theories have been advanced in the past, each with its own problems. The simplest explanation is that the moon formed at the same time as Earth from the coalescing debris of the early solar system. However, measurements of the moon's density, less than that of Earth, don't support this "little twin" scenario. Moreover, rocks brought back from manned missions lack the water-bearing minerals found in Earth's stones. Another theory holds that Earth captured the wandering moon as it floated past. This, however, would have been a very difficult task given the moon's size. A third scenario

Astronauts from Apollo 17, the last manned mission to the moon, photographed a boulder field at their Taurus-Littrow landing site in 1972. The Apollo missions brought back hundreds of kilograms of moon rocks, which are still being studied.

has the spinning Earth somehow spitting out the material that made the moon, leaving the Pacific Ocean Basin as evidence. But this, too, is physically implausible.

The current model of the moon's origin is a dramatic one, but it accounts for the similarities and differences between Earth and its satellite. In this hypothesis, a giant object the size of Mars struck Earth a glancing but titanic blow about 4.5 billion years ago, soon after its formation. In the heat of the catastrophic impact, the impactor's metallic core merged with that of Earth, while great chunks of Earth's crust and mantle were ejected into space. The intense heat vaporized water and most volatile elements from the cast-off material, which clumped together and re-formed into the orbiting moon within about a century. Because it contained relatively little iron, the moon had a small iron core and was therefore less dense than the earth. The reeling Earth, meanwhile, had been knocked askew by some 23 degrees. It was not long before the mother and daughter worlds settled into the close orbital partnership we know today.

LANDS AND SEAS

The dark and light patches on the full moon's piebald surface may look like a man's face, if you're not too picky about faces, but even a small telescope will reveal the bright mountains, dark plains, and thousands of giant craters that tell of a long history of violence against Earth's satellite.

Galileo, viewing the moon through his newly built telescope in 1609, likened its variable surface to that of Earth and named the lighter areas *terrae,* "lands," and the smooth dark regions *maria,* "seas." His maria labels survive on some of the moon's largest features: Mare Tranquillitatis, for instance, the Sea of Tranquillity, or Mare Imbrium, the optimistically named Sea of Rains.

The notion of watery seas on the moon did not survive into modern times, and thanks to space age studies and the Apollo missions, we know that the terrae are actually lunar highlands, ancient regions of the moon's crust containing rocks dating back almost to the moon's birth, roughly 4.5 billion years ago. These heavily cratered, mountainous areas cover more than 80 percent of the moon's surface. The maria, two to five kilometers (1.2 to 3 mi) lower than the highlands, are huge impact basins filled with cooled dark lava.

CRATERS

Craters abound on the moon, at least 30,000 of them boasting a diameter greater than one kilometer (0.6 mi). When rocky bodies smacked into the moon, the energy of their motion was converted into heat and sent out shock waves through the crust. The impacting object was vaporized upon collision, while the pulverized debris from the surface was ejected outward into a circular rim 10 to 20 times as large as the impactor. The bigger craters typically have a central peak, where the crater floor rebounded after the shock of the impact, as well as surrounding carpets of debris known as an ejecta blanket. Many of the ejected rocks were large enough and fell from the sky hard enough to create their own craters in turn. The biggest craters of all are called impact basins. The moon's far side has the solar system's largest impact basin, the continent-size South Pole-Aitken Basin, which measures about 2,500 kilometers (1,500 mi) across.

THE MOON'S HISTORY

In the very earliest stages of its formation, roughly 4.5 billion years ago, the satellite was largely molten. Denser, heavier material sank inward toward its center, while its lighter elements rose to form the surface crust as the moon gradually cooled.

Then, sometime between 3.9 billion and 3.8 billion years ago, debris from the early solar system bombarded the poor moon, huge rocks blasting crater after crater out of its surface with the force of multiple hydrogen bombs. Even as this bombardment began to slow, volcanism took over. Heat from the decay of radioactive elements within the moon pushed molten rock through the thin crust beneath the biggest impact basins, where it spread out and cooled to form the lunar maria.

LUNAR ANATOMY

Beneath its soil, called regolith, is the lunar crust, thinner on the near side, particularly under impact basins, and thicker on the far side. Under that is a mantle, cool, dense, and semi-rigid, surrounding a partially molten zone. The moon's iron-rich core is small, perhaps 700 kilometers (430 mi) in diameter, reflecting the moon's birth from the Earth's lighter, outer regions.

The moon is relatively cold and quiet, geologically speaking. It does, however, experience moonquakes, relatively gentle but long-lasting tremors that ring the moon like a bell.

TIDES

The pull of the moon on Earth causes its tides, a fact known since ancient times. But why are there high tides on the side of Earth opposite the moon?

It helps to think of the Earth as a solid ball surrounded by a liquid envelope. The moon's gravitational force pulls strongly on the water nearest it, causing it to bulge toward the moon in a high tide. It also tugs on the center of Earth itself, pulling the solid Earth toward it within its watery covering and away from the ocean's surface on the far side. That "left-behind" water thus raises a high tide.

The sun's gravitational pull contributes to tides as well, although more weakly than that of the moon. When the sun, moon, and Earth are aligned at the new and full moons, Earth experiences unusually high, or "spring" tides.

The moon's gravitational pull is also slowing Earth's rotation and lengthening our day by two milliseconds per century. To balance the gravitational equation, the energy lost as Earth's momentum slows is gained by the moon, which moves slowly away from Earth. Laser measurements from Earth's surface confirm that our satellite is pulling away by 3.8 centimeters (1.5 in) per year.

EARTH'S MOON: ONCE & FUTURE MISSIONS

On December 11, 1972, the U.S. spacecraft Apollo 17's lander touched down in the Taurus-Littrow region of the moon. Astronauts Eugene Cernan and Harrison Schmitt (a geologist) were aboard. After the flurry of landing instructions and thruster noise, the first thing Cernan noticed, he said later, was the silence.

"Boy, when you said shut down, I shut down and we dropped, didn't we?" said Cernan to Schmitt.

"Yes, sir! But we is here," Schmitt replied.

"Man, is we here," said Cernan.

They were the last to be there. After a spectacular three-year run of lunar visits, beginning with Apollo 11 in 1969, the manned spaceflight program was shut down indefinitely after 1972. The program had contributed a wealth of information not only about the moon, but also about Earth and the origins of the solar system. Just as important was the heartening message that humans could, in fact, set foot on other planets.

EARLY MISSIONS

Planetary scientists can thank the Cold War for much of the lunar knowledge they've gained. The Soviet Union's launch of Sputnik I on October 4, 1957, shocked the United States into turbocharging its own infant space program. President John F. Kennedy declared in 1961 that the United States would "send a man to the moon and return him safely to Earth" before the end of the decade.

The 1960s then saw a torrent of unmanned missions to the moon from both the Soviet Union and the United States: the Luna series, the Ranger series, Surveyors, Lunar Orbiters, and more. Many failed, missing the moon altogether or crashing (unintentionally) into its surface. But those that succeeded sent back detailed images of the moon's surface and news about its geology. When the first spacecraft orbited the moon, for instance, scientists were surprised to find that the crafts' orbits were deflected by unexpectedly dense gravitational pulls from the moon's maria. These mass concentrations, or mascons, are now thought to be high-density mantle rocks that rose toward the surface in the wake of giant impacts. Craft that reached the surface sent back the reassuring information that landers would not vanish beneath a deep layer of lunar dust, as some had feared. Instead, a lunar soil called regolith (from the Greek for "blanket of stone"), a thin, fine powder with scattered rocks, carpets the moon's surface. With no atmosphere to disturb it, the regolith holds imprints, such as astronauts' footprints, indefinitely.

NEIL ARMSTRONG

ONE SMALL STEP

Cool-headed test pilot Neil Armstrong (1930–) earned his first steps on the moon the hard way. He flew 78 combat missions in Korea and had experience in aircraft ranging from gliders to the supersonic X-15. In 1966 he commanded the Gemini 8 mission that performed the first successful docking of two vehicles in space.

Armstrong's first words on the moon—"That's one small step for man, one giant leap for mankind"—were not quite what he intended. In the heat of the moment, he omitted the "a" before "man." But the millions listening knew exactly what he, and the moment, meant.

APOLLO

NASA's Apollo program began with tragedy when three Apollo 1 astronauts died in a fire on the launchpad in 1967. But by the end of 1968 the manned spacecraft Apollo 8 had circled the moon and returned. And on July 20, 1969, Apollo 11 landed in Mare Tranquillitatis, with Neil Armstrong and Buzz Aldrin emerging to become the first humans to walk on another world. With their bouncing steps broadcast to a worldwide television audience, they explored the lunar surface for 21 hours and collected 20 kilograms (44 lb) of samples before returning to the orbiting spacecraft and departing for Earth.

Apollo 17 astronaut Jack Schmitt leaves the lunar rover to collect samples from the moon's surface.

Over three eventful years NASA sent seven more Apollo missions to the moon. One, Apollo 13, popularized in the 1995 Oscar-winning movie *Apollo 13,* had to abort its landing and return perilously to Earth after a near-fatal onboard explosion. The other missions were hugely successful, eventually exploring more and more of the moon's surface in golf cart-style rovers and delivering 382 kilograms (842 lb) of moon rocks to Earth. This geologic treasure trove confirmed the igneous and waterless nature of the moon's crust. Astronauts also left behind scientific instruments that returned data to Earth about moonquakes and the solar wind. A small array of silica reflectors, left on the moon's surface by Apollo 11 astronauts, still serves scientists looking to accurately measure the moon's distance from Earth. Laser beams directed through optical telescopes from Earth bounce off the reflectors and return, yielding information a few photons at a time and confirming that the moon is receding at the rate of 3.8 centimeters (1.5 in) per year.

THE SEARCH FOR ICE

The Apollo missions halted in 1972 for a host of political and financial reasons. To the great disappointment of scientists and space enthusiasts worldwide, no other piloted missions have been launched toward the moon or toward any other planet so far. However, with an eye to eventually returning in force, various unmanned missions have scrutinized the moon's polar regions, where water ice may lurk in the shadows.

One was Clementine, a NASA/Department of Defense orbiter launched in 1994, which sent back 1.6 million digital images of the lunar surface and may have detected ice in a dark crater on the moon's south pole. NASA's little Lunar Prospector, launched in 1998, orbited the moon for 18 months and detected hydrogen at the moon's south pole, perhaps indicating the presence of water. NASA even sacrificed the orbiter to the cause of ice, crashing it at the end of its mission into a crater in hopes of stirring up water vapor—but none was detected. Hopes for ice were dimmed in 2006 when radar studies conducted from Puerto Rico's Arecibo Observatory indicated that the earlier readings may have come from other sources, such as young, scattered rocks and the solar wind.

RETURN TO THE MOON?

Why such interest in ice? As NASA's space shuttle program winds down, plans for people to return to the moon as well as venture to Mars have ratcheted up. In 2004, President George W. Bush announced an ambitious proposal to return humans to the moon by the year 2020. The vision was later expanded to include a possible moon base on the rim of the Shackleton Crater near the lunar south pole. The solar-powered base would need to be as self-sufficient as possible, including retrieving ice if available from nearby deposits. Valuable not only as water, the ice could be processed into oxygen and hydrogen.

The United States is hardly the only country with an interest in the moon. Space agencies from Europe, Japan, and China have all sent spacecraft to the moon in the 21st century as they developed their space programs. India launched its first mission to another solar system body with Chandrayaan-1, which entered lunar orbit in November 2008. The craft is training a cluster of sensors toward the moon to map its mineral elements and gain some insight into the processes that went into its creation. Any such knowledge will inevitably help us understand the birth of Earth, as well.

APOLLO MISSIONS

Nine spacecraft and 27 astronauts reached lunar orbit during the storied Apollo program, with first launch in December 1968. The last mission concluded four years later.

Apollo 8—Launched December 21, 1968; returned to Earth December 27, 1968.

Apollo 10—Launched May 18, 1969; returned to Earth May 26, 1969.

Apollo 11—Launched July 16, 1969; landed on moon July 20, 1969; returned to Earth July 24, 1969.

Apollo 12—Launched November 14, 1969; landed on moon November 19, 1969; returned to Earth November 24, 1969.

Apollo 13—Launched April 11, 1970; returned to Earth April 17, 1970.

Apollo 14—Launched January 31, 1971; landed on moon February 5, 1971; returned to Earth February 9, 1971.

Apollo 15—Launched July 26, 1971; landed on moon July 30, 1971; returned to Earth August 7, 1971.

Apollo 16—Launched April 16, 1972; returned to Earth April 27, 1972.

Apollo 17—Launched December 7, 1972; landed on moon December 11, 1972; returned to Earth December 19, 1972.

THE RED DESERT
MARS

The chilly red planet next door has some striking similarities to Earth and remains the closest extraterrestrial body with a source of water—and maybe even evidence of life—in the solar system.

Mars was well known to ancient cultures, one of the five "wandering stars" tracked by ancient sky-watchers. With an orbit outside Earth's, Mars seems to skate in a great loop across the night sky, sometimes back-tracking in a retrograde motion before starting forward again. When the planet's at opposition (when Earth is between it and the sun) and at perihelion (the closest point in its orbit to the sun) it can come as close to Earth as 55,800,000 kilometers (34,700,000 mi). Such a close opposition happens only three times a century. • The planet is our closest neighbor after Venus and is, in some ways, surprisingly similar to our own planet. For example, Mars rotates on its axis every 24.6 hours, giving a day much like ours. Its axial tilt is similar to Earth's as well, inclined at about 24°, which gives it seasons as it orbits around the sun. In addition, it has its own atmosphere, clouds, and polar caps. However, it is considerably smaller than our planet, measuring just 6,794 kilometers (4,222 mi) in diameter, giving it a correspondingly smaller mass only 11 percent that of Earth's.

Symbol: ♂
Discovered by: Known to the ancients
Average distance from sun: 227,936,640 km (141,633,260 mi)
Rotation period: 1.026 Earth days, or 24.62 hours
Orbital period: 1.88 Earth years, or 687 Earth days

Equatorial diameter: 6,794 km (4,222 mi)
Mass (Earth=1): 0.107
Density: 3.94 g/cm³ (compared with Earth at 5.5)
Surface temperature: -87°C to -5°C (-125°F to 23°F)
Natural satellites: 2

SKYWATCH

- The best time to observe Mars is at opposition, when the planet is closest to Earth, which happens every two years for about two to four months.

AMAZING FACT The soil on Mars can get as warm as 27°C (81°F) in the summer, but the air rarely exceeds 0°C (32°F).

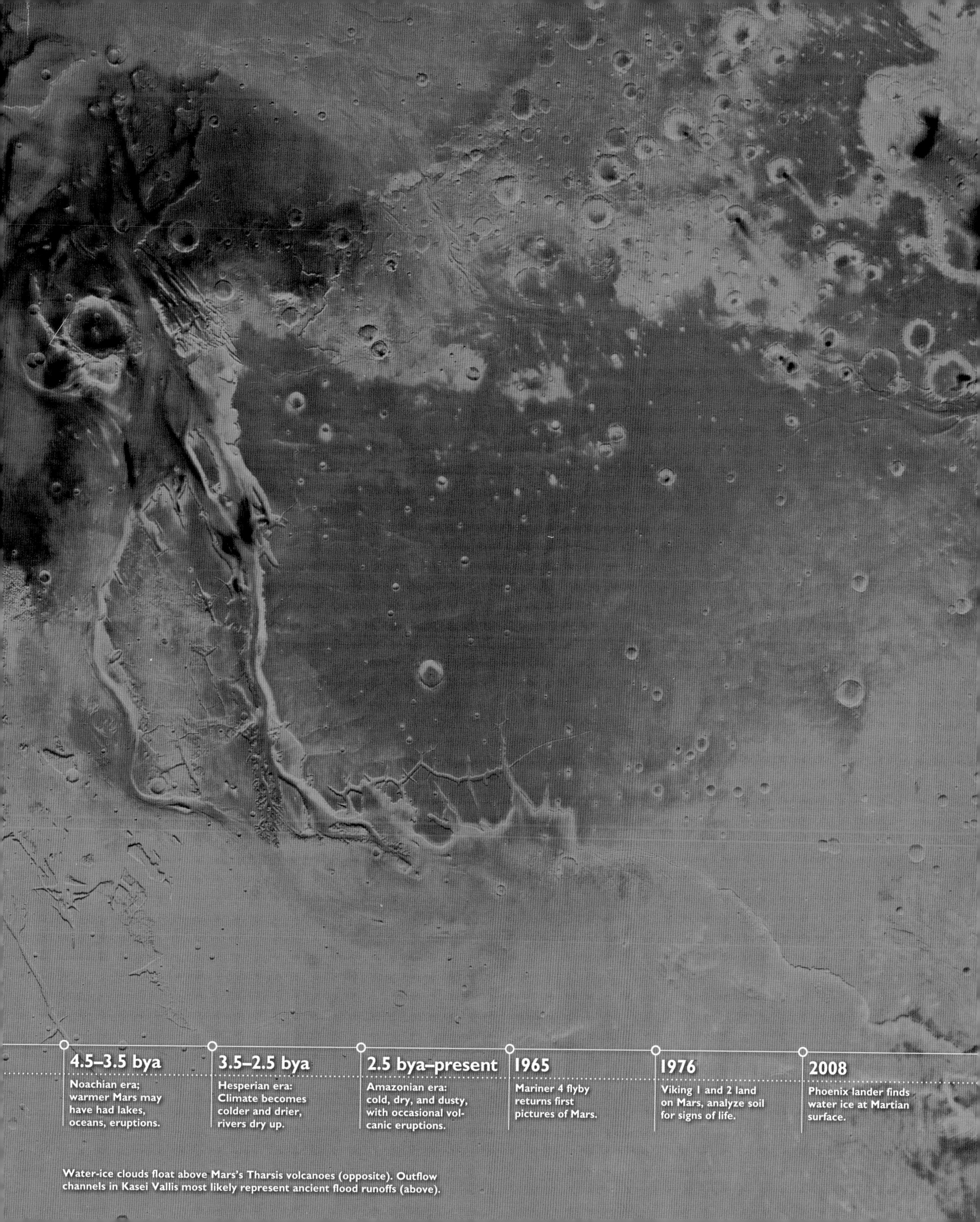

4.5–3.5 bya

Noachian era; warmer Mars may have had lakes, oceans, eruptions.

3.5–2.5 bya

Hesperian era: Climate becomes colder and drier, rivers dry up.

2.5 bya–present

Amazonian era: cold, dry, and dusty, with occasional volcanic eruptions.

1965

Mariner 4 flyby returns first pictures of Mars.

1976

Viking 1 and 2 land on Mars, analyze soil for signs of life.

2008

Phoenix lander finds water ice at Martian surface.

Water-ice clouds float above Mars's Tharsis volcanoes (opposite). Outflow channels in Kasei Vallis most likely represent ancient flood runoffs (above).

MARS: GOING TO EXTREMES

Crisscrossing winds have shaped twisted dunes of basaltic sand in this false-color image of a crater floor in Arabia Terra. The landscape is tinted orange and yellow where the surface is composed of warmer, consolidated sediments, and blue where there are colder drifts of dust and fine-grain sand.

Shining rust red, Mars has long been associated with blood and destruction. Babylonians knew it as Nergal, the star of death, and the Greeks and Romans named it after their god of war. Its baffling retrograde motion was the despair of many an early astronomer, including a pupil of Copernicus's who became so enraged by attempting to calculate it that he pounded his head against the wall. It remained for the great astronomer Johannes Kepler to elucidate the planet's orbit in his *Commentaries on the Motions of Mars* just as the first telescopes were invented in 1608.

The new device fueled both discovery and speculation about Mars. In the 1600s, Dutch astronomer Christiaan Huygens and then Italian astronomer Giovanni Cassini observed the regular appearance of a dark feature on the planet (now called Syrtis Major) that allowed Cassini to estimate Mars's rotation at 24 hours 40 minutes (just two and half minutes off the actual time). Hopes soon arose for the existence of life, intelligent life, on Mars. Astronomers noted that bright and dark patches on the planet's surface changed with time and the seasons, while Mars's bright white polar caps shrank and grew. It wasn't unreasonable to think that the dark spots were vegetation, fed by meltwater from the poles. Indeed, wrote astronomer William Herschel in the 18th century, "the inhabitants probably enjoy a situation similar to our own."

MOONS

No speculation was involved when American astronomer Asaph Hall discovered Mars's small, lumpy, inconspicuous moons in August 1877, following a persistent night-after-night search. Hall named them for the god of war's attendants: Phobos (Fear) and Deimos (Panic). Deimos, the outer moon, is only 15 by 12 kilometers (9 by 7 mi) in size and orbits Mars every 30 hours at a distance of 23,459 kilometers (14,577 mi). Phobos, the inner moon, is slightly larger at 27 by 22 kilometers (17 by 14 mi) and whizzes around the planet three times a day only 9,378 kilometers (5,827 mi) from the surface. Its spiraling orbit brings it 1.8 meters (6 ft) closer to Mars every century, which ensures that it will either crash onto Mars's surface or break up into a ring within about 50 million years. The origins of these odd little moons are unknown, but most scientists believe they are captured asteroids.

THE CANALS THAT WEREN'T

In the same year, when the planet was particularly close during its favorable opposition, Italian astronomer Giovanni Schiaparelli was able to make a good set of maps showing Mars's surface features. His drawings included quite a few straight, crisscrossing lines that Schiaparelli called *canali,* meaning "channels." No sooner had this word been mistranslated as "canals" than the idea of constructed waterways on Mars gripped the public imagination. Wealthy American amateur Percival Lowell wrote three popular books around the turn of the 20th century that envisioned a dying world whose inhabitants were heroically fighting for survival. "A mind of no mean order would seem to have presided over the system we see," he wrote. "Certainly what we see hints at the existence of beings who are in advance of, not behind, us in the journey of life."

Lowell's flights of fancy inspired science fiction writers and annoyed serious astronomers, who saw no signs of artificial canals on Mars, not to mention water or vegetation. Indeed, most 20th-century observations pointed in the other direction. Analysis of light from Mars showed that its thin atmosphere contained carbon dioxide and nitrogen, with virtually no oxygen. Vast, planetwide dust storms swirled across the planet's surface. Without a thick atmosphere to trap solar heat, temperatures averaged well below the freezing point of water. In short, the environment was hostile to life as we know it.

And yet, and yet—it was far from impossible that the planet could hold water in some form, even if frozen beneath its surface. And water is the most important prerequisite for life. Beginning in the 1960s, spacecraft began visiting the nearby planet with the goal of learning about its geography, its atmosphere, and the possibility of water.

NORTH AND SOUTH

More than a dozen flybys, orbiters, landers, and rovers, as well as countless earthbound instruments, have scanned the Martian landscape. Thanks to them, we know the surface of Mars better than that of any planet save our own—though that doesn't mean we completely understand it.

Perhaps the most striking overall aspect of the Martian terrain is the distinct difference between the northern and southern hemispheres. The north has low, smooth, rolling plains with relatively few craters. Scientists think the surface there is young, geologically speaking, having been resurfaced by lava a few hundred million years ago. The south is higher (its average elevation 5.5 kilometers/3.2 mi above the north) and rougher, with an older, heavily cratered surface. Straddling the equator is the Tharsis Bulge, a region the size of North America that rises 10 kilometers (6 mi) above the surrounding landscape. Opposite it, on the far side of the planet, is the Hellas Basin, a huge impact feature and Mars's lowest point at 8.2 kilometers (5.1 mi) below the average surface level.

DEEPER AND HIGHER

Enormous canyons radiating out from the east side of the Tharsis Bulge were created by a cracking of the planet's crust. The biggest, Valles Marineris, called the Grand Canyon of Mars, puts the Terran Grand Canyon to shame. Running some 4,000 kilometers (2,500 mi) around the Martian equator, in places it is 120 kilometers (75 mi) wide and at its deepest more than 8 kilometers (5 mi). Earth's entire Grand Canyon could fit into one of its smaller side cracks.

Just as spectacular as Mars's huge canyons are its towering volcanoes, the biggest in the solar system. The Tharsis Bulge holds four, including the largest, Olympus Mons. Rising 21.3 kilometers (13.4 mi) high and spreading out for 600 kilometers (370 mi), it is three times as tall as Earth's Mount Everest and roughly as wide as Colorado. Its caldera, the summit crater, is 80 kilometers (50 mi) across. Olympus Mons and its giant companions are shield volcanoes like those in Hawaii, broad domes with wide, sloping flanks. They grew not over plate boundaries but over hot spots, where magma broke through the planet's crust. Their immensity is caused largely by Mars's low gravity, which allowed them to build themselves up without collapsing under the weight of accumulating lava. They were active up to at least 100 million years ago; whether Mars's volcanoes are still active is not yet known.

DID MARS GET WHACKED?

Mars's lopsided form—low on top and high on the bottom—may have been caused by some unevenness in the planet's early internal dynamics. But an alternative theory proposes that a giant collision shaped the planet. Computer models show that a Pluto-size body could have hit the northern hemisphere of the young planet at a shallow angle, stripping off its crust. The prodigious impact would have created an elliptical impact basin about the size of Mars's depressed area. Topographical maps below depict the south pole at left, north pole at right, and the contrast between northern and southern hemispheres (bottom).

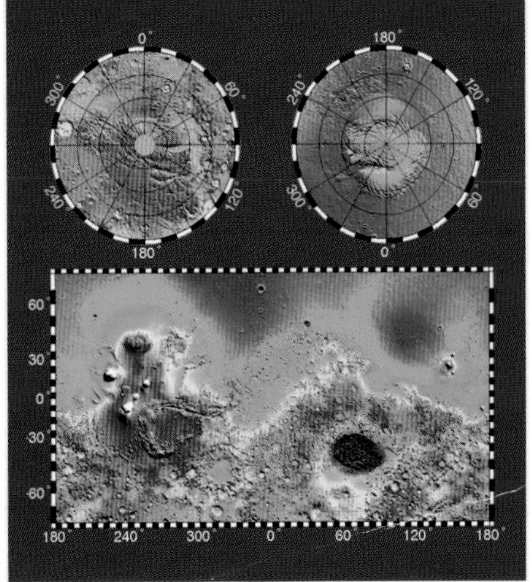

MARS: AIR & WATER

Large impact craters pock the surface of Mars, with many more in the southern highlands than in the north. The reason for this difference is that the southern craters were probably created during the era of heavy bombardment some four billion years ago, while those on the younger northern plains are more recent. Small impact craters are few. These may have been rubbed out by the planet's thin, scouring winds.

Mars's surface features and what we know of its rocks tell us that the planet formed from the same kinds of rocky materials as the other terrestrial planets, but cooled quickly. The planet's crust is stiff and lacks plate tectonics—that is, it isn't broken up into the moving plates that distinguish Earth's geology. Mars probably has an iron-sulfide core between 2,600 and 3,000 kilometers wide (1,600 and 1,800 mi), possibly some of it molten. Despite some magnetized spots in the crust, possibly left over from a period of magnetism early in the planet's history, Mars has no global magnetic field now.

Beginning in the 1970s, we have enjoyed close-up views of small pieces of the Martian surface, thanks to landers such as Viking 1 and 2 and rovers such as Opportunity and Spirit. Revealing a red, rocky, sandy, desert world, images and samples from these missions have confirmed that the soil contains a high percentage of iron. Small amounts of free oxygen in the atmosphere oxidize the metal, turning it the rust color visible even from Earth.

ATMOSPHERE

The first successful missions to Mars confirmed earlier, Earth-based observations that Mars's atmosphere is composed mainly of carbon dioxide (95 percent), with small amounts of nitrogen, argon, oxygen, carbon monoxide, and water vapor. It's thin, with atmospheric pressure only about one hundred fiftieth that of Earth's at sea level, though that pressure varies with the seasons. Icy clouds float above Mars's surface, those made of water ice closer to the ground, with carbon-dioxide-ice clouds above them.

Given its orbit and thin atmosphere, Mars is naturally quite cold, but temperatures vary considerably from day to night and from summer to winter, ranging from -87°C (-125°F) to a relatively balmy -5°C (23°F). As temperatures rise during the Martian day, thin winds stir up the dusty soil, dying down again at night. Dust devils, such as those seen in the American deserts, spin across the dry ground, while in the southern summer, when temperatures are highest, huge dust storms can sweep dust and sand into the atmosphere for months at a time, veiling the entire planet and sculpting sand dunes on the surface.

MARS ON EARTH

Humans haven't yet been able to walk on Mars, but they have found places on Earth that make a pretty good substitute. Among them is Chile's Atacama Desert, the driest spot on our planet (seen below are its salt flats and the aptly named Valley of the Moon). The virtually sterile area is a good testing ground for developing life-hunting instruments for Martian expeditions. The Dry Valleys of Antarctica can also stand in for Mars, geologically. Extreme cold, long periods of darkness, and ice-based cracking of the soil there all have their counterparts on Mars. And in perhaps the most elaborate Mars simulation, researchers have set up a base on Canada's Devon Island in the high Arctic, where a rocky polar desert and a well-preserved impact crater allow scientists to live and study in a Mars-like environment.

Mars's axis is tilted at 24°, giving it seasons, like Earth. Because of the planet's eccentric orbit, the southern summer is considerably warmer than the northern summer. The most noticeable effect of this seasonal variation is seen at the Martian poles. Each has a residual cap, ice that remains all year, but around it is a seasonal cap that grows and shrinks, adding so much carbon dioxide to the atmosphere in the summer that it increases atmospheric pressure by up to 30 percent. Although the polar caps are made primarily of carbon dioxide ice—dry ice—scientists have found that the inner, residual caps hold water ice as well.

Why is the atmosphere on Mars so different from the balanced, humid air of Earth or the torrid crush of Venus? Scientists are eager to learn the answer to this question. Its original atmosphere was probably outgassed, released from the planet's interior, like those of the other terrestrial planets. In its early years, it may have been denser than it is now, allowing for blue skies and puffy clouds. But in the billions of years since then, the atmosphere has largely disappeared. Much may have leaked into space, released by Mars's weak gravity, and some may even have been expelled by huge impacts during the era of bombardment. Much atmosphere was also lost when the dynamo creating the magnetic field collapsed, allowing solar radiation particles to erode the atmosphere, atom by atom. If Mars once had liquid water, that water may have absorbed carbon dioxide from the air, reducing the greenhouse effect and cooling the planet, which led to greater absorption of carbon dioxide, rapidly accelerating the thinning of the air and cooling of the planet in a reverse greenhouse effect. But a warmer early climate might also have been nurtured by heat from impacts, volcanic sulfur, or methane.

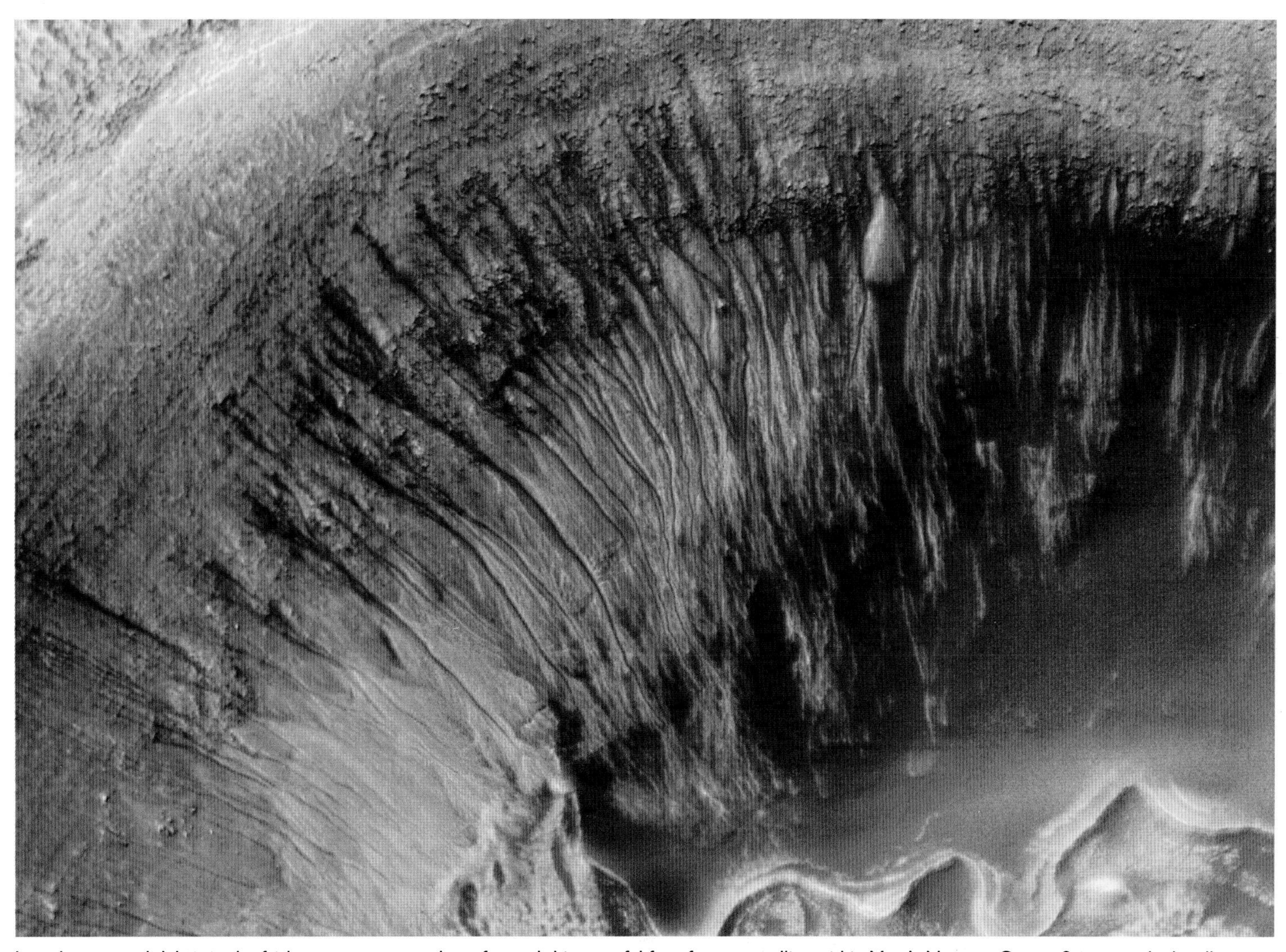

Liquid water and debris in the fairly recent past may have formed this graceful fan of narrow gullies within Mars's Newton Crater. Scientists think gullies may have been eroded as snow melted and ran down the crater walls, carrying debris to the bottom before evaporating or freezing and leaving fingerlike lobes.

WATER, WATER, ANYWHERE

Questions about Mars's history, its atmosphere, and particularly the possibility of life on its surface are all bound up in the question of water: How much did Mars have? How much was liquid? Only in recent years have we begun to answer those questions, thanks to a small armada of spacecraft launched by the U.S. and the European Space Agency.

Yes, Mars has water—frozen into the soil at latitudes higher than about 50° in both hemispheres, and likely abundant in the soil elsewhere at depths of several to many meters as a kind of permafrost. In the summer of 2008, NASA's Mars Phoenix lander dug a trench into the soil at its landing place in Mars's high northern latitudes, revealing a shiny patch of ice a couple centimeters down. Samples of the ice, scooped up and deposited into the lander's gas-analyzing oven, confirmed that the ice was good old H_2O. Not long afterward, as winter approached, the lander also detected snow falling from the clouds, although the flakes evaporated before they reached the ground.

Did liquid water ever exist on the planet's surface? Sinuous features on the planet's surface and geological studies of its rocks suggest that it did, but the jury is still out. Interesting evidence from orbiting spacecraft includes detailed images that have shown two different kinds of channels. Runoff channels in the south wind down from the mountains into the valleys, looking exactly like dry riverbeds on Earth. And outflow channels spread from the southern highlands into the northern plains, very much like the remains of ancient floods or river deltas. These channels imply that Mars once was warmer and wetter, with rain falling from the sky and coursing across the ground. Perhaps it even had lakes and oceans in the lower, northern lands or in the Hellas basin. Analysis of Martian soil also supports the notion that liquid water once soaked into the ground. Landers have spotted rounded pebbles, possibly eroded by water, and possibly calcium carbonate and clays, which form only in the presence of water. And cameras on the orbiting Mars Odyssey have detected salt deposits in the southern hemisphere that may also be relics of standing pools of water.

MARS: THE SEARCH FOR LIFE

Did life ever emerge on Mars? If so, what was its fate? The search for answers to these questions has been a driving force in Martian exploration. Hopes for life have risen and fallen over the years, reaching a peak in Percival Lowell's fantasies of an advanced Martian race, and plummeting when the first visiting spacecraft found a desiccated, lonely world. A more credible source was 20th-century American astronomer Carl Sagan (1934–1996). A science fiction fan from boyhood, he was entranced by the idea of life on other planets; as an adult, he became an expert in planetary climates and exobiology. Sagan was among the first to suggest that early Mars could have had a warmer, wetter climate conducive to life. Like him, many modern scientists still believe the planet, so similar to ours in many ways, could have harbored living organisms at least in its early history.

THE VIKING EXPERIMENTS

Looking for existing life on Mars has usually meant two things: searching for liquid water and searching for microbes. In 1976, NASA's Viking 1 and Viking 2 landers touched down on Mars and conducted three kinds of remote-controlled experiments. The pyrolitic-release experiment incubated a Martian soil sample in a simulated CO_2 atmosphere that was radioactively tagged. After five days, the sample was heated and analyzed to see if any microorganisms had absorbed the gas. A second, gas-exchange experiment added organic compounds to a soil sample and tested it after 12 days to see if any organisms had eaten and metabolized the compounds, releasing telltale gases. The third test, the labeled-release experiment, was similar: Compounds containing radioactive carbon-14 were added to a moistened soil sample, incubated for ten days, and heated to see if organic gases would be given off.

Thrillingly, all three experiments produced positive responses. But the first excitement soon faded as it became clear that inorganic chemical reactions, not microbes, were probably responsible for the results. For instance, samples heated to intense, sterilizing temperatures and then retested gave off the same results, although no Earthlike life could have survived the heat.

However, Mars-watchers have not lost hope of detecting the chemical signatures of life. In 2009, astronomers using the Mauna Kea Observatory announced that they had found the spectral lines for methane in the Martian atmosphere. Methane can be produced by microorganisms or by certain geological processes. The discovery suggests either that there is some form of life currently living below the Martian surface or that previously unknown chemical processes in Mars's rocks were releasing the gas. More definitive results await future missions.

NAME GAME

Giovanni Schiaparelli, the first person to map Mars in detail, was also the first to apply classical, historical, and biblical names to the planet's features: Syrtis Major, Tharsis, Nix Olympia (now known as Olympus Mons). The International Astronomical Union has retained many of Schiaparelli's names as well as his categories. On Mars, large craters are named for scientists who studied Mars or for those who contributed to its legends, ranging from Galileo to Edgar Rice Burroughs. Small craters take the names of smaller cities or towns, such as Princeton, Annapolis, or Porvoo (Finland). Large valleys, or "valles," are named for the words for Mars in various cultures: Auqakuh Vallis, for instance, from the Quechua for "Mars," or Mawrth Vallis, from the Welsh.

LIQUID WATER

Other searches for existing life on Mars focus on finding liquid water, a prerequisite for any life as we know it. We know that Mars has water ice—but so far we have not detected water in liquid form. It's possible that it exists in pockets of groundwater not far below the surface near the equator, or in the form of meltwater from ice in the sunlight. But conditions on the Martian surface are tough. Very low atmospheric pressure and frigid temperatures make it almost impossible for liquid water to survive on the surface. Furthermore, ultraviolet light from the sun streams virtually unimpeded through Mars's thin atmosphere, practically sterilizing the ground. Exposed life would have a difficult time. On the other hand, life on Earth exists around geothermal vents and in underground pockets, so subsurface life on Mars is still possible.

More recent searches for life have turned toward the stronger possibility that life once existed on ancient Mars, though it may be extinct today. Mars probably had a thicker, warmer, wetter climate over 3.5 billion years ago. Many features seen on the planet's surface today support the idea that water once flowed freely over Martian ground (see p. 113). However, Mars's climate cooled rapidly as its atmosphere thinned, and so any microbial life may not have been able to evolve into larger organisms as it did on our planet. It's also possible that the old runoff channels were formed by lava or that the first findings about the soil will not hold up. But if we can prove that Mars once had abundant liquid water, it will take us further toward understanding the past and future of our own planet and Mars.

METEORITE CONTROVERSY

One important source of information about Martian history is right here on Earth. At least 30 meteorites in current collections have fallen to Earth from Mars. We know this because analysis shows similar typical Martian chemical ratios, unlike any on Earth, the moon, or asteroids. The most famous—or infamous—of these meteorites is ALH48001, discovered in Antarctica in 1984. The sole Martian meteorite to date back more than 3.5 billion years, it holds carbon-containing chemicals that could have formed only in a watery environment. But does it hold more than that? In 1996, a group of scientists announced that they had discovered fossilized remains of microbial life inside the meteorite. Tiny, rodlike structures reminiscent of bacteria do appear in the rock. However, the structures are far smaller than bacteria on Earth. Most scientists today believe they are some kind of nonbiological mineral formation. Samples found on Earth also face the issue of contamination—over millions of years, Terran organisms could have been incorporated into the meteorite. The meteorite

TERRAFORMING MARS

The process of changing the climate of an entire planet to support human life is called terraforming, and it exists at the intersection of science and science fiction. Mars is a popular candidate. A common scenario goes like this: First we'd heat the planet by focusing sunlight on its surface with giant orbiting mirrors. Polar caps would melt, releasing carbon dioxide gas to thicken the atmosphere. Water frozen in the soil would melt into lakes, rivers, and oceans (below). Genetically engineered green plants, flown in from Earth, could then take hold and begin their task of translating the carbon dioxide into breathable oxygen. A few millennia later: Green Mars.

remains controversial, but most scientists are skeptical of claims for fossilized life.

The existence of ancient Mars rocks on Earth raises another question. Could life have piggybacked to Earth aboard Martian meteorites? It is far from impossible that rocks knocked off Mars during its warmer, wetter era could have reached our planet, possibly bringing organic molecules.

FUTURE MISSIONS

Scientists are eager to apply modern methods to the search for life in Martian soil. NASA is now weighing a proposal to send up an astrobiology field laboratory to look for evidence, the first such experiments since Viking. Another proposal calls for a sample return mission that would collect rocks for study. Lunar rocks collected by Apollo astronauts have been a treasure trove of information about the early solar system. Martian rocks might hold the key to life itself.

Water ice is clearly visible on the floor of a small trench scraped out by NASA's Phoenix Mars lander. Icy spots around the trench represent frost deposited in the morning.

CHAPTER 4

ROCKY DWARF PLANETS

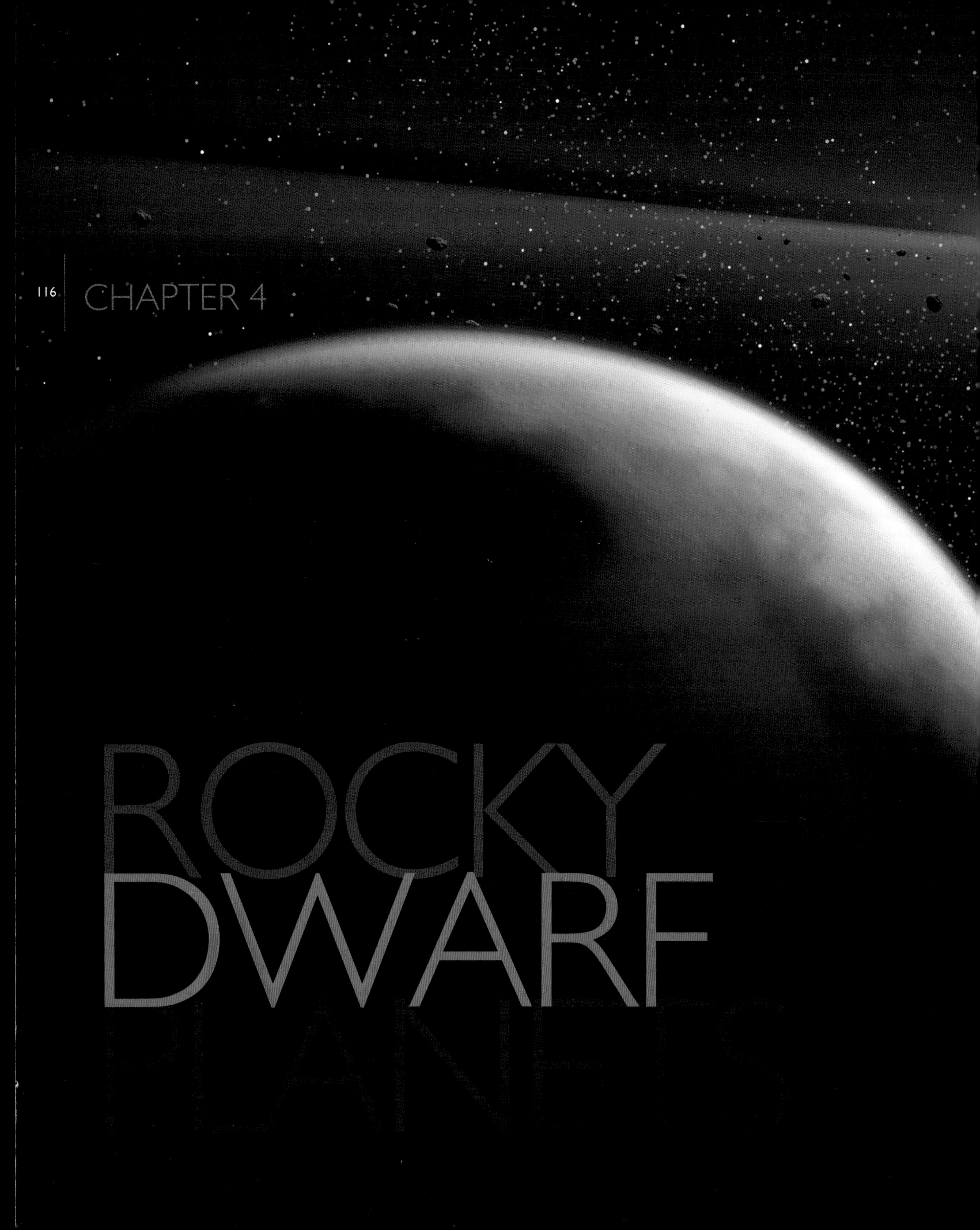

ROCKY DWARFS

The eight major planets, as they formed, became massive enough to attract most of the rocky debris in their orbital paths. But elsewhere in the solar system vast fields of chilly rubble remain. Countless fragments from the system's early history orbit between Mars and Jupiter, beyond Neptune, and in a shell at the extreme edge of the sun's gravitational reach. When they were first discovered in the early 19th century, astronomer William Herschel suggested that they be called asteroids, from the Greek root word for star, *aster,* because their tiny specks of light looked like little stars. Although European astronomers called them small or minor planets, in modern times the term *asteroid* has once again gained favor.

Astronomers once made a clear distinction between asteroids and comets. Asteroids were rocky and inactive, while comets were icy and volatile. We know now that these objects differ mainly by their location and less so in their substance. All began their existences as chunks of rock and ice. Most of the ice on asteroids evaporated quickly in the heat of the inner solar system, while in the outer solar system, comets hold on to their ice until they travel close to the sun, where it vaporizes in a stream of gas and dust. With enough trips through the inner solar system, any comet will come to resemble an asteroid as its volatile elements become buried under a thick rind of rocky dust.

Nor are there fundamental differences between these objects and planets, save for size. Dwarf planet Ceres, previously known as the largest asteroid, is a rocky sphere similar to Mercury, but smaller. In an effort to clarify solar system nomenclature and to recognize the existence of newly discovered planetlike bodies in

3.8 bya
Asteroids, comets, dwarf planets remain in orbit after major planets formed.

65 mya
Asteroid 10 kilometers (6 mi) wide strikes Earth, leads to mass extinctions.

50,000 ya
A 50-meter-wide (64-ft) meteorite strikes the North American continent.

A.D. 1492
First meteorite fall recorded and studied, Ensisheim, Alsace.

A.D. 1801
Giuseppi Piazzi discovers first asteroid, Ceres.

1802–07
Next three asteroids—Pallas, Juno, and Vesta—are discovered.

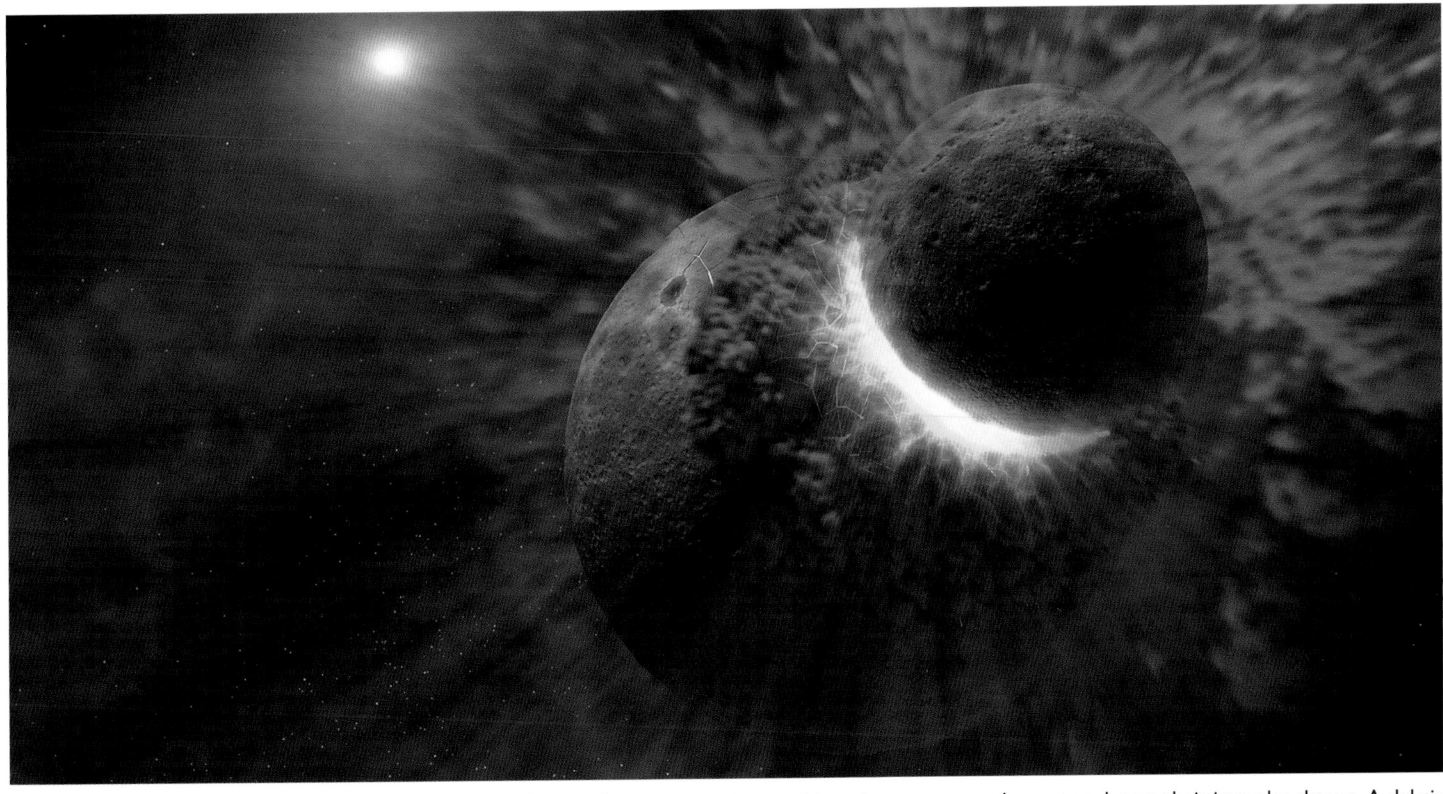

Most current asteroids are probably fragments left over from collisions among larger objects in our young solar system, hence their irregular shapes. A debris belt like the sun's has been seen around the bright star Vega, visible in the background, indicating that asteroids may be common among other systems.

the Kuiper belt, members of the International Astronomical Union (IAU) created the new category of dwarf planet in 2006. Major planets orbit the sun, are spherical, and have cleared their gravitational lanes of debris. Dwarf planets also orbit the sun and are roughly spherical, but they lack the mass to clear the debris out of their orbital neighborhoods. The first members of the dwarf planet category were Ceres, Pluto, and Eris. Later, in 2008, Pluto, Eris, and other trans-Neptunian dwarf planets gained the name plutoids, which are a subcategory of dwarf planet (see p. 166).

The bad news for Pluto was good news for Ceres. The newly christened dwarf planet was the first asteroid discovered by Giuseppe Piazzi on January 1, 1801. The Italian astronomer, working at the southern observatory in Palermo, was systematically compiling a star catalog when he noticed a new star—"a little faint, and of the colour of Jupiter, but similar to many others which generally are reckoned of the eighth magnitude," he wrote later. "Therefore I had no doubt of its being any other than a fixed star. In the evening of the 2d I repeated my observations, and having found that it did

1906
Max Wolf discovers first Trojan asteroid, Achilles, near orbit of Jupiter.

1908
Asteroid or comet explodes over Siberia, with the force of a hydrogen bomb.

1993
NASA's Galileo discovers first moon, Dactyl, to orbit an asteroid, Ida.

2001
NEAR Shoemaker spacecraft makes first landing on an asteroid, Eros.

2006
Ceres named a dwarf planet.

2007
Dawn mission launched toward Vesta (arrival 2011) and Ceres (2015).

not correspond either in time or in distance from the zenith with the former observation, I began to entertain some doubts of its accuracy. . . . The evening of the third, my suspicion was converted into certainty, being assured it was not a fixed star."

Within a couple of weeks, Piazzi wrote a letter to a friend in which he says that "I have announced this star as a comet, but since it is not accompanied by any nebulosity and, further, since its movement is so slow and rather uniform, it has occurred to me several times that it might be something better than a comet. But I have been careful not to advance this supposition to the public." Other astronomers were not so circumspect, and Piazzi wrote later that many "were instantly of the opinion that it was a new planet; and settled nearly the same elements of its orbit, as I have done." But within a few years several other such little "planets" had been found in similar orbits, and Ceres became just another asteroid, although the largest of a ragtag bunch.

And yet for an asteroid, it was remarkably big and round and planetlike. With a diameter of 940 kilometers (584 mi), Ceres is almost twice as large as the next biggest asteroid (Vesta). It is the only member of the asteroid belt massive enough to maintain a spherical shape. As small as it is, just one ten-thousandth the mass of Earth, it still represents one-third of the mass of the main asteroid belt. Like other terrestrial planets, it is quite likely that it has a differentiated interior, with a rocky core, an icy mantle, and a cratered surface. It may even have a tenuous atmosphere. And so when astronomers were forced to say just what defined a planet, Ceres found itself boosted into the new dwarf planet category.

These rocky bodies attract astronomers for several reasons. They represent material left over from the solar system's formation and can thus tell us much about how our planets were born and how planets and smaller bodies may have migrated from orbit to orbit in the solar system's early days. And there's a less scholarly but perhaps more pressing reason for studying them. Some of these circling rocks, knocked out of their orbits by Jupiter's gravity or passing bodies, come very close to Earth. Asteroid impacts in the past have been catastrophic, causing worldwide climate change and mass extinctions, and another one could happen at any time. Even a modestly sized asteroid would cause an impact equivalent to many hydrogen bombs.

JOSEPH-LOUIS LAGRANGE
CELESTIAL MATHEMATICIAN

Joseph-Louis Lagrange (1736–1813), whose name is given to key gravitational points in the solar system, was a modest genius who in his day contributed to many fields of mathematics and mechanics. Born in Italy to a French family, he was teaching mathematics by the age of 19, and his publications about number theory and gravitation quickly gained him a reputation as the greatest mathematician in Europe. First in Berlin, and then in Paris, Lagrange published paper after groundbreaking paper about prime numbers, differential equations, probability, and celestial mechanics, including the equations that describe the location of many satellites and asteroids today.

Several large sky-searching organizations scan for near Earth asteroids, but the discovery of asteroids is one area in which amateur astronomers shine. In Piazzi's days, finding the dim objects was a slow and painstaking process, involving drawing star fields by hand and comparing each night's stars with previous charts in hopes of finding a wanderer. Today, amateurs who invest in high-end telescopes equipped with charge-coupled devices and the relevant software can forgo much of the tedious eye-straining work — the computer programs will do their comparisons for them. Their instruments will record the sky several times, a few minutes apart each time, and identify any moving objects. The amateurs can then notify the Minor Planet Center, which will make the observations available to professional astronomers able to track their orbits.

In this way, amateurs dedicated to their hobbies of watching and charting the heavens have found many thousands of the small travelers. For example, in 2008, an amateur astronomer in England operated a telescope in Australia via the Internet and spotted the fastest spinning asteroid every discovered: a compact near Earth object (NEO) that rotates every 42.7 seconds. NASA and other agencies ask amateurs to help them follow up on the orbits of newly discovered asteroids, particularly NEOs. Each year, a privately funded cash prize is awarded to an amateur who discovers a significant NEO.

When first discovered, an asteroid is given a temporary designation that includes the year of discovery, a letter assigned to each half month, and a letter indicating how many asteroids have been reported previously in that half-month (although the letter *I* is not used). Asteroid 2009 DB, for instance, would be the second asteroid discovered in the second half of February 2009. Once an asteroid's orbit is tracked and confirmed, it can be given a permanent name. The IAU, which governs the naming of astronomical objects, has unusually relaxed rules for asteroids. The small bodies can be named for almost anyone or anything, except recent politicians or military figures. (Pets' names are also discouraged, but they've been used in the past.) Names must be 16 characters or fewer and not confusing or obscene. Among the named asteroids are Zappafrank, Jabberwock, Purple Mountain, and Dioretsa (a backward-orbiting asteroid).

LUIS AND WALTER ALVAREZ
HUNTING THE DINOSAUR KILLER

The father-son duo of Luis Alvarez (1911–1988) and Walter Alvarez (1940–) was the right combination to put forth daring new ideas. Luis (above, left) was a famed scientist who earned the Nobel Prize in physics in 1968; Walter (above, right, examining a slide containing iridium), a geologist who trained at Princeton. After Walter Alvarez discovered a thin clay layer containing high levels of the rare metal iridium and dated it back 65 million years, he, his father, and colleagues made an extraordinary connection. The metal, more common on asteroids, had been spread across the world in a dust cloud following a giant asteroid impact that led to the demise of the dinosaurs. Controversial at the time, the theory has since become widely accepted.

ASTEROIDS
WHERE THEY ARE

The asteroid belt marks a broad dividing line between the orbits of Mars and Jupiter, but asteroids can also be found at other key orbital points. Each asteroid, no matter how small, is its own little world, following its own orbit around the sun.

Sky-watchers have spotted and officially named more than 100,000 asteroids, with more added to the list each year. Almost all can be found in the main belt at distances ranging from 2.1 to 3.3 AU between the orbits of Mars and Jupiter. Another group—called Trojans because the first one discovered was named Achilles—lie farther out in Jupiter's orbital path. One group of the Trojans orbits 60° ahead of Jupiter, and another group 60° behind Jupiter. And smaller classes also populate the outer and inner regions of the solar system: the Centaurs, for instance, cometlike bodies beyond the orbit of Saturn, or the Amor, Apollo, and Aten asteroids, speeding close to Earth. • Numerous as they are, the total mass of asteroids in the main belt is far less than that of Earth's moon. This paltry mass, and the fact that some classes have distinctly different compositions, shoots down the old notion that asteroids represent the remains of a destroyed planet between Mars and Jupiter. Most likely, they are solar system leftovers, pulled about by Jupiter's gravity and unable ever to coalesce into even a petite planet.

Ceres discovered by: Giuseppe Piazzi, 1801
Named for: Roman goddess
Avg. dist. from sun: 414,628,870 km (257,645,470 mi), 2.76 AU
Orbital period: 4.6 Earth years
Equatorial diameter: 940 km (584 mi)

Mass: 9.43×10^{20} kg
Density: 2.1 g/cm^3 (0.38 x Earth)
Rotational period: 9 hours
Axial tilt: 3°
Surface temperature: -106°C (-158°F)

SKYWATCH

* Asteroids look like dim stars. With a backyard telescope, look for starlike points that move against the background stars over the course of a few nights.

AMAZING FACT Small, spinning asteroids can have a day and night just five minutes long.

1801
Piazzi identifies first asteroid, marking the belt between Jupiter and Mars.

1898
Gustav Witt discovers first near Earth object.

1906
Max Wolf locates Achilles, first Trojan asteroid, orbiting in Jupiter's path.

1932
Eugene Delporte observes 1221 Amor, first Amor asteroid.

1932
Karl Wilhelm Reinmuth discovers 1862 Apollo, first Earth-crossing asteroid.

1990
David Levy and Henry Holt observe asteroid following Mars's orbit.

Asteroids and comets tracked by the Minor Planet Center show up in this solar system map (opposite) in green (main-belt asteroids), red (near-Earth asteroids) and blue (comets). Most asteroids are irregular and smaller than 160 kilometers (100 mi) in diameter (above).

ASTEROIDS: ANCIENT FRAGMENTS

About 60 years after the first asteroids were discovered, American astronomer Daniel Kirkwood discovered several relatively empty orbital paths within the asteroid belt. These "Kirkwood gaps," as they came to be called, are evidence of the gravitational patterns that control the solar system. At certain distances from the sun, asteroid orbits fall into resonances with Jupiter's orbit, the asteroids receiving a regular gravitational tug from Jupiter as they circle between it and the sun. These repeated strains eventually pull the asteroids out of their old orbits into long, eccentric paths that take them close to Mars or Earth or knock them out of the solar system altogether.

The Jupiter-sun gravitational connection also controls the Trojan asteroids. Five spots in the solar system are known as Lagrangian points, after the French mathematician Joseph-Louis Lagrange, who worked out the orbital mathematics in the 18th century. Objects at any of the points should always orbit in synchrony with Jupiter. Three of the points, directly between Jupiter and the sun, are somewhat unstable, so that bodies at those locations can easily be pulled away. The other two spots, 60° ahead of and behind Jupiter in its orbital path, are stable points around which the Trojan asteroids have collected. Astronomers have found almost 3,000 Trojans to date. Some have also been discovered near similarly stable points in the orbits of Mars, Saturn, and Neptune.

Even light can change an asteroid's orbit, an idea first suggested by 19th-century Russian theorist I. O. Yarkovsky. As they spin, the rocky bodies absorb sunlight and reradiate it in the infrared spectrum as heat. Those infrared photons carry a tiny but perceptible momentum, each one pushing the asteroid just a little. The warmer, afternoon side of a rotating asteroid will get a bigger photon boost than the cool, morning side; eventually this "Yarkovsky effect" will alter the asteroid's path and spin.

Even the closest asteroids can be difficult to see from Earth, small and dark as they are.

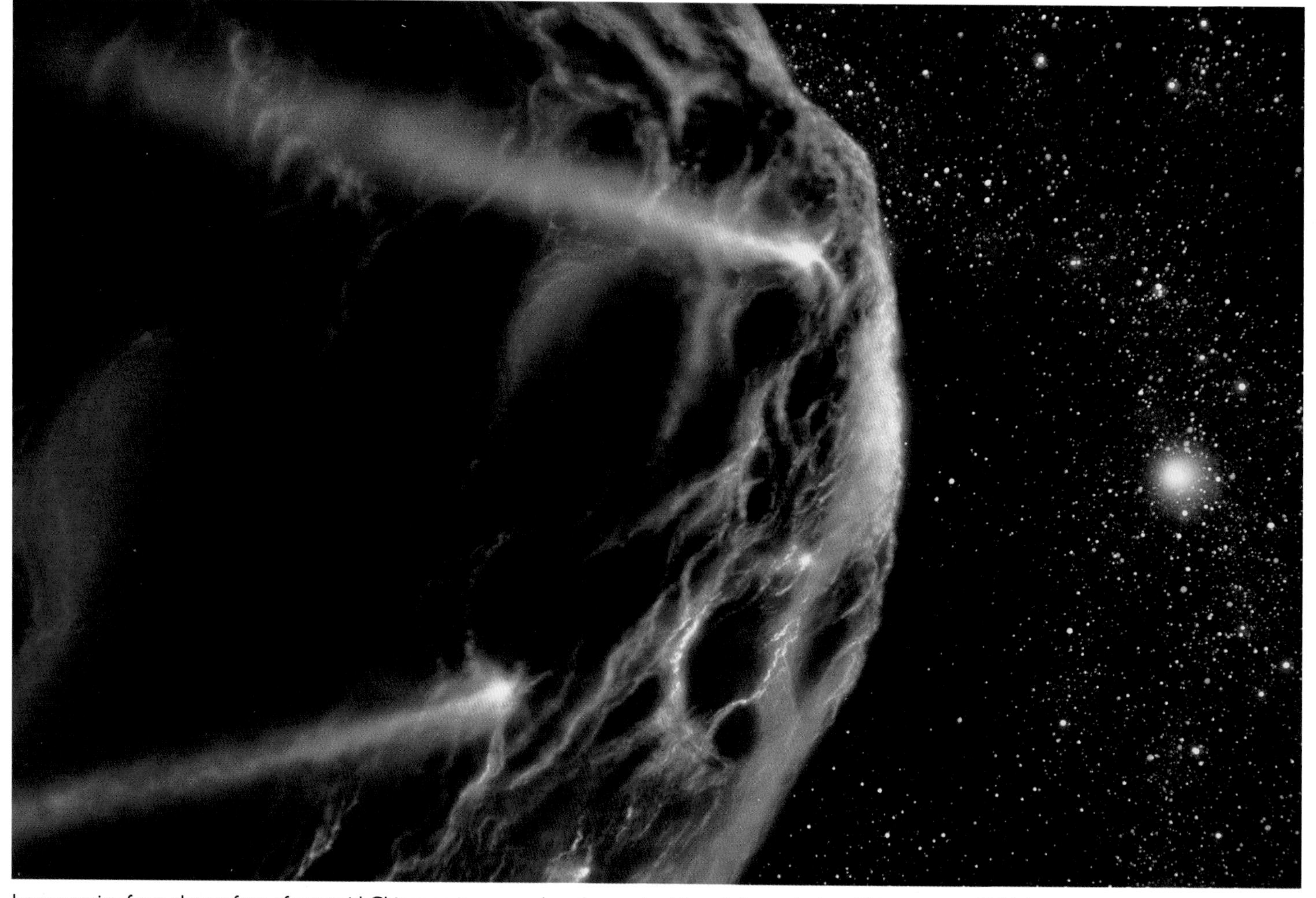

Ices vaporize from the surface of asteroid Chiron as it approaches the sun in this artist's rendering. Chiron, some 300 kilometers (185 mi) wide, belongs to an unusual class of asteroid/comets called Centaurs located past the orbit of Saturn. It may be an escaped moon.

(Only 25 have a diameter greater than 200 kilometers/125 mi.) Even when they have spotted an asteroid, astronomers may need years to decipher its orbit and glean an idea of its dimensions. Occasionally they will see one as it occults (crosses in front of) a star, which gives them a good estimate of its size. More often, they estimate sizes from the amount of reflected sunlight, although this can be misleading in the case of surprisingly dark, but large, asteroids. Changes in the object's light intensity can also give observers an idea of its spin.

Despite the observational challenges, astronomers have learned enough about asteroid composition through meteorites and through spectroscopy—analysis of their reflected light—to be able to divide them into more than a dozen classifications. Most common are the dark C-type, carbonaceous asteroids, containing large amounts of carbon. Shinier S-type asteroids hold more silicates, while M-types have nickel and iron. The S-types are found mostly in the inner regions of the asteroid belt, whereas C-types become more common with increasing distance from the sun. The more distant bodies probably represent material relatively unchanged from the earliest days of the solar system. Inner, brighter asteroids may have undergone heating and differentiation like the terrestrial planets.

MISSIONS TO ASTEROIDS

Spacecraft began to visit asteroids in the 1990s. The Galileo mission, cutting through the belt on its way to Jupiter, passed by asteroid 951 Gaspra in 1991, and 243 Ida in 1993. Both appeared to be fragments of larger bodies, broken up long ago in collisions. Ida was heavily cratered and seemed to be much older than Gaspra. The big surprise came when close inspection of the Ida images revealed a tiny moon, later named Dactyl, orbiting Ida. The 1.5-kilometer-wide (1 mi) body was the first confirmed satellite of an asteroid. To date, more than 100 such asteroid moons have been discovered, most in binary systems, but at least two in triple systems and one in a quadruple system.

In 1997, NASA's Near Earth Asteroid Rendezvous (NEAR—later named NEAR Shoemaker) spacecraft flew past asteroid 253 Mathilde and then went on to orbit 433 Eros. Heavily cratered, Mathilde had such a low density that it was apparently made of rubble. Eros seemed solid, though fractures and craters on its surface also testified to the violent life of asteroids. Although NEAR was not designed to land on the body—it had no landing gear—at the end of its mission controllers decided to give it a shot anyway. Descending gently at 1.5 meters a second (3.6 mph), NEAR successfully touched down on Eros's surface, the first craft to land on an asteroid. Through 2001, it returned data and images to Earth. Among other things, the images showed boulders and regolith, loose rocky material, covering the asteroid's surface, confirming that even the modest gravitational pull of the small body was sufficient to hold on to this debris. Two other NASA missions, Deep Space 1 and Stardust, also scooted past asteroids in 1999 and 2004 on their way to study comets, while Japan's Hayabusa craft landed on gravelly asteroid 25143 Itokawa in 2006. In the next few years, NASA's Dawn spacecraft will swing by dwarf planet Ceres and the large asteroid Vesta.

WHITTLED TO THE BONE

Mapped by Earth-based radar, asteroid 216 Kleopatra amused and intrigued astronomers when it turned out to look very much like a spinning dog bone. But Kleopatra might shed some more serious light on asteroid formation. The shiny, metallic body with the unusual shape could well be the remains of the dense core of a larger asteroid, destroyed in a collision billions of years ago.

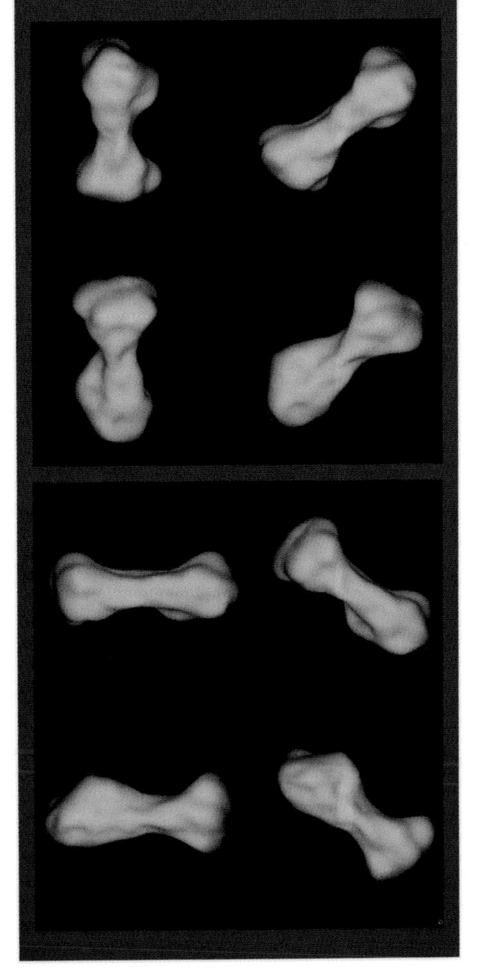

FROM RUBBLE TO RESOURCE

What we've learned so far about asteroids tells us a lot about the solar system's early years. Models of solar system formation suggest that the asteroid belt area should originally have contained enormous amounts of asteroidal debris. The gravitational influences of Jupiter and Saturn must have stirred up the rocky material, smashing the asteroids into each other and ejecting the great majority of them from the belt altogether. Some remaining fragments pulled back together, resulting in porous asteroids like Mathilde. Other ejected asteroids may have been captured as moons—Mars's Phobos and Deimos, for instance, and some of the irregular satellites of the gas giants. (New Mars Express data suggest that little Phobos is pretty rubbly throughout, too.) More than a few of the discarded asteroids may also have whacked the Earth during the era of bombardment, possibly bringing water and organic chemicals to the young planet.

Asteroid studies also help us understand the perils and the possibilities of the solar system's small rocky bodies. We have a pressing interest in learning how to forestall future collisions with Earth-crossing asteroids (see pp. 126–29). But asteroids represent rich resources, as well. Future builders of space stations, even orbiting hotels and factories, might profitably mine the asteroids for metals such as iron or platinum, silicate minerals, and water, usable as a propellant in spacecraft. It's not inconceivable that the costs of space mining might drop below that of lifting the massive materials into orbit, making it a business opportunity of the future.

FOR MARS EXPRESS DATA SUGGESTING RUBBLE ON PHOBOS: HTTP://WWW.ESA.INT/SPECIALS/MARS_EXPRESS/SEM8MUSG7MF_0.HTML

EARTH'S VISITORS
DOOMSDAY ROCK

Collisions are a fact of solar system life, and Earth is not immune. In recent years, scientists and governments have become more alert to those dangers, scrambling to set up programs that will help us track Earth-skimming objects.

On June 17, 2002, astronomers from the Lincoln Near-Earth Asteroid Research (LINEAR) program discovered a new asteroid, designated 2002 MN. The 100-meter-wide (330-ft) object had passed only 120,000 kilometers (74,000 mi) from Earth, just one-third of the distance to the moon—by astronomical standards, a very close shave. Perhaps more alarming was the fact that the discovery occurred three days *after* the asteroid flew by. • An asteroid 100 meters across is small and difficult to detect. Even so, if it had struck Earth, the impact would have been at least 100 times as powerful as the bomb that destroyed Hiroshima. An asteroid roughly half that size flattened forests around Tunguska, Siberia. Larger objects can hit with energies greater than a worldwide nuclear war. And history tells us that such impacts are not just possible but, over time, inevitable. • Such science fiction-made-real scenarios have spurred the development of near Earth object search programs around the world. Observatories in North America, Europe, and Japan now dedicate time to finding and tracking all large NEOs, those with diameters greater than 1 kilometer (0.6 mi). At the beginning of 2009, more than 1,000 potentially hazardous asteroids (PHAs) had been found.

NEOs TO WATCH
- **2007 VK184:** 0.13 km (0.08 mi) wide, next pass 2048
- **99942 Apophis:** 0.27 km (0.17 mi) wide, next pass 2036
- **2004 XY130:** 0.50 km (0.31 mi) wide, next pass 2009
- **2008 AO112:** 0.31 km (0.19 mi) wide, next pass 2009
- **1994 WR12:** 0.11 km (0.07 mi) wide, next pass 2054
- **1979 XB:** 0.68 km (0.42 mi) wide, next pass 2056
- **2000 SG344:** 0.037 km (0.022 mi) wide, next pass 2068
- **2006 QV89:** 0.030 km (0.018 mi) wide, next pass 2019
- **2008 CK70:** 0.031 km (0.019 mi) wide, next pass 2030

SKYWATCH

- Track NEOs using the Minor Planet NEO page, *www.cfa.harvard.edu/iau/NEO/TheNEOPage.html*, and the JPL NEO site, *neo.jpl.nasa.gov/*.

AMAZING FACT Astronomers who find that asteroids get in the way of their observations sometimes call them "vermin of the skies."

1908
An asteroid levels 2,000 square km (772 sq mi) of Siberian forest.

1970
Searches begin for near Earth asteroids, only some of which cross Earth's orbit.

1980
Luis Alvarez finds large asteroid was responsible for mass extinction 65 mya.

1999
The IAU endorses the Torino Scale, created by Professor Richard Binzel.

2007
Computer simulations show Tunguska asteroid smaller than originally thought.

2009
As year begins, 6,046 NEOs are known. Of these, 1,018 are classified as PHAs.

Small metal-and-stone meteorites (opposite) hit Earth often, larger asteroids rarely, but when bigger objects do strike the Earth (artwork, above) they can release the energy of thousands of atomic bombs.

EARTH'S VISITORS: DANGER AND DEFENSE

Today's world might be ruled by saurians were it not for an asteroid that whacked Earth 65 million years ago near the Yucatán peninsula. At an estimated 10 kilometers (6 mi) in diameter, it hit the ground with an energy far greater than all of the world's modern weaponry detonated at once. The resulting fireball, tsunamis, wildfires, and climate change resulting from airborne debris most likely caused the mass extinction of about half of all animal and planet species on Earth, including the dinosaurs.

IMPACTS OF THE PAST

Impacts are a fact of life for all solar system bodies. Earth has not experienced a major collision in human history, but impact craters around the planet tell us that it is not immune. Plate tectonics and erosion erase most craters from Earth's surface over time. Arizona's sharp-edged 1.2-kilometer-wide (0.75-mi) Meteor Crater, the result of an impacting body about 60 meters (200 ft) across, is a youngster at 50,000 years old.

Most space rocks smaller than about 40 meters (130 ft) in diameter will break up in the atmosphere. The largest of these can still do damage, as evidenced by the Tunguska explosion of 1908. Scientists now believe that the Tunguska object may have been a fractured asteroid no more 40 meters across. Even so, traveling at supersonic speeds, its shock wave created a fireball that smashed the landscape even though the asteroid itself disintegrated in the air.

On average, an object the size of the Tunguska asteroid will hit Earth once every thousand years. Big rocks perhaps 2 kilometers (1.2 mi) wide would arrive about once or twice in a million years, and the catastrophically large visitors, 10 kilometers (6 mi) across or bigger, would intersect Earth's orbit about once every 100 million years. Of course, these are averages taking in long swaths of time. The world could go for millions of years without any impacts, or it could suffer three massive blows in three succeeding years without invalidating the statistics.

Category	Level	Description
No Hazard	0	The likelihood of collision is zero, or is so low as to be effectively zero. Also applies to small objects such as meteors and bolides that burn up in the atmosphere, as well as infrequent meteorite falls that rarely cause damage.
Normal	1	A routine discovery in which a pass near the Earth is predicted that poses no unusual level of danger. Current calculations show the chance of collision is extremely unlikely, with no cause for public attention or public concern. New telescopic observations very likely will lead to reassignment to Level 0.
Meriting Attention by Astronomers	2	A discovery, which may become routine with expanded searches, of an object making a somewhat close but not highly unusual pass near the Earth. While meriting attention by astronomers, there is no cause for public attention or public concern, as an actual collision is very unlikely. New telescopic observations very likely will lead to reassignment to Level 0.
	3	A close encounter, meriting attention by astronomers. Current calculations give a one percent or greater chance of collision capable of localized destruction. Most likely, new telescopic observations will lead to reassignment to Level 0. Attention by the public and public officials is merited if the encounter is less than a decade away.
	4	A close encounter, meriting attention by astronomers. Current calculations give a one percent or greater chance of collision capable of regional devastation. Most likely, new telescopic observations will lead to reassignment to Level 0. Attention by the public and public officials is merited if the encounter is less than a decade away.
Threatening	5	A close encounter posing a serious but still uncertain threat of regional devastation. Critical attention by astronomers is needed to determine conclusively whether a collision will occur. If the encounter is less than a decade away, governmental contingency planning may be warranted.
	6	A close encounter by a large object posing a serious but still uncertain threat of a global catastrophe. Critical attention by astronomers is needed to determine conclusively whether a collision will occur. If the encounter is less than three decades away, government contingency planning may be warranted.
	7	A very close encounter by a large object, which if occurring this century poses an unprecedented but still uncertain threat of global catastrophe. For such a threat in this century, international contingency planning is warranted, especially to determine conclusively whether a collision will occur.
Certain Collisions	8	A collision is certain, capable of causing localized destruction for an impact over land or possibly a tsunami if close offshore. Such events occur on average between once per 50 years and once per several thousand years.
	9	A collision is certain, capable of causing unprecedented regional devastation for a land impact or the threat of a major tsunami for an ocean impact. Such events occur on average between once per 10,000 years and once per 100,000 years.
	10	A collision is certain, capable of causing a global climatic catastrophe that may threaten the future of civilization as we know it, whether impacting land or ocean. Such events occur on average once per 100,000 years or less often.

The Torino Impact Hazard Scale categorizes the impact hazard level associated with potentially dangerous asteroids or comets on a ranking of zero to ten. Zeros and ones represent routine finds extremely unlikely to cause damage. Eights through tens represent certain collisions of increasing magnitude.

SEARCH PROGRAMS

These alarming numbers, highlighted by the impact of the Shoemaker-Levy comet on Jupiter in 1994, spurred increased scientific action beginning in the 1990s. NASA supports several programs that search for NEOs, comets or asteroids whose orbits come within 45 million kilometers (28 million mi) of Earth's. Particularly interesting are potentially hazardous asteroids (PHAs), defined as those coming within 0.05 AU (about 4.65 million mi) and larger than approximately 150 meters (500 ft). Search programs in Japan, Italy, and Germany also track nearby asteroids and comets.

Although comets can certainly hit Earth with devastating results, relatively few pass near the Earth in their elongated orbits. Of greater interest are near Earth asteroids (NEAs). These generally

Propelled by ion-drive thrusters, a spacecraft uses the gravitational attraction of its mass to gently tug an asteroid out of an Earth-impacting trajectory in this illustration of a possible rescue scenario. Gravity tractors have not yet been built, but they are entirely feasible.

come from three groups, known as the Apollo, Aten, and Amor asteroids. Most are small and still invisible to telescopes. By the beginning of 2009, searchers (mainly Lincoln Near-Earth Asteroid Research—or LINEAR—and the Catalina Sky Survey) had discovered 767 NEAs 1 kilometer (0.6 mi) in diameter or larger. Altogether, the programs have identified more than 1,000 PHAs to date.

These statistics include some good news. Astronomers believe that about 1,100 NEOs larger than 1 kilometer (0.6 mi) exist in total, so the search programs are closing in on identifying all of them. Even better is the news that none so far is on a direct collision course. None of the PHAs identified so far rates more than a 1 on the 10-point Torino impact scale.

And the bad news? Well, there are those undiscovered big impactors. Furthermore, scientists estimate that one million NEAs larger than 40 meters (130 ft) are still out there in the dark. A direct hit from rocks ranging from this size up to one kilometer would not do catastrophic damage. Even so, such an impact could take out a major city, create massive tsunamis, or more. Search programs today are not designed to warn of imminent impacts but are simply meant to identify NEAs and track their orbits. These identifications may take place only when an asteroid has already skimmed past us, as happened with asteroid 2002 MN. According to NASA, "The most likely warning today would be zero—the first indication of a collision would be the flash of light and the shaking of the ground as it hit."

DEFENSE

Once an asteroid's orbit is known, earthbound instruments can track it and, in theory, warn us of a future impact. The next step is to devise a strategy for deflecting a dangerous asteroid. Blowing it up makes good cinema but bad practice—we'd just be trading one big impact for many smaller ones.

Other suggestions involve gently deflecting the object. A plasma rocket, for instance, might land on the asteroid and act as a space tug to push it into a different orbit. Or such a rocket might work as a "gravity tractor" instead, hovering near the object and pulling it out of a dangerous trajectory using the gravitational attraction between them. We might even be able to employ sunlight to nudge an asteroid, using gossamer-thin solar sails or a change to the reflectivity of the rock itself to alter its orbit. But the larger the asteroid, the harder it is to shove it aside—and all these solutions require advance planning and years of notice before the NEO arrives. So far, plans for asteroid defense are purely theoretical.

EARTH'S VISITORS: METEORITES

Far from being empty, solar system space is laced with rocks, pebbles, and dust. Earth passes through a grainy mist of pulverized asteroids and comets as it orbits; scientists estimate that 1,000 tons of such material enters the atmosphere every day.

Pieces of space rock too small to be considered asteroids are called meteoroids. Traveling at speeds between 10 kilometers and 70 kilometers per second (20,000 and 150,000 mph), almost all of those reaching Earth burn up about 80 kilometers (50 mi) high in the atmosphere. The brief, fiery trails their fragments leave in the sky—reminiscent of certain types of fireworks—are sometimes known as shooting stars, but are more properly called meteors.

Men look over a two-ton meteorite embedded in the sand. Deserts are good hunting grounds for meteorites; the relatively unchanging, waterless environment preserves them where they can be spotted.

About once a month, larger meteoroids hit the atmosphere with a bang, flaming out in a fireball and creating a sonic boom tracked by military satellites (programmed to distinguish between these explosions and those of actual bombs). Thankfully for Earth dwellers, only about one in a million meteoroids, the most massive specimens, makes it all the way to the ground. Once it reaches Earth's surface, it is known as a meteorite.

Worshipped as sacred sky stones for millennia, meteorites have attracted scientists and collectors since the 19th century, when their extraterrestrial nature was first accepted. Until the space age, they were the only samples we possessed of other planetary bodies.

KINDS OF METEORITES

Meteorites fall into three categories: stony, iron, or stony-iron. About 94 percent are stony, made mostly of silicates, but with flakes of iron and nickel. Five percent are iron—actually mixtures of iron and nickel. And one percent are stony-irons, roughly half and half silicates and the nickel-iron mix. Though far more common, stony meteorites are harder to spot than irons, since they blend in with rocks on Earth. Metal detectors will find the irons, shiny and distinctive.

Stony meteorites are further subdivided into chondrites and achondrites. Chondrites contain chondrules, tiny spheres of rock surrounded by grains of other minerals. Ordinary chondrites make up the bulk of this sort of meteorite. Enstatite chondrites contain the mineral enstatite, and carbonaceous chondrites, less dense than the others, have significant amounts of carbon and water. Meteorites' composition tells scientists how and when they were formed. All of the chondrites represent primitive materials of the early solar nebula. The little chondrules they contain were formed when the rocky materials melted into droplets and then cooled again during intense, hot phases of the solar system's history. Carbonaceous chondrites may be the most primitive. Achondrites, on the other hand, are igneous rocks that appear to have been completely melted and re-formed within larger objects. And iron meteorites may be the remains of the cores of differentiated bodies, those in which the dense, metallic materials sank to the center.

These findings, as well as other evidence, tell us that almost all meteorites are remnants of rocky bodies, asteroids or planetesimals, that accreted in the first days of the solar system. At some point the bodies broke apart, probably in collisions, and their fragments flew into space. Those that landed on Earth are the oldest rocks on our planet, dating back 4.6 million years. Some contain materials created even before the solar system was formed.

Some meteorites, such as Australia's Murchison meteorite, even carry organic molecules, suggesting that such chemicals may have been brought to Earth in its early history.

LUNAR AND MARTIAN METEORITES

A few valuable meteorites appear to come not from early asteroids, but from elsewhere in the solar system. Thirty-one are almost identical to rocky samples brought back from the moon by Apollo astronauts, and very different from typical meteorites. These must be lunar meteorites, blasted by impacts off our satellite's surface. Scientists treasure them because they represent a truly random sample of the moon's surface. At least one of them shows a basaltic composition unlike any of the rocks brought back by astronauts.

At least as interesting are 34 meteorites from Mars that apparently broke off from six to eight large meteorite impacts long ago. Tiny amounts of gas trapped within them match the profile of the Martian atmosphere. One, dubbed ALH 84001, caused (and continues to cause) quite some controversy when a group of scientists announced the discovery of tiny bacteria-like fossils within it (see p. 114). Most experts now believe the structures do not represent any form of life. However, ALH 84001 is still prized as representing a particularly ancient piece of the Martian surface. Studies show that it may be about 4.5 billion years old, having been blasted from Mars's heavily cratered southern highlands some 17 million years ago.

It is remotely possible that meteorites might have reached Earth from other planets or moons, flying off bodies such as Mercury that have escape velocities similar to that of Mars. However, any bits knocked off more remote planets would probably have been gravitationally attracted to objects near them, rather than to Earth.

METEORITE INJURIES

Though accounts in various cultures tell of people being struck and killed by stones from the sky, few verifiable records exist of meteorite injuries. One exception is the case of Ann Hodges of Sylacauga, Alabama, who was hit and badly bruised by a 4-kilogram (8.5-lb) stony meteorite that crashed through her roof in 1954.

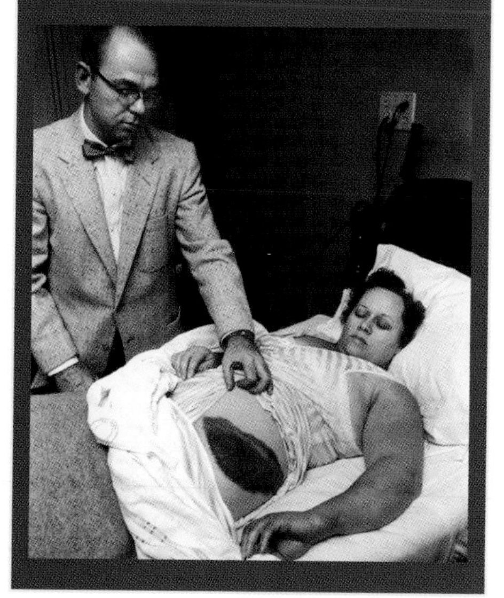

WHERE THEY LAND

Meteorites can and do land anywhere on Earth. Even if their fall is witnessed, they can be hard to find, and over time stony meteorites weather away and look much like any other rock. They come in all sizes, from pebble to boulder, such as the 60-metric-ton iron monster found on a farm in Namibia in 1920. Some terrains are better for discovering the visitors. The ALH 84001 meteorite was discovered in Antarctica, as were more than 15,000 others in recent years. Antarctica's dry, frigid environment, where the rocks are rapidly encased in ice, preserves the meteorites and keeps them from weathering. For similar reasons, quite a few meteorites have been found in deserts. Not only are the conditions apt for preservation, but the meteorites, their surfaces darkened by a fusion crust generated by the heat of their flight through the atmosphere, stand out from the surrounding light-colored rocks (these known informally as meteorwrongs).

Given the large numbers of meteorites entering our atmosphere, it's remarkable that so few have hit people or human structures. One such was the Peekskill meteorite, which appeared as a spectacular fireball over the eastern U.S. in October 1992. After soaring with a crackling sound over various Friday-night football games, it smashed in the trunk of a Malibu sedan in Peekskill, New York. (The car was subsequently taken on a world tour.) A meteorite fragment badly bruised a woman in Alabama in 1954; another, in 1992, struck a Ugandan boy in the head. Even an unfortunate Egyptian dog was reportedly killed by a meteorite in 1911—and by a rare Martian meteorite, at that. Many more meteorites have been recovered harmlessly and taken into private collections or sold, bit by bit. Tiny fragments of Martian meteorites go for about 20 times the price of gold. From a scientific point of view, the true value of meteorites lies in the information they convey about the origins and behavior of our solar system, brought to us on Earth in a convenient package as "the poor man's space probe."

The Holsinger meteorite is the largest piece remaining of the massive meteorite that blasted out Arizona's Meteor (also called Barringer) Crater some 50,000 years ago. When explorers first discovered the crater, it was surrounded by 30 tons of iron fragments.

FOR INFORMATION ON METEORITES AND LINKS TO ASSOCIATIONS AND COLLECTORS, GO TO HTTP://METEORITE.ORG

CHAPTER 5

RINGED PLANETS

RINGED PLANETS

The four outer worlds are often called the Jovian planets, after Jupiter, their biggest representative. Or they're known as gas giants, from their sizes and atmospheres. They can also be divided into gas giants (Jupiter and Saturn) and ice giants (Uranus and Neptune). But they could with equal justice be known as the ringed planets, because all four are distinguished by the remarkable halos of rings that surround them.

Jupiter and Saturn have been known to sky-watchers since antiquity. Jupiter is the brightest object in the sky after the sun, moon, and Venus. Its steady white light moves slowly across the constellations over the course of almost 12 years, a grand journey that may have earned it its title as King of the Planets long before we knew about its immense size. Saturn, pale gold, is also prominent in the night sky. Its slow motion against the stars, given an orbit of more than 29 years, may account for its ancient association with Father Time and venerable old age in the person of Cronus, the Greek god who fathered Zeus.

Uranus and Neptune, far more distant, are relatively recent astronomical discoveries. German-English astronomer William Herschel, a towering figure in 18th-century astronomy, spotted Uranus in 1781 through his beautifully handcrafted telescope. The mathematics of gravitation led to Neptune's discovery in 1846; English and French mathematicians figured out where the planet should be, and a German astronomer confirmed it.

The four ringed planets are as different from the inner, or terrestrial, worlds as—well, as gas is from rock. Their remote and widely spaced orbits, for instance, contrast with the relatively crowded inner solar system. Mercury, Venus,

250–50 B.C. Babylonians are able to predict the longitudinal positions of Jupiter.

A.D. 1660 Jean Chapelain proposes Saturn's rings are made up of many small satellites.

1846 William Lassell discovers the first of Neptune's moons, Triton.

1971 NASA launches Pioneer 10, the first spacecraft to reach the outer planets.

1977 Uranus's rings discovered. Before this, Saturn was the only known ringed planet.

1977–89 NASA's Voyagers 1 and 2 provide information on all four ringed planets.

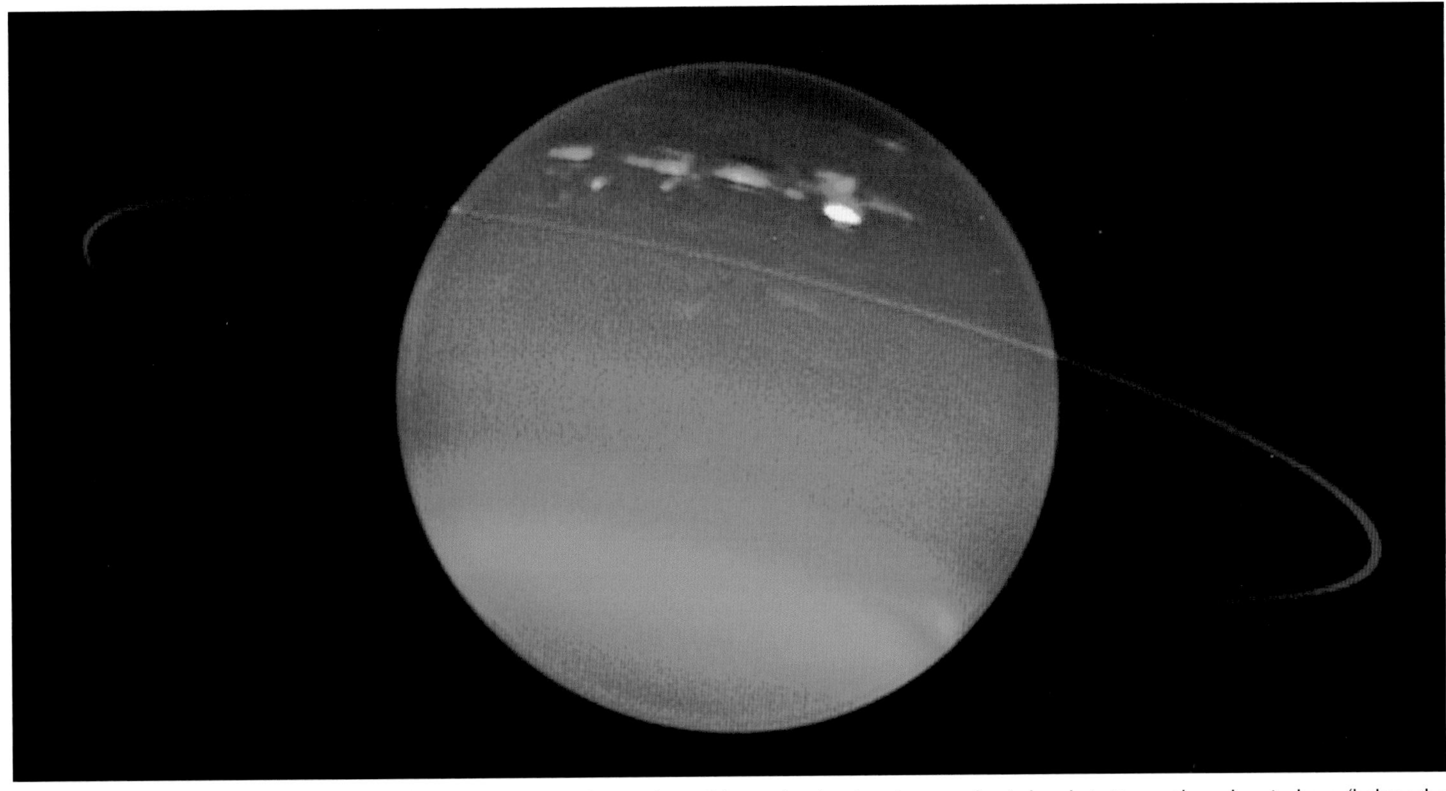

An infrared image of Uranus picks out clouds in its southern hemisphere (above the rings) and atmospheric bands in its northern hemisphere (below the rings). The planet's slender, widely spaced, very flat rings were discovered in 1977.

Earth, and Mars orbit within 1.5 astronomical units (AU) of the sun. Jupiter orbits at more than 5 AU, Saturn at 9.5, Uranus at 19, and Neptune at 30, in the frigid reaches where the sun is no more than an unusually bright star.

The outer planets also differ sharply in size and composition from their terrestrial cousins. All are gaseous, consisting largely of hydrogen and helium atmospheres that grow increasingly dense with depth until they reach a rocky core. They have no real surfaces to speak of, and almost all of our observations are of the upper levels of their atmospheres.

Despite their gaseous natures, the Jovian planets are massive. Jupiter alone contains more than twice the mass of all the other planets combined. Its huge gravitational pull is second only to that of the sun in the solar system. Saturn has less than one-third Jupiter's mass, but it's still a giant that could swallow more than 700 Earths. Uranus and Neptune are somewhere in between these two monsters and the terrestrial planets in size. They are made mainly of gas, like their bigger brethren, but are icier and denser. All four have powerful magnetic fields, apparently generated by electrically conducting liquids under their

1979
Jupiter's moon Io seen as most geologically active body in solar system.

1989
Voyager 2 discovers Uranus's Great Dark Spot, like Jupiter's Great Red Spot.

1994
Pieces of comet Shoemaker-Levy 9 impact Jupiter, cause clouds, fireballs.

2005
Cassini spacecraft finds that Saturn's ring system has its own atmosphere.

2006
Cassini spacecraft discovers potential water on moon Enceladus.

2009
Cassini imaging teams discover new, small moon orbiting in Saturn's G ring.

atmospheres; the fields of Uranus and Neptune are strangely askew.

Unlike the inner planets, the outer planets are richly endowed in moons and rings. The four big planets are orbited by 164 icy moons, from giant Ganymede to petite Pasithee. Some are worlds unto themselves. Saturn's Titan has lakes and a thick atmosphere; Jupiter's Europa possesses a moon-enveloping ice-capped sea.

Rings circle each big planet as well. More properly, they should be called ring systems, consisting as they do of untold millions of rock and ice particles following orbits shaped by the complex web of gravitational forces among the parent planets and their moons. The origins of these rings are still mysterious. They seem to be younger than their parent planets, on the whole. A prevailing theory holds that they are the pulverized remains of moons that were dragged toward their massive planets and pulled to pieces by gravitational tides. Some rings also appear to be constantly replenished by fragments knocked off existing moons. But these theories don't completely explain why Saturn's rings are so spectacular and the others so faint. It's possible that they are younger and fresher than others in the solar system. If and when astronomers can observe rings around extrasolar planets, that explanation may come forth.

Why are the gas giants all in the solar system's farthest suburbs? The one-word answer is "temperature." As the infant solar system condensed from gas and rocky dust grains into planetesimals, the temperatures around the growing sun were too high for water ice; metals and rocky grains survived close in and melded into rocky planets. From about five AU outward — the "snow line" — the region was cool enough for some of the solar system's most common gases, such as hydrogen, helium, ammonia, and methane, to condense into icy grains right away. According to one theory, solid protoplanets formed in this way eventually became massive enough to pull in abundant hydrogen and helium directly from the solar nebulae as their atmospheres. Another theory holds that this formation scenario takes too long; when the young sun's solar wind ignited and blew away the solar system's gases, it would have blown away the giant planet's burgeoning atmospheres as well. This second theory proposes that the big planets formed not by accretion, but by condensing

IS JUPITER A FAILED STAR?

Jupiter is sometimes called a failed star because of its size and hydrogen/helium composition. If it had picked up more gas from the solar nebula, would fusion have begun in its core?

The answer is probably "no." However the planet originally took shape in the chilly outer reaches of the solar nebula, it ended up with an icy/rocky core inside its gas envelope. The sun, on the other hand, condensed directly from a gas cloud and never possessed a rocky core. True failed stars can be seen in the substellar objects known as brown dwarfs. These dense, lukewarm gaseous bodies, between 13 times and 80 times the mass of Jupiter, are like a missing link between a gas giant planet and a small star.

directly from unstable areas in the solar nebulae, compacting very rapidly into the planets we see today. This scenario is supported by evidence that young extrasolar systems already host giant planets.

The terrestrial planets settled into their current orbits fairly quickly, but the outer planets may have migrated in and out for a while in huge sweeping paths of gravitational disruption. The theory of giant-planet migration proposes that Uranus and Neptune originally formed near the current orbits of Jupiter and Saturn, and that all four big planets were close to the inner edge of a huge belt of planetesimals. Gravitational interactions among the planetesimals and the four planets pulled the planets inward at first, and then pushed Saturn, Uranus, and Neptune outward toward the planetesimal belt. The big planets' gravity scattered many of the planetesimals, shooing them completely out of the solar system or inward on long elliptical orbits. Jupiter was pulled in slightly to its current orbit. And many of the smaller bodies, knocked from their orbital shelves, fell toward the inner solar system and pummeled the terrestrial planets (and our moon) during the period of Late Heavy Bombardment about 700 million years after the planets formed.

The migration theory might explain how Uranus and Neptune could have become so massive. If they had originally formed in the far reaches of the system, the relative paucity of primordial material would have restricted their growth. It may also explain how Jupiter has retained certain gases in its atmosphere that are more typical of elements in the outer solar system.

Studies of extrasolar systems will help us to understand how our own planets formed. So would additional missions to the outer solar system, but these will take a few years. Saturn and its moons are coming into clearer focus in the beginning of the 21st century with the Cassini mission, which landed a probe on Titan's surface. Beginning in 2016, two more spacecraft may visit the big planets, including NASA's Juno, a Jupiter polar orbiter, and the joint NASA-ESA Europa Jupiter System Mission, designed to search out life on the moons Europa and Ganymede. In the absence of spacecraft, planetary scientists are also studying the outer planets with such advanced observatories as the Hubble Space Telescope and the Keck telescopes.

GODS, NYMPHS, AND LOVERS

As soon as the telescope began to reveal the many moons of the outer solar system, astronomers began to argue over what to call them. Galileo wanted to name the four big Jovian moons he discovered after his patrons, the Medici, and so they were known as the Medicean stars for years. Astronomer Simon Marius suggested naming them after other planets—the Saturn of Jupiter, the Jupiter of Jupiter, the Venus of Jupiter, and the Mercury of Jupiter—an idea that was wisely ignored. But another scheme he proposed, originally suggested by Johannes Kepler, caught on, and the four big moons were named for the god Jupiter's objects of desire: Io, Europa, Callisto, and Ganymede.

Thus the convention was born of naming the outer planet's satellites after mythological figures associated with the parent world. Jupiter's moons are named after Jupiter's lovers and descendants. Saturn's satellites are identified with Greco-Roman titans, their descendants, and other mythological giants. Greek or Roman nymphs or deities of the sea were used for Neptune's moons. Only Uranus is different. Following a fanciful system established by William Herschel's son John, its moons bear the names of characters from Shakespearean plays or Alexander Pope's *The Rape of the Lock*.

JUPITER
KING OF PLANETS

Enormous Jupiter dominates the planets, far bigger and more massive than any other world in the solar system. Its family of satellites includes surprisingly diverse and intriguing moons, including some that might harbor oceans.

Jupiter, first of the giant gas planets, orbits more than 778 million kilometers (483 million mi) from the sun, on average, separated from Mars by the rocky interface of the asteroid belt. At that distance, it needs almost 12 years to circle the sun. But despite its ponderous size, it is a whirling dervish of a planet, rotating once every 9.9 hours. So fast does it spin that its gaseous body is less a sphere than an egg, bulging at the equator. • Jupiter is big enough to contain 1,400 Earths, but it's far less dense than our planet. Composed mostly of hydrogen and helium, like the sun, it has no real surface, but a deep and windy atmosphere over a liquid hydrogen ocean. Long-lived, Earth-size hurricanes breach its upper layers in swirling ovals of red and white. The planet may have a solid, rocky core, but so far we've been able to see its cloud tops only at the one-bar (Earth surface pressure) level. • Jupiter isn't just a planet, but a planet-moon system. Its many satellites, huge and tiny, interest scientists almost as much as the parent planet. So do its delicate, dusty, fascinating rings, discovered—to the surprise of many—by the Voyager 1 mission in 1979.

Symbol: ♃
Discovered by: Known to the ancients
Average distance from sun: 778,412,020 km (483,682,810 mi)
Rotation period: 9.925 Earth hours
Orbital period: 11.86 Earth years

Equatorial diameter: 142,984 km (88,846 mi)
Mass (Earth=1): 317.82
Density: 1.33 g/cm³ (compared with Earth at 5.5)
Effective temperature: -148°C (-234°F)
Natural satellites: 63

SKYWATCH

* Jupiter is easy to spot with the naked eye. The second brightest planet after Venus, it moves through one constellation of the zodiac per year.

AMAZING FACT The pressure at Jupiter's center is 70 million times that of Earth's atmosphere at sea level.

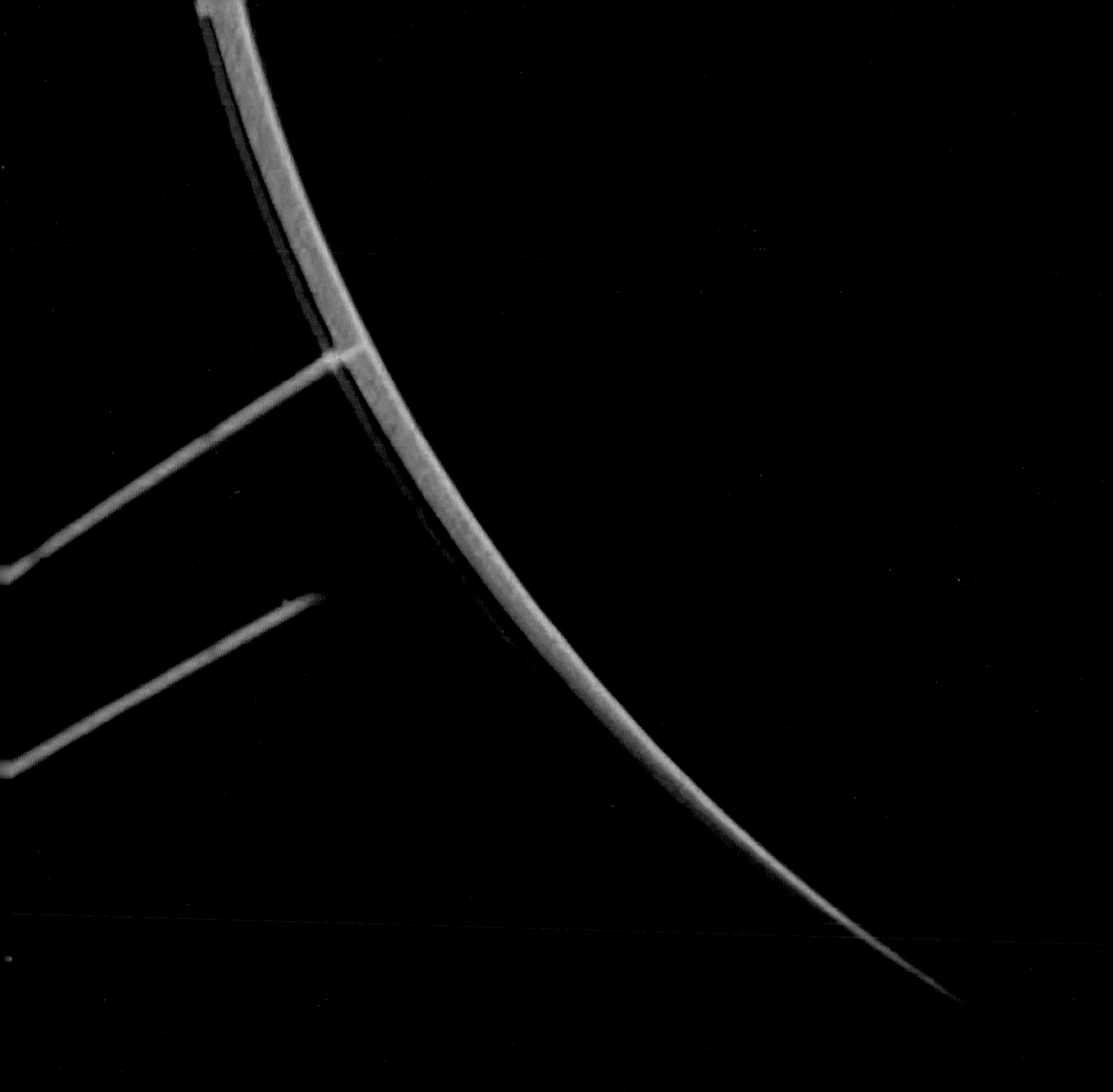

757
Chinese astronomers record an occultation of Jupiter by Venus.

1610
Galileo discovers four largest moons: Io, Ganymede, Europa, and Callisto.

1664
Robert Hooke uses an early telescope to discover Jupiter's Great Red Spot.

1892
Edward Emerson Barnard discovers Jupiter's fifth moon, Amalthea.

1979
Scientists discover Jupiter's ring system and active volcanism on Io.

2003
Galileo spacecraft purposely impacts Jupiter to avoid collision with Europa.

Jupiter's swirling winds sweep across the planet in bands and zones (opposite). Viewed from within Jupiter's shadow, the giant planet's rings look particularly slender (above).

JUPITER: WORLD OF STORMS

Unlike the inner planets, Jupiter managed to hold on to the atmosphere it pulled in during the primordial days of the solar system. Best estimates put its composition at about 86 percent hydrogen and 13 percent helium, with traces of other gases, such as ammonia, methane, and water vapor. Like the sun and other giant planets, it shows differential rotation, the gases at its equator rotating a little faster than those at its poles.

Jupiter's atmosphere appears to be layered, like Earth's, with a cold, smoggy haze at the top. Below the haze, to a depth of about 40 kilometers (15 mi), is a region of white, crystalline ammonia clouds. Deeper in the atmosphere, the white clouds give way to golden clouds, probably a mix of ammonia and sulfides, overlying a blue layer of water ice crystals. Temperatures increase with depth, from 110K (-163°C/-261°F) at the top of the atmosphere to 400K (127°C/260°F) about 150 kilometers (93 mi) down. Because Jupiter lacks a solid surface, there is no clear demarcation between atmosphere and interior, but most of its weather occurs in the top 200 kilometers (124 mi) of thin gas.

STORMY WEATHER

And what weather it is! On Earth, high- and low-pressure zones form fronts and systems, but on fast-spinning Jupiter they are stretched in even strips around the entire planet. Even small telescopes from Earth reveal the strikingly beautiful colored bands of Jupiter's cloud tops, stirred into whorls and scallops by storms. Bright zones above uprising warm gas are interspersed with darker belts of cooler, sinking gas. Powerful winds in the bands alternate between east and west, blowing at up to 530 kilometers an hour (330 mph).

Earth's weather is powered by solar heat, but not so on Jupiter. The big planet actually gives off more heat than it receives from the sun. Jupiter's warmth most likely flows from gravitational compression at its core, the heat

In 2008, the Hubble Space Telescope found that a third red storm (at left in image) had joined the Great Red Spot and Red Spot Jr. in Jupiter's atmosphere. Previously it had been white; some unknown chemical reaction has changed its color recently.

LIQUID METALLIC HYDROGEN

Common on Jupiter and Saturn, liquid metallic hydrogen has existed only momentarily on Earth in laboratory conditions. It comes into being when extreme pressure crushes hydrogen atoms together until they are stripped of their electrons. The resulting sea of unbound protons and electrons now behaves like a metal, conducting heat and electricity.

originally created when the planet's interior was squashed by its enormous overlying mass, now very slowly leaking into space.

Huge oval spots, white, red, and brown, break up the symmetry of Jupiter's bands. These enormous storms, made of circling hurricane-like winds, appear and disappear over the years, but the largest are remarkably stable—unlike storms on Earth, they aren't broken up by underlying landmasses. The most famous is the Great Red Spot (GRS), seen by Robert Hooke in the 17th century and still going strong. About twice as wide as Earth (though its size varies), the GRS rotates with the planet, possibly powered by energy from below. Spinning counterclockwise, it seems to be rolled between two alternating bands of winds. But exactly what causes it, and the reason for its rust-red color, are still in question. Other oval spots, white and brown, have also been seen in Jupiter's atmosphere, along with flashes that may be lightning. Three such white storms merged into a larger white oval in 2000, then surprised astronomers five years later by turning red, due to some sort of chemical reaction with ultraviolet light. "Red Spot Jr." still rides Jupiter's currents, roughly half the diameter of its bigger counterpart. In 2008, it was joined by yet a third red oval, smaller but in the same turbulent band. These recent appearances have led some scientists to think that Jupiter is undergoing climate change.

A METAL OCEAN

Beneath the roiling clouds, Jupiter's gaseous interior is increasingly dense with depth. After a few thousand kilometers, intense pressures have turned the gas into a hot, almost liquid substance. About 20,000 kilometers (12,400 mi) below the cloud tops temperatures reach 7000K (6727°C/12,140°F), hotter than the sun's face, and pressures bear down three million times that of Earth's surface. Here, the gas probably undergoes a phase transition to liquid metallic hydrogen, an electrically conductive hydrogen soup with the density of water. Most of Jupiter probably consists of this strange, hot liquid, which may encompass about 50,000 kilometers (31,000 mi) of the planet's radius. At its core, Jupiter may contain a dense nugget of rocky materials in the range of five Earth masses. Temperatures here could be as high as 20,000K (19,727°C/35,540°F), hotter than the center of the sun.

JUPITER'S MAGNETOSPHERE

The planet's vast metallic ocean, combined with its rapid rotation, give it the strongest magnetic field in the solar system. The area covered by this field, its magnetosphere, is larger than the sun. On the sunward side, it is compressed by the solar wind to about 3 million kilometers (1.9 million mi) from Jupiter's cloud tops. Away from the sun, its magnetotail stretches at least 700 million kilometers (435 million mi), past the orbit of Saturn. A plasma of charged particles, most probably emitted by the volcanic moon Io (see p. 142) speeds along Jupiter's magnetic field lines, emitting radio waves detectable on Earth. This plasma is a potent hazard to any spacecraft that enters its field. These same charged particles sweep down the magnetic field lines to produce shimmering polar auroras, like the aurora borealis on Earth but much more powerful.

MISSIONS

Almost all of the measurements given above are estimates based on observations of Jupiter's upper atmosphere and on our knowledge of the behavior of hydrogen and helium under pressure. Several NASA missions beginning in the 1970s contributed much of what we know. Pioneers 10 and 11 flew by Jupiter in 1973 and '74, surveying the planet's clouds and registering surprisingly strong magnetic readings. Voyagers 1 and 2 reached the giant planet in 1979, discovering Jupiter's thin rings, new moons, and volcanoes on Io.

The Galileo orbiter, arriving in 1995, also sent back detailed information about Jupiter's moons. The probe it dropped into Jupiter's clouds survived for an hour until crushed by pressure. Although it returned valuable data, as luck would have it, the probe entered through a rare, cloudless hole in the upper atmosphere, and so couldn't tell scientists much about the composition of those layers. The Cassini-Huygens and New Horizons missions flew by the planet on their way to other targets, yielding sharp new images; the upcoming Juno orbiter, planned to reach Jupiter in 2016, will study the planet from polar orbit.

RADIO JUPITER

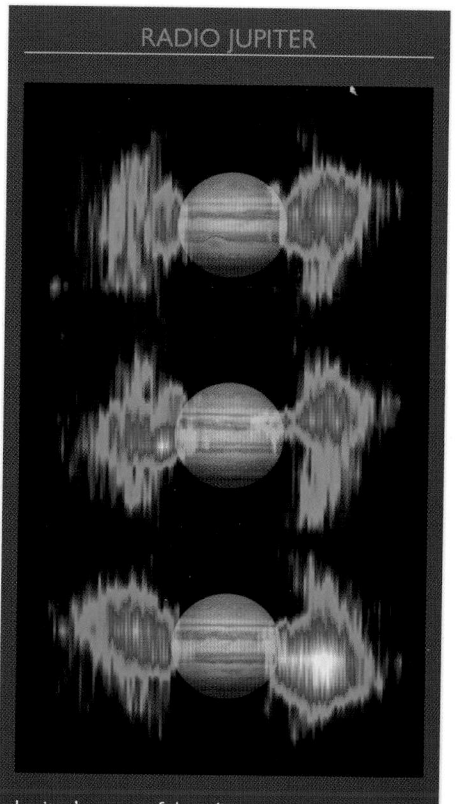

Jupiter's powerful radio waves, emitted by its magnetic field, can be detected on Earth's surface. NASA's Cassini orbiter captured images of the intense waves from space.

An image of Jupiter (above) is superimposed on false-color radio images, taken three times during one of the planet's rotations. Whirling rapidly in its ten-hour day, Jupiter flings charged particles in a current sheet along its magnetic equator. The radiation belts in these three images seem to tilt back and forth because Jupiter's magnetic axis is tilted by about ten degrees relative to its rotation axis. The strength of the magnetic field is 20,000 times that of Earth; radiation within its harsh bath of charged particles would kill an unprotected human.

JUPITER: RINGS AND MOONS

Jupiter, the most sunlike of the planets, has its own mini-planetary system. Orbiting the gas giant are at least 63 moons. Four of them, known as the Galilean moons because they were discovered by Galileo in 1610, are as large as or larger than Earth's moon and might qualify as planets themselves if they orbited the sun. Twelve others were discovered between 1894 and 1980 by astronomers or by the Voyager mission team. But all the rest, most of them quite small, have been found since 1999 by ground-based telescopes using special software that allows them to survey large areas of the sky for tiny moving objects. More tiny Jovian satellites will undoubtedly be added to the list in the years to come.

The Galilean moons—Io, Europa, Ganymede, and Callisto—are among the most fascinating objects in the solar system. These four probably formed at the same time as Jupiter, condensing from the debris of the early solar system. Their nearly circular orbits lie along Jupiter's equatorial plane, and such is the big planet's gravitational grip that all four are locked into synchronous rotation, with one face always pointing toward Jupiter. The three innermost have also settled into orbital periods that follow a simple 4:2:1 mathematical ratio; Io orbits four times for every two Europa orbits and each single Ganymede orbit. But all the moons tug on each other as well, so the gravitational tussling between massive Jupiter and its many satelllites means that the bodies of its inner moons are pulled and flexed, heating up in dramatic ways.

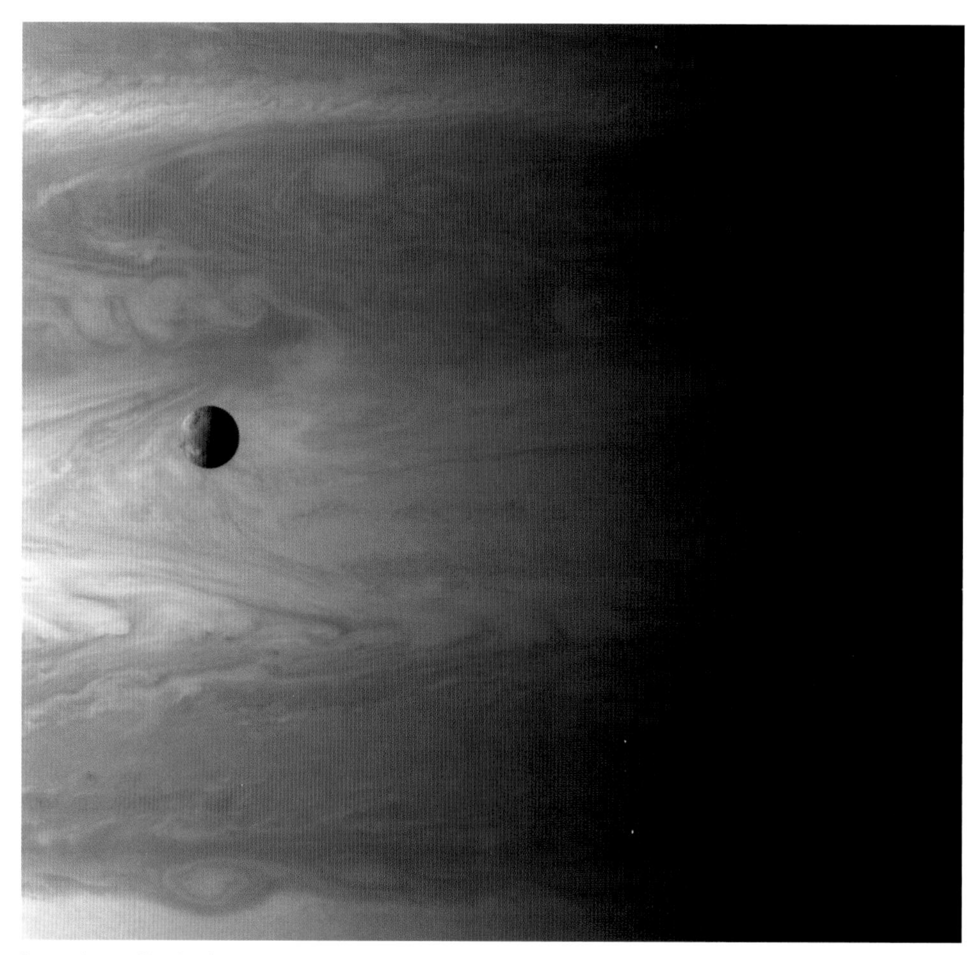

Jupiter's satellite Io floats 350,000 kilometers (217,000 mi) above the planet's cloud tops. The size of Earth's moon, Io is dwarfed by the gas giant.

IO

When Voyagers 1 and 2 first returned images of Io in 1979, scientists were astonished to see active volcanoes on its surface shooting plumes far into space. In a region of space long viewed as dead and frigid was the most geologically active body in the solar system. In 1995, the visiting spacecraft Galileo found even more active volcanoes; to date more than 80 have been discovered. Loki, the largest, pours out more than 1,000 square meters (10,700 square yd) of lava per second during its most energetic bursts. Other volcanoes have been seen spewing sulfur dioxide gas up to 290 kilometers (180 mi) above the moon. Smoothed by repeated outpourings of lava, Io's surface is a pizzalike mélange of white, orange, red, yellow, and brown deposits. Surprisingly high nonvolcanic mountains also rise from Io's plains, among them Boosaule Mons, twice the height of Mount Everest. The erupting moon even has an atmosphere, albeit a very thin one, consisting of sulfur dioxide gas from its volcanoes.

All this heat and activity is generated by the gravitational tug-of-war among Jupiter, Io, and its neighboring moon, Europa. Tidal stresses tear at Io's rocky body, pulling and squeezing it so violently that much of its interior stays hot and molten. The moon is also a significant player in Jupiter's magnetosphere. Charged particles from its volcanoes enter the planet's magnetic field lines and are swept up into an orbiting plasma torus, a belt of deadly radiation.

EUROPA

Icy Europa, next in line of the four big satellites, shows none of Io's volcanism but has its own fascinations. Like terrestrial planets, Europa had a rocky interior and iron core. Its surface, however, appears to be a fractured, striated layer of water ice. Between the rock and the surface ice may lie an ocean of salty liquid water 100 kilometers (62 mi) deep—bigger than the one on our own planet.

Images returned by Voyager and Galileo show Europa's relatively uncratered surface crisscrossed with trenches, ridges, fractures, and even chunks that look like icebergs. Like the Arctic Ocean in winter, the ice seems to have repeatedly pulled apart, its cracks filled in with upwelling liquid from below. This leads scientists to think that the surface ice is fairly shallow, perhaps a few kilometers thick, over a deep ocean. Evidence for a saltwater ocean is also bolstered by the fact that Europa has a weak magnetic field, which can be generated by electrically conducting salt water.

If Europa proves to have such an ocean, it will rise toward the top of the list of candidates for life outside our Earth (see pp. 180–95). Astronomers will be turning renewed attention to this frozen moon in years to come.

GANYMEDE

Ganymede, third of the Galilean satellites, is the biggest moon in the solar system. Its diameter of 5,262 kilometers (3,280 mi) makes it larger than the planet Mercury, although less dense and less massive. Like Europa, it is ice-covered, but Ganymede's ice lies thick over an interior of rock, with an iron core.

Ganymede looks much like a larger version of our own moon. Impact craters cover its mottled surface. A huge, dark area, Galileo Regio, represents an ancient region of dusty ice—other portions of the moon's surface are younger and lighter. A surprising magnetic field, one percent as strong as Earth's, may be produced by the moon's iron core interacting with some sort of liquid. Ganymede may even turn out to have some liquid water sloshing about under its ice.

CALLISTO

Callisto, outermost of Jupiter's Galilean moons, is the second-largest of the Jovian satellites. More than 4,800 kilometers (2,985 mi) in diameter, it is also the third biggest moon in the solar system, after Ganymede and Saturn's Titan. Although it is similar to Ganymede in composition—rock covered with ice—its distance from Jupiter protects it from the tidal stresses that helped to fracture the larger moon. Instead, its surface is relatively unchanged since its infancy some four billion years ago. Heavily cratered, it features Valhalla basin, a huge crater surrounded by concentric ridges, ripples from an ancient impact with an asteroid or comet.

Ganymede, the largest moon in the solar system, is marked by the large dark region Galileo Regio.

Callisto is the most heavily cratered body in the solar system.

Volcanic Io is famous for its eruptions and its pizzalike coloration.

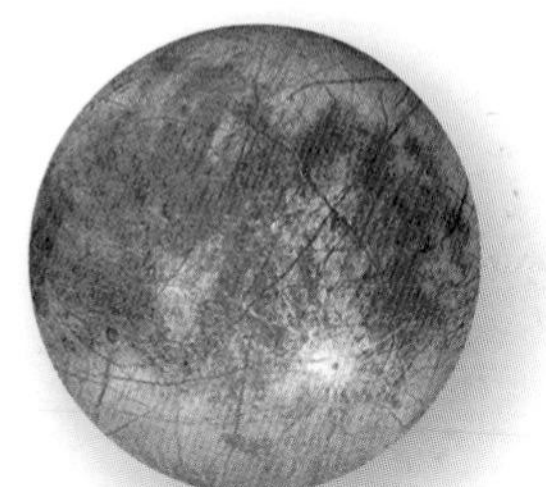

Icy Europa's cracked surface may hide a vast, salty sea.

RINGS

Jupiter's delicate rings were unknown until Voyagers 1 and 2 visited the planet. Far narrower and darker than Saturn's rings, they consist of a main ring about 7,000 kilometers (4,300 mi) wide surrounded by a dusty inner halo ring and a pair of tenuous outer gossamer rings. Judging by their location and gritty composition, the rings seem to have been made from fragments chipped off four tiny, irregular moons in the same orbits: Adrastea and Metis for the main ring, Amalthea and Thebe for the outer rings.

OUTER MOONS

A swarm of small, lumpy satellites, most just a few kilometers wide, make up Jupiter's outer moon system. These tiny bodies were probably snatched up by the planet's powerful gravity in its early days. Some orbit in groups, sometimes retrograde to the other moons, indicating that they may be the remains of larger objects.

THE GALILEAN MOONS

Jupiter's four big moons are the size of small planets.

IO
Distance from Jupiter: 422,000 km (262,000 mi)
Diameter: 3,660 km (2,270 mi)
Mass (Earth's moon=1): 1.22

EUROPA
Distance from Jupiter: 671,000 km (417,000 mi)
Diameter: 3,130 km (1,945 mi)
Mass (Earth's moon=1): 0.65

GANYMEDE
Distance from Jupiter: 1,070,000 km (665,000 mi)
Diameter: 5,262 kilometers (3,280 mi)
Mass (Earth's moon=1): 2.02

CALLISTO
Distance from Jupiter: 1,883,000 km (1,170,000 mi)
Diameter: 4,800 km (2,985 mi)
Mass (Earth's moon=1): 1.46

SATURN RINGWORLD

Saturn's bright, expansive rings, stormy atmosphere, and cloudy moon Titan, dotted with methane lakes, make it a popular planetary target for amateur and professional astronomers alike.

Glorious Saturn, floating within shining rings, is the iconic planet. The farthest of the worlds known to the ancients, its strange, protruding belt of icy moonlets astonished the first observers to view it through a telescope. Today, thanks to visiting spacecraft, we know much more about the giant planet and its ring system, although new findings bring as many questions as answers. • At 1.3 million kilometers (838,000 mi) from the sun, this gas giant planet is far more distant than its big brother Jupiter. Its body could contain 763 Earths, but its density is less than that of water. In fact, Saturn would float in, say, a Jupiter-size tub of water. Saturn needs more than 29 Earth years to orbit the sun, but only a little over ten hours to rotate on its axis. The dizzying spin of its light, gassy body pulls it into an oblate shape, bulging at the equator. Ferocious winds circle the planet and spiral into high-speed vortices at each pole. • Billions of icy particles form Saturn's impressive ring system, which is sculpted into multiple bands by the gravity of some of Saturn's many moons. Its big moon, Titan, is the only satellite in the solar system to possess a thick atmosphere and, apparently, liquid lakes on its surface, possible havens for life. Bright Enceladus may have liquid oceans under its ice, while Iapetus is a study in black and white: its leading hemisphere dark and sooty and its trailing half bright.

Symbol: ♄
Discovered by: Known to the ancients
Average distance from sun: 1,426,725,400 km (885,904,700 mi)
Rotation period: 10.656 Earth hours
Orbital period: 29.4 Earth years

Equatorial diameter: 120,536 km (74,900 mi)
Mass (Earth=1): 95.16
Density: 0.70 g/cm³ (compared with Earth at 5.5)
Effective temperature: -178°C (-288°F)
Natural satellites: 61

SKYWATCH

• Saturn looks like a bright, pale yellow star to the naked eye. Through any good, small telescope its rings and largest moon, Titan, are clearly visible.

AMAZING FACT The transmitter that the Titan probe Huygens used to signal Earth used no more power than a cell phone.

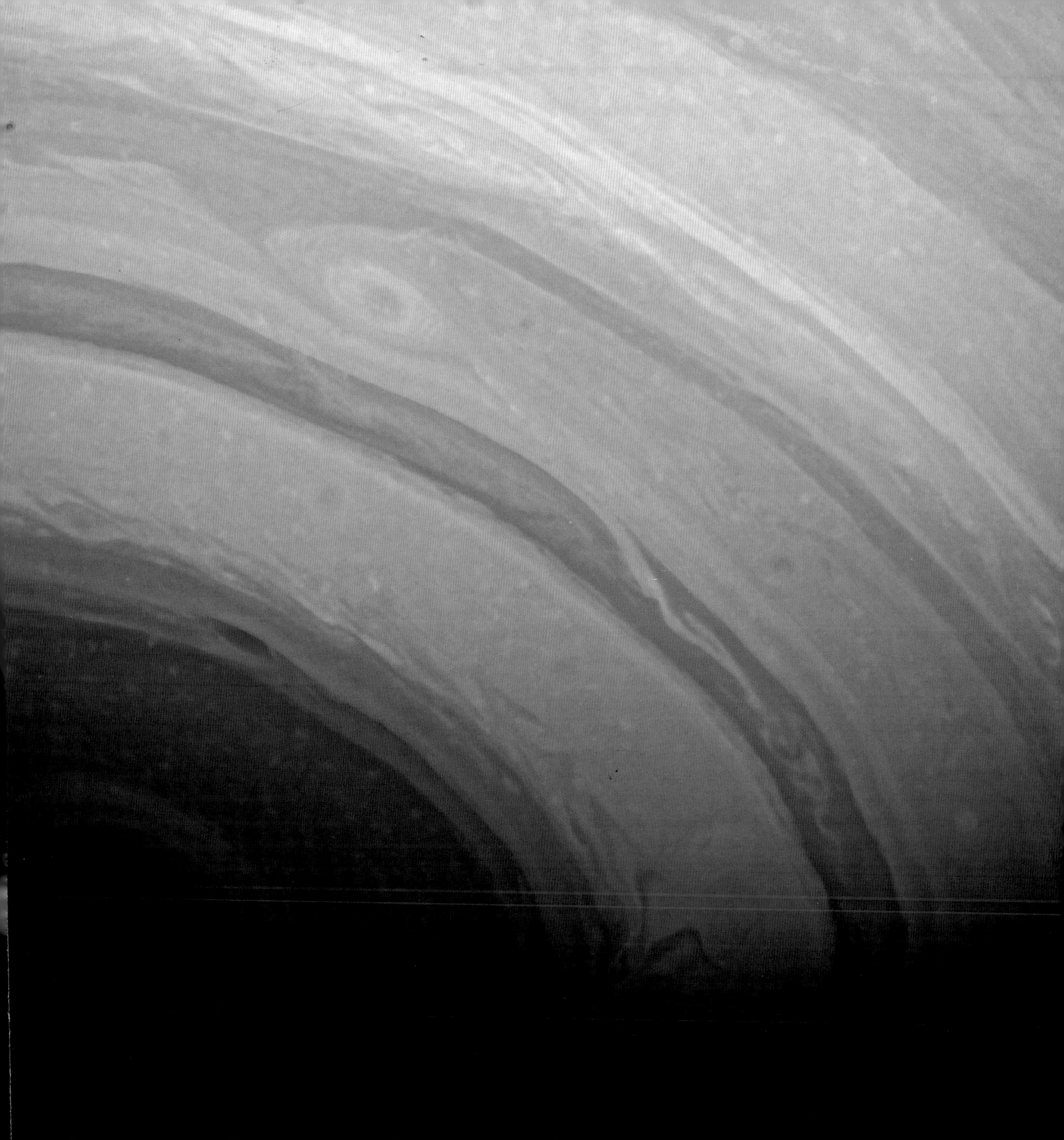

1027
Chinese astronomers record an occultation of Saturn by Mars.

1610
Galileo observes Saturn's rings but believes them to be two separate bodies.

1655
Christiaan Huygens discovers the first of Saturn's moons, Titan.

1659
Christiaan Huygens suggests that Saturn is surrounded by a flat, solid ring.

1883
The first photograph of Saturn's rings is taken.

2007
Cassini spacecraft determines Saturn's outermost A ring contains moonlets.

Saturn has the most spectacular rings in the solar system (opposite). A close view of its southern atmosphere shows oval storms and meandering clouds (above).

SATURN: WILD WINDS

Saturn's rings have always been a source of fascination, but until the space age, the planet itself revealed little to astronomers' eyes. In the 1790s, William Herschel tracked markings on the rings and in the planet's atmosphere to guess at a rotation period of 10 hours 16 minutes, quite close to today's estimate of 10 hours 45 minutes (based on radio signals). What lay under the atmosphere was anyone's guess, and early speculation ranged from a solid surface covered with molten lava to pure gas through and through. Spectroscopy in the 1930s put those theories to rest when hydrogen, methane, and ammonia were identified in its atmosphere, making it look like a twin to the better-known Jupiter.

Unlike Mars, which seems to put a curse on many of the missions that approach it, Saturn has welcomed several highly successful robotic missions over the years. Much of what we know now about the planet comes from these spacecraft and their instruments. Pioneer 11 flew by the planet in 1979, charting a risky path through a gap in the rings at 113,000 kilometers an hour (70,000 mph) and narrowly dodging a newly discovered little moon. Scooting to within 21,000 kilometers (13,000 mi) of the planet's cloud tops, it sent back information about Saturn's atmosphere and magnetic field before continuing toward the boundaries of the solar system.

Voyagers 1 and 2, flying past the planet in 1980 and '81, returned tens of thousands of images of Saturn. The atmosphere was stormier and more complex than it had seemed from Earth, with bands and scallops like the clouds of Jupiter. The spacecraft also studied Saturn's surprisingly complex rings and moons, discovering three new satellites in the process. Voyager 1 also took a turn past Titan, hoping to lift the veil on the solar system's most tantalizing moon. Alas, Titan hid its secrets well beneath a dull and impenetrable orange haze.

More than 20 years elapsed before the next mission reached the ringed planet. The Cassini orbiter (designed by NASA) and Huygens probe (designed by the ESA) were created specifically to study Saturn and to take another shot at its stubbornly mysterious moon Titan. After reaching the planet in 2004, Cassini settled into a looping, elliptical orbit, soaring above the cloud tops, checking out the planet's stormy poles, studying the magnetosphere, and watching the seasons change as the northern hemisphere began to enter summer. As it swings around Saturn, Cassini is also programmed to make close-up flyby visits to some of the planet's more interesting moons, such as Enceladus and Hyperion.

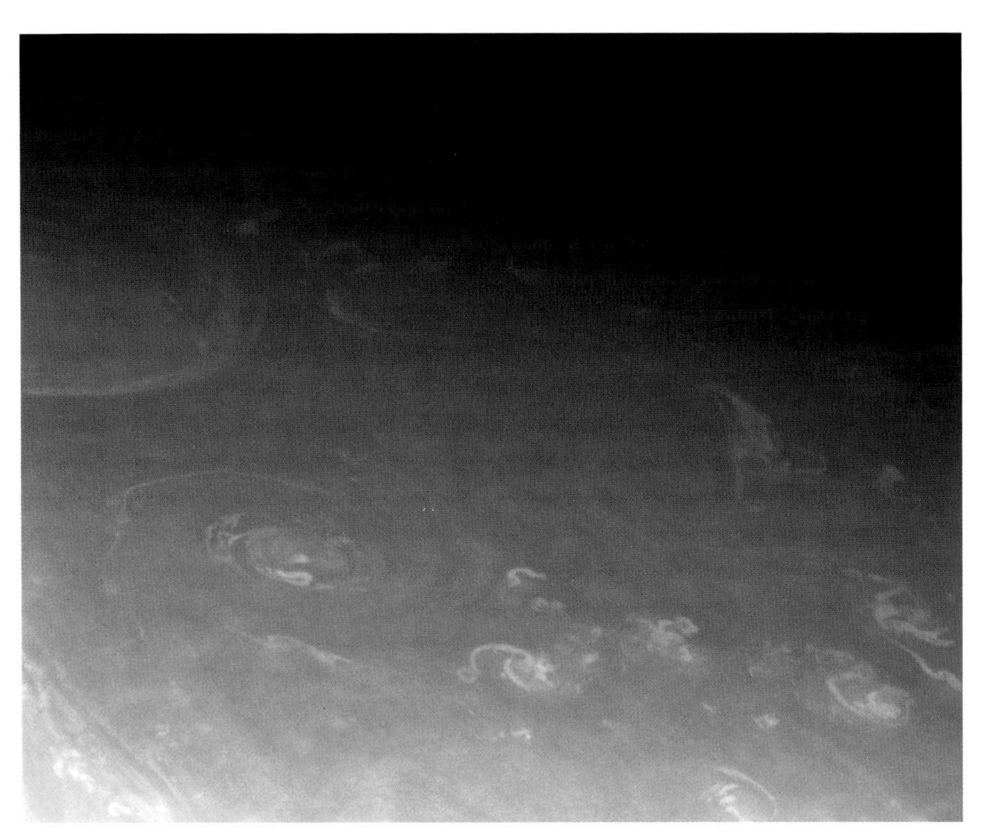

Spiraling clouds, like scattered scribbles, decorate the top of Saturn's busy atmosphere in this visible-light image from the Cassini orbiter.

Of particular interest to this recent mission was the smoggy, mysterious moon Titan, possible cradle of life. Soon after Cassini entered into orbit around Saturn, it launched one of the more dramatic missions in space exploration history. Early in 2005, it released the Huygens probe, which parachuted through Titan's dense atmosphere and revealed never-before-seen details of a strange, lake-filled landscape (see pp. 150–51). To date, Cassini remains in orbit around Saturn, its mission extended and more flyby visits to Titan planned.

A GOLDEN ATMOSPHERE

These fruitful missions have added many details to the basic portrait of distant Saturn portrayed by ground-based telescopes. Like Jupiter, Saturn has no solid surface: The clouds and winds of its atmosphere merge into the planet's vast, gaseous body. Its gases are a lightweight mix, 92 percent molecular hydrogen and 7 percent helium, with small but pungent amounts of methane and ammonia. Beneath a layer of hydrocarbon haze, three layers of clouds run 200 kilometers (124 mi) deep. The ringed planet's buttery color comes from the thick top layers of these clouds, golden ammonia ice over reddish ammonium hydrosulfide; below them is a seldom-viewed layer of blue water ice. The Cassini orbiter found that Saturn changes color during its slow seasons. In winter, the northern hemisphere becomes distinctly bluer as the overlying haze thins out. The planet is, of course, distinctly chilly year-round.

Receiving only one percent as much solar energy as Earth, Saturn registers only 95K (178°C/289°F) at the top of its atmosphere.

Saturn's clouds stretch around the planet in bands, like those of Jupiter, but without the dramatic shifts in color or visible swarms of storms. But the outwardly bland appearance is deceptive. Ferocious winds blow through the atmosphere from east to west at speeds up to 1,600 kilometers an hour (1,000 mph) near the equator, slowing and shifting from west to east near the poles. At each pole, winds spiral downward in a vortex larger than the Earth, a counterclockwise hurricane with wind speeds of 550 kilometers an hour (350 mph) and a deep central eye. (As comparison, the highest wind speed ever measured on Earth was 372 kilometers an hour/231 mph.) Multiple thunderstorms with tall cumulus clouds swirl within the hurricanes, the heat from their condensing liquid powering their motion. A strangely angular hexagon of clouds surrounds the north pole's cyclone. This even, six-sided figure is 25,000 kilometers (15,000 mi) across and has been visible since the Voyager spacecraft visited the planet. Its orgins are unknown.

Occasional white oval ammonia-ice storm clouds breach the cloud tops across the planet, as they do on Jupiter. Many more storms are thought to rumble below the clouds: country-size hurricanes lashed by water and ammonia rain, producing giant lightning strikes millions of times more powerful than those on Earth. Though Saturn has no equivalent of Jupiter's Great Red Spot, it did produce the elegantly convoluted "Dragon Storm" in 2004. Forming in a particularly tempestuous region of the planet known informally as storm alley, the twisting cloud pattern gave off regular bursts of radio waves. Scientists think these probably came from lightning flashing beneath the violent storm below the cloud tops.

Below its icy clouds, Saturn's outer layer of hydrogen and helium gas extends downward for some 30,000 kilometers (18,600 mi), becoming hotter and denser with depth. Raining through these gassy layers are droplets of liquid helium. As the helium rain falls through the gas envelope, increasing pressure squeezes and heats it enough to warm up the entire planet by an extra 20K (253°C/424°F). (In a billion years or so the helium rain will stop and the planet will cool.)

When temperatures in the depths reach about 8000K (7727°C/13,940°F), hydrogen gas is transformed into a sea of liquid metallic hydrogen, as on Jupiter. The electrically charged liquid, about 15,000 kilometers (9,300 mi) deep, surrounds a dense, solid core of heavy elements (or so scientists believe) with a mass ten to twenty times that of Earth.

MAGNETOSPHERE

The electrical currents in Saturn's liquid metallic hydrogen sea, rotating on a rapidly whirling body, produce a powerful magnetic field. Though weaker than that of Jupiter, it is nonetheless a thousand times stronger than Earth's magnetic field. Saturn's magnetic field's axis is lined up exactly with the planet's axis of rotation, the only such case in the solar system. The immense bubble of Saturn's magnetosphere extends past its rings and inner moons, deflecting the solar wind around the planet. Radiation trapped within the rings forms a ring around the planet; the magnetic field also interacts with Titan's atmosphere and Enceladus's icy plumes in complex ways.

Charged particles from the solar wind, racing down Saturn's magnetic field lines, create glowing auroras at its poles. The auroras can be widespread and complicated, radiating in the infrared and ultraviolet. At times they cover an entire pole, brightening and darkening and varying in size according to some so-far-unknown rules.

Probably fueled by the solar wind, an aurora circles Saturn's south pole.

MISSIONS TO SATURN

SPACECRAFT	LAUNCH DATE	ARRIVAL	SOURCE	TYPE OF MISSION
PIONEER 11	April 5, 1973	September 1, 1979	NASA	successful Saturn flyby
VOYAGER 2	August 20, 1977	August 26, 1981	NASA	successful flyby
VOYAGER 1	September 5, 1977	November 12, 1980	NASA	successful flyby
CASSINI-HUYGENS	October 15, 1997	July 1, 2004	NASA and ESA	orbiting Saturn; Huygens probe dropped January 14, 2005

SATURN: THE RINGS

A view from just above Saturn's ring plane reveals two of its shepherd moons. Pan, 28 kilometers (17 mi) wide, floats within the Encke gap at left. Prometheus, 86 kilometers (53 mi) wide, hugs the inside edge of the outer F ring.

We now know that all the outer planets have rings, but Saturn's outclass the others in size and beauty. With each new generation of telescopes and spacecraft we have seen the rings in more detail and have learned more about their intricate structure, even while their origins remain a mystery.

Galileo saw the rings in 1610, but in his simple telescope they looked like two planets bulging from Saturn's waist. In a letter, he noted: "the star of Saturn is not a single star, but is a composite of three, which almost touch each other, never change or move relative to each other, and are arranged in a row along the zodiac, the middle one being three times larger than the lateral ones, and they are situated in this form: oOo."

If they were stars, they didn't behave like them, seeming to vanish and reappear over the years. In 1655, Dutch astronomer Christiaan Huygens (see p. 32) made the intuitive leap: Saturn's elliptical appendages were "a thin, flat ring, nowhere touching, and inclined to the ecliptic." The realization that the ring (which Huygens believed to be a single, solid object) was inclined to the ecliptic cleared up another mystery. It seemed to appear and disappear because it was tilted with respect to the Earth. Over Saturn's 29-year orbit, it would be seen from below, then edge-on and vanishingly thin, and eventually from above.

Closer observations soon broke the single ring into two and then three concentric hoops. In 1676 Italian-born astronomer Giovanni Cassini spotted a dark gap two-thirds of the way out along the ring, a band now called the Cassini division, which split the rings into the A (outer) and B (inner) ring. In 1837, German astronomer Johann Encke found another gap within the A ring, known now as the Encke gap. Thirteen years later a third, thin ring was found inside the B ring and named, predictably, the C ring.

Although most 19th-century astronomers took the view that the rings were solid, rocky planes, in 1857 Scottish physicist James Clerk Maxwell pointed out that such an orbiting disk would be pulled to pieces by gravitational stresses. He suggested, and later observations confirmed, that the rings were in fact composed of countless tiny particles orbiting independently, like an infinitude of tiny moons.

ICY RINGLETS AND SHEPHERD MOONS

Today, ground-based telescopes, close observation by the Voyager spacecraft, and even closer scrutiny by the Cassini orbiter have filled in our pointillist picture of Saturn's rings. They are apparently composed of billions of highly reflective particles, mostly water ice and some rock, ranging mainly from dust-size to boulder-size. The ice balls cluster into tens of thousands of intricate ringlets, guided and restrained by

Saturn's smaller moons. Seven main rings and two major gaps are recognized: From inside to out, they are labeled D, C, B (the brightest), the Cassini division, A, the Encke gap (actually within the A ring), F, G, and E (the last two very faint and thin). Though the rings span 282,000 km (175,000 mi), they are gossamer thin, on the order of 10 meters (30 ft) thick from top to bottom. Stars shine through them.

Viewed close-up, the seemingly smooth rings are revealed as thousands of individual ringlets clustered together in wavy or corrugated planes, with peaks and valleys forming ridges that spiral inward. Some ringlets wiggle or crisscross. All these rings, waves, and gaps seem to be shaped by the complex gravitational interplay of Saturn and its many inner moons. Saturn's moon Mimas, for example, orbits outside the A ring; the outer edge of the B rings orbits in a two-to-one resonance with the little moon, whose influence also ejects most of the orbiting particles from the Cassini division. Tiny moon Pan circles within the Encke gap, while even smaller Daphnis clears out the Keeler gap in the A ring. In at least one case, a pair of moons works together to keep a ring confined. The narrow, braided-looking F ring, just 100 kilometers (60 mi) wide, is flanked by two little moons, Prometheus and Pandora. Orbiting about 1,000 kilometers (621 mi) on either side of the ring, these shepherd satellites keep the ring's particles from straying too far. The shapes of other narrow rings suggest that shepherd moons may also be found near them in the future.

Voyager and Cassini spotted other odd ring features that have yet to be completely explained. Dark spokes cross the rings in places, circling in tandem with Saturn's rotation, appearing and dissolving repeatedly. Scientists speculate that these are formed from dust hovering just above the rings and gripped by Saturn's magnetic forces.

THE MAXWELL GAP

The Cassini mission cameras were able to focus on ring details down to 4 or 5 kilometers (2.5 or 3 mi) wide. In 2008 they captured this image of the C ring's Maxwell gap and its odd, twisted structure, the Maxwell ringlet, shaped by forces yet unknown.

ORIGINS OF THE RINGS

So where do Saturn's rings come from? Why didn't these particles clump together into moons long ago? The answer—at least a partial one—to these questions invokes the Roche limit. Named for the 19th-century French mathematician Édouard Roche, who first explained it, the Roche limit is the boundary around a planet within which moons cannot form. Inside the Roche limit, a planet's gravitational field exerts too much stress on orbiting objects for them to hold together. Any moon more than ten kilometers (six mi) wide caught venturing within the limit will be torn apart by tidal forces. Any pieces already circling inside the limit will be too perturbed ever to coalesce into a larger satellite. Saturn's Roche limit lies at 144,000 kilometers (90,000 mi) from the planet's center, and all of the major rings lie within its boundaries.

This explains why no large moons orbit close to Saturn. But how did all that icy debris get there in the first place? Small moons, smacked by micrometeoroids, do provide a little of the rings' substance. Anthe and Methone, for instance, produce their own mini-rings, arcs of pulverized matter ahead and behind them in their orbits. Ice spewed out of Enceladus adds material to the E ring.

But when it comes to the bulk of the rings, several theories compete. One says that the rings represent material left over from Saturn's creation 4.6 billion years ago, just as the asteroids are debris left over from the solar system's birth. Another theory suggests that the rings are the remains of a small planetesimal, perhaps 250 kilometers (155 mi) wide, that originally orbited the sun, then wandered too close to Saturn and was torn apart by tidal forces within the Roche limit. Finally, a third scenario proposes that the rings began as many small moons that were either blasted apart by comets or other impacting objects or were pulled apart by Saturn's gravity within the Roche limit. Some tiny moons would remain to shepherd the rings into their multiple components.

Boulder-size debris spotted by the Cassini orbiter within Saturn's A ring and the generally dynamic nature of the rings in general make the second and third theories more likely than the primordial rings thesis. Saturn's rings are also icy and shiny, unlike the dusty, dark rings of Neptune and Uranus. So the rings seem to be relatively young by astronomical standards, perhaps 50 million to 100 million years old, and still in an unstable pattern of collisions. Over time, the constant impacts will wear the rings down to dust. Mutual collisions among ring particles seem to steadily drain orbital velocity from them, driving them inward until they eventually spiral into Saturn. Unless another body ventures too close to the big planet and is pulled to pieces to give birth to a gorgeous new band of icy debris, we may be seeing Saturn's rings in the temporary height of their glory.

GIOVANNI CASSINI

FRENCH-ITALIAN ASTRONOMER

Giovanni Domenico Cassini (1625–1712) left his mark in several areas of solar system astronomy, but perhaps most permanently on the rings of Saturn, where the prominent dark gap between the A and B rings is named for him. Born in Italy, Cassini studied the sun, determined the rotations of Jupiter and Mars, and worked out the positions of Jupiter's satellites. After becoming director of the newly founded Paris Observatory, he discovered Saturn's moons Iapetus, Tethys, Rhea, and Dione, as well as the ring division that now bears his name. Though he resisted revolutionary ideas, such as Newton's theory of gravitation, Cassini was a meticulous observer and is considered one of the finest astronomers of his era.

SATURN: ACTIVE MOONS

Saturn's extended family of moons shows just how diverse natural satellites can be, even when orbiting the same planet. At least 61 icy moons circle the ringed world. They include one whopper—Titan, the second biggest moon in the solar system—six midsize moons—Rhea, Iapetus, Dione, Tethys, Enceladus, and Mimas—and 54 smaller bodies. More are doubtless waiting to be found. Early observers, including Christiaan Huygens, Giovanni Cassini, and William Herschel, discovered the larger satellites, but modern telescopes and the up-close scrutiny of the Voyager and Cassini spacecraft have picked out the petite ones down to just a few kilometers wide. Many exhibit curious features; at least two, Titan and Enceladus, have the potential for life.

TITAN

Titan, discovered by Huygens in 1655, is one of the most remarkable bodies in the solar system. At 5,150 kilometers (3,200 mi) wide, it's bigger than Mercury and far more Earthlike. Even before the space age, the spectrum of its hazy orange light told astronomers it had a thick atmosphere. So dense is the haze at its upper levels that it frustrated the attempts of the Voyager spacecraft to peer through its clouds, but the Cassini orbiter and its probe Huygens, with imagers specially tuned to smog-penetrating wavelengths, have had better luck revealing its surface.

Titan's atmosphere reaches 600 kilometers (370 mi) into space and is dense enough that air pressure at the moon's surface is one and a half times that of Earth. Standing on Titan's surface would feel like standing at the bottom of a swimming pool. Its atmosphere consists of about 95 percent nitrogen and 5 percent methane, but its upper layers are chemically complex. Powered by distant sunlight, molecules here split apart and recombine into organic compounds containing carbon, hydrogen, and sometimes oxygen or nitrogen. Hydrocarbon molecules such as ethane, propane, and carbon monoxide create an orange smog, thick with suspended droplets of liquid.

Troughs, fractures, and craters disrupt the surface of Saturn's close-in moon Enceladus. The bright, reflective satellite may hold a liquid ocean beneath its surface.

Big methane raindrops occasionally drift down from Titan's clouds, scouring the landscape and collecting into lakes and ponds. So much methane is in the atmosphere that if it held much free oxygen, a single flame could set the moon on fire.

Absent such a heat source, Titan is intensely cold. Its average temperature of -179°C (-290°F) keeps the water ice on its surface as hard as rock. These low temperatures may have helped to shape its atmosphere. Forming so far from the sun, Titan's icy body was able to absorb and hold on to methane and ammonia from the gases of the early solar nebulae. Later, warmed by internal heat, the gases began to escape into the atmosphere but were prevented from vanishing into space by the moon's substantial gravity. Because methane is broken down by sunlight in the atmosphere, researchers think Titan may continually replenish it somehow—perhaps from ice volcanoes.

The Huygens probe, which landed on Titan's surface in 2005, was the first craft to land on a solar system outer body. Able to survive in the harsh environment for only a few hours, as it descended it returned startling images of drainage channels and a shoreline that would not have been out of place on Earth. The ground around its landing site was a hazy scene of scattered, eroded-looking "rocks"

MISSION TO TITAN

Built by the European Space Agency, the Huygens probe piggybacked on the Cassini orbiter for seven years on the way to Saturn. On December 25, 2004, during its second flyby of Titan, Cassini released the probe. Three weeks later, the probe reached an altitude of 1,270 kilometers (790 mi) above Titan's surface and began its descent. Parachuting through the dim haze, it turned on a light part of the way down before making a soft landing on a frozen surface littered with rounded bits of ice. Huygens continued to broadcast to Cassini, and from there to mission controllers on Earth, for a little more than an hour from the solar system's most atmospheric moon.

An artist's conception of a lake region on Titan shows a smooth hydrocarbon lake beneath smoggy skies. Recent observations of the moon support the presence of ethane-containing lakes near its north and south poles, some appearing only after rainstorms.

made of hardened ice. Radar from the Cassini orbiter has added further details to Titan's surface, including dramatic portraits of land near Titan's north pole liberally pockmarked with smooth regions, almost certainly seasonal lakes of liquid methane. If this proves to be so, it will make Titan the only body outside of Earth, that we know of, to possess open liquid on its surface. Low-lying areas were revealed to have rippling dunes, probably sculpted from hydrocarbon particles. Strong winds blowing mainly east to west not only shape the dunes, they also shift the entire surface of the satellite around as a single piece. They can accomplish this because Titan's surface seems to float atop a submerged ocean of liquid water.

MIDSIZE MOONS

Saturn's middling moons are a varied lot. Enceladus, smaller (512 km/318 mi wide) than Titan and closer to Saturn, is so bitterly cold at -201°C (-330°F) that any activity there seems unimaginable. The moon is so icy that its surface reflects almost 100 percent of the sunlight that reaches it. Therefore Voyager's scientists were astonished to see evidence of volcanic activity, of all things, on the frigid moon. Admittedly, the plumes of material ejected into space from Enceladus appear to be icy, not hot, but the fact that they exist at all argues for liquid water, warmth, and motion beneath the moon's surface.

The Cassini orbiter, skimming close to the satellite, captured images of a world scarred by sinuous ice ridges. Large areas of its surface are uncratered, indicating that they have probably been resurfaced with freezing water in the fairly recent past. Plumes of water vapor intriguingly mixed with organic materials jet out of vents in the "tiger stripe" fractures in the planet's southern hemisphere, spreading into space and contributing material to Saturn's E ring. Like Jupiter's Io (see pp. 142–43), Enceladus may get its subsurface heat from the flexing motion of tidal stresses.

MYSTERIOUS MOTIONS

Other moons also hold mysteries. Slow-moving Iapetus is almost pitch-black on its forward side and bright and shiny on its trailing hemisphere. Like a spherical dust rag, it could be gathering up gritty material cast off from its sister Phoebe as it sweeps through its orbit. Cratered Rhea may be surrounded by its own thin ring. Irregular Hyperion tumbles about in a chaotic orbit, tossed this way and that by the gravity of nearby Titan. Repeated impacts have gouged away its spongy surface. Some of the smaller moons share an orbit: Little Telesto and Calypso orbit 60° ahead and behind Tethys. Janus and Epimetheus even swap orbits. Every four Earth years, the inner moon catches up to the outer one and the two switch places. Yet other moons circle Saturn in a retrograde motion, counter to Saturn's own rotation, indicating that they may be captured objects. Where the moons come from, what they're made of, and why they behave as they do will occupy scientists for many years to come.

URANUS AND NEPTUNE
ICE GIANTS

The cold, blue giants, Uranus and Neptune, were discovered in the 18th and 19th centuries. Much about them remains mysterious–Uranus is tipped on its side and Neptune is strangely warm—and their moons are as interesting as the planets.

Ice giants Uranus and Neptune are the most distant members of the primary solar family and the only major planets discovered in modern times. They are often lumped together as planetary twins, but they are far from identical. • Both are considered giant planets, but they're much smaller than their enormous cousins, Jupiter and Saturn. Uranus has about 63 times the volume of Earth and Neptune 58—big, but modest compared with Jupiter, which could hold 1,300 Earths. Uranus, the closer of the two ice giants, orbits 2.9 billion kilometers (1.8 billion mi) from the sun on average, 19 times as far as the Earth. It needs 84 Earth years to complete one orbit. Neptune is far more distant at 4.5 billion kilometers (2.8 billion mi), 30 times Earth's distance from the sun. With one orbit lasting 165 years, it is just finishing up one Neptunian year since its discovery. Both planets have a rich complement of quirky moons, including Neptune's massive Triton and Uranus's gouged-up Miranda. In the late 20th century, astronomers were surprised to discover that each planet had a ring system as well, albeit slender, dusty, and dark ones compared with the rings of Saturn.

URANUS
Symbol: ⛢
Discovered by: William Herschel in 1781
Average distance from sun: 2,870,972,200 km (1,783,939,400 mi)
Rotation: -17.24 Earth hours (retrograde)
Orbital period: 84.02 Earth years
Diameter: 51,118 km (31,764 mi)
Mass (Earth=1): 14.37
Density: 1.3 g/cm³ (Earth is 5.5)

NEPTUNE
Symbol: ♆
Discovered by: Adams, Le Verrier, Galle in 1846
Average distance from sun: 4,498,252,900 km (2,795,084,800 mi),
Rotation period: 16.11 Earth hours
Orbital period: 164.79 Earth years
Diameter: 49,528 km (30,776 mi)
Mass (Earth=1): 17.15
Density: 1.76 g/cm³ (Earth is 5.5)

AMAZING FACT Despite their frigid exteriors, Neptune and Uranus are as hot as the sun's surface in their dense cores.

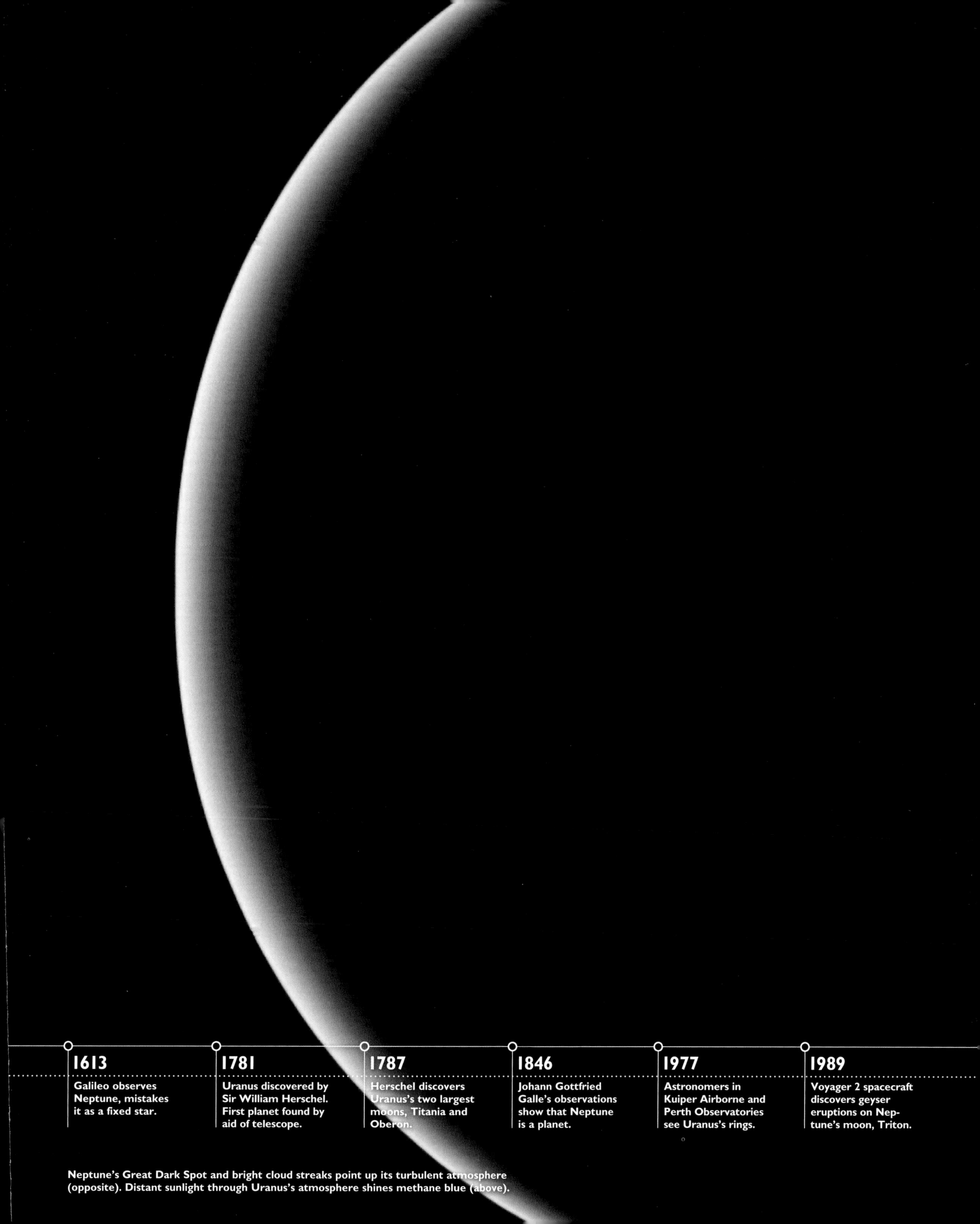

1613
Galileo observes Neptune, mistakes it as a fixed star.

1781
Uranus discovered by Sir William Herschel. First planet found by aid of telescope.

1787
Herschel discovers Uranus's two largest moons, Titania and Oberon.

1846
Johann Gottfried Galle's observations show that Neptune is a planet.

1977
Astronomers in Kuiper Airborne and Perth Observatories see Uranus's rings.

1989
Voyager 2 spacecraft discovers geyser eruptions on Neptune's moon, Triton.

Neptune's Great Dark Spot and bright cloud streaks point up its turbulent atmosphere (opposite). Distant sunlight through Uranus's atmosphere shines methane blue (above).

ICE GIANTS: URANUS

When British astronomer William Herschel discovered the planet Uranus through his handcrafted reflecting telescope in 1781, it marked the first new planet added to the planetary pantheon in recorded history (see p. 18). Though Herschel's discovery astonished his contemporaries, it turned out that Uranus had been there in the sky all along. Tiny and dim, barely visible to the naked eye, it had been recorded as a star by various early astronomers. It took Herschel's careful observation of the object's movement relative to background stars to reveal it as a planet.

To the eye, through the telescope, and even from the 1986 vantage point of Voyager 2, Uranus was for a long time a singularly featureless world. A smooth, opaque ball, its color a rich blue-green, in the past it revealed almost no atmospheric markings. In recent years, however, clouds and storms have begun to bloom on Uranus, possibly because the planet is entering its long, warmer spring.

Uranus is called an ice giant, rather than a gas giant, because it contains relatively large amounts of icy methane and water, rather than hydrogen gas like Jupiter and Saturn. About 14 times more massive than the Earth, it's midway in mass density between the Earth and Saturn, a finding that tells us it probably has a largish rocky core. Temperatures at its cloud tops are a superchilled -216°C (-357°F), making it the coldest of the major planets. Unlike every other major planet, Uranus seems to have no internal source of heat. Why it is so different from its siblings is not clear.

WILLIAM HERSCHEL

ORGANIST TO ASTRONOMER

Perhaps the amateur astronomer William Herschel (1738–1822) acquired his habits of persistence and perfectionism through long hours of organ practice as a child. The son of a German musician, Herschel worked for years as an organist in England before astronomy began to take over his life. He and his sister Caroline built numerous fine telescopes, one of which helped him detect a new planet, later called Uranus, on March 13, 1781. King George III named him Royal Astronomer in 1782. Herschel went on to study the nature of nebulae and theorize that stars are organized into "island universes" (now called galaxies).

PLANET SIDEWAYS

Bland as it may look, Uranus has some bizarre attributes. Perhaps the most startling is its sideways axis of rotation. All the other planets have axes that are more or less perpendicular to the ecliptic, the plane of the solar system. Uranus spins on its side (at 98° from the perpendicular), with its poles pointing toward and away from the sun. Its moons and rings circle about its equator, flat-on to the sun, so that the Uranian system looks like a planetary bull's-eye. With its tilt at a little more than a right angle, so that its pole is below the solar system's orbital plane, its 17-hour rotation is considered to be retrograde. It seems unlikely that the planet formed this way. The best explanation is that a massive impact during Uranus's early days knocked it on its keister.

Spinning on its side, Uranus has odd seasons. During the 21-year northern summer, the north pole points directly at the sun. A well-insulated observer in Uranus's northern latitudes would never see the sun set in the height of summer—it would merely circle around the sky's northern point every 17 hours. At the opposite pole, the sun would never rise. At the equator, spring and fall are the warmest seasons, while winter and summer would see the sun barely rising above the horizon.

THE TURQUOISE ATMOSPHERE

Uranus shines a rich blue-green due to methane molecules in its atmosphere. (Methane absorbs the redder portions of the spectrum, so sunlight bouncing off Uranus's clouds looks blue when it returns to space.) However, the atmosphere consists mainly of molecular hydrogen (84 percent) and helium (14 percent). Methane makes up a remaining 2 percent or so. Like Jupiter and Saturn, Uranus has no solid surface. Its upper atmosphere appears to be layered, with methane haze overlying methane clouds, and below them possibly water- and ammonia-ice clouds. Close examination of the atmosphere at various wavelengths reveals that it flows around the planet in bands in the same westward direction as the planet's rotation, reaching speeds between 100 and 600 kilometers an hour (60 and 400 mph). Unlike on other planets, its winds blow fastest near the poles, but this is probably due to the sideways polar orientation. However, there is little difference in heat between the poles and the equator, which implies that some mechanism below the clouds is spreading heat efficiently around the planet.

When Voyager flew by in 1986, Uranus revealed very few clouds or storm features, giving it the unfair reputation of Mr. Boring Planet. In the 21st century that began to change. The Keck II telescope began to see both storms and clouds appearing and disappearing through the atmosphere. One large storm, called the Great Spot at 37° south, has persisted for years, sometimes climbing to high altitudes. Bands of clouds 18,000 kilometers (11,000 mi) long swell and fade in the northern hemisphere. The increased activity may be caused by seasonal changes in solar heating.

DOWN TO THE CORE

Uranus is well hidden beneath its clouds. Scientists can only theorize about its internal structure based on its density and visible gases. Its hydrogen-rich atmosphere may extend fairly deep into the planet, encompassing perhaps 80 percent of its radius. Under that, Uranus may possess a dense, slushy mix of water, methane, and ammonia. If these materials were not under extreme pressure, they would be frozen, and astronomers refer to them as ices—however, the heat inside Uranus keeps them densely liquid. Like the bigger gas giants, it may have a rocky core—in the case of Uranus, one about the size of Earth, but denser.

Something about Uranus's materials or its history gives it strange, off-kilter magnetism. When Voyager 2 flew past the planet it found a strong magnetic field, about 48 times as powerful as Earth's. Uranus's magnetosphere traps charged particles, electrons and protons, that travel along magnetic field lines and give off radio waves. But the field itself is tilted 60° from the planet's rotational axis (which, as noted earlier, is roughly perpendicular to the ecliptic). Furthermore, the magnetic axis isn't aligned with the planet's axis, but is shoved to the side, offset by about one-third of the planet's radius. It's unlikely that the force that knocked Uranus on its side also sent its magnetic field reeling. Something about Uranus's internal structure, perhaps in its electrically conducting icy chemicals, seems to be responsible for its odd magnetic axis, but we don't know yet what that is.

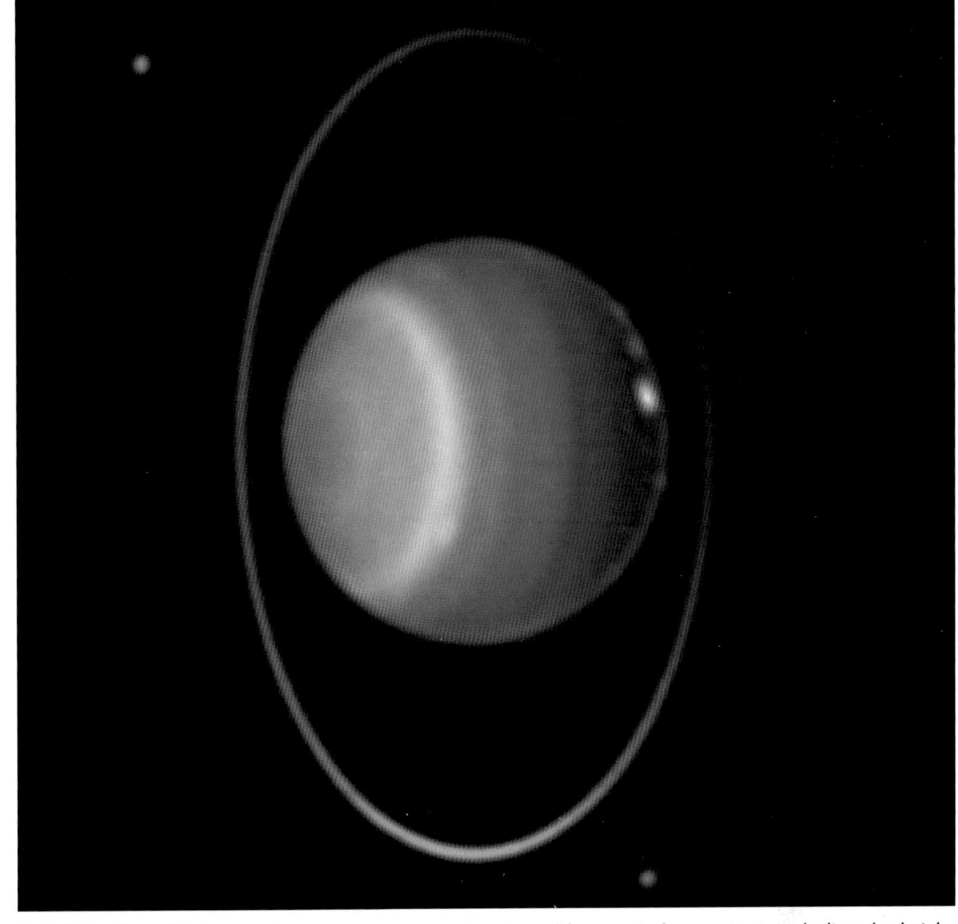

A Hubble Space Telescope image shows storm clouds on Uranus, its larger rings, including the bright epsilon ring, and several of its moons.

THE 21-YEAR SUMMER

A false-color image of Uranus compiled from Voyager 2 data highlights atmospheric banding at the planet's south pole during its long summer season. The sun shines directly on the planet's poles in summer and winter.

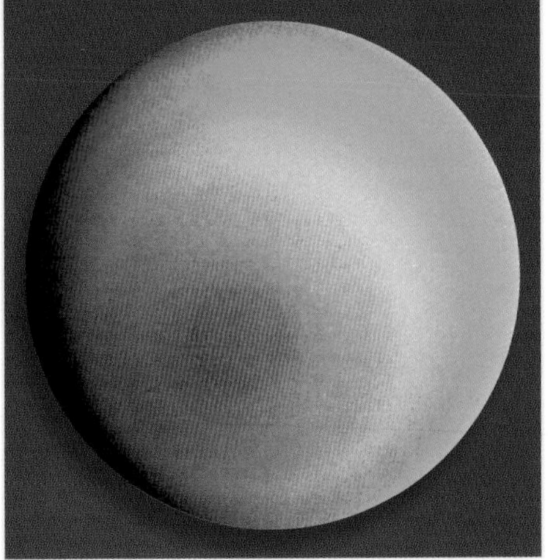

RINGS

In 1977, astronomers hoping to learn more about Uranus's atmosphere were studying the planet as it passed in front of a distant red giant star. To their surprise, about 40 minutes before the planet blocked out the star, the star's light flickered repeatedly. Something very narrow and very close to Uranus was crossing in front of the star—and thus were Uranus's rings discovered. They were the first planetary rings found since Galileo spotted Saturn's appendages. The earthbound observers counted nine slender rings. Voyager 2 spotted two more in 1986, and the Hubble Space Telescope brought the total to 13 in 2003, although these last 2 were not recognized as such until 2005. In order from inside to outside, they bear the unromantic labels 1986 U2R, 6, 5, 4, alpha, beta, eta, gamma, delta, lambda, epsilon, R2, and R1. Uranus's rings circle its equator (though some are tilted slightly). This means that like their parent planet, they are perpendicular to the plane of the solar system and at times face the Earth flat-on. All are formed of small, dusty, dark particles, similar in size to those in Saturn's rings but not as shiny. Most are relatively close to the planet and quite shallow and narrow, less than 10 kilometers (6 mi) wide, with occasional bands of dust between them. The two rings discovered recently are more distant and wider, like broad trails of dust.

The narrowness of most of Uranus's rings implies that they are constrained by the gravitational pressures of shepherd moons (see pp. 148–49). And in fact, two such moons, Cordelia and Ophelia, have been discovered flanking the epsilon ring. The outermost ring discovered by Hubble is distinctly blue and appears to consist of icy particles shed by Uranus's moon Mab. Its color may be a result of the way its extrafine ice particles reflect blue wavelengths.

ICE GIANTS: NEPTUNE

Distant Neptune marks the boundary of the traditional solar system, the realm of the eight planets. At 30 AU from the sun, or 4,498,252,900 km (2,795,084,800 mi), it is half again as far as Uranus. Observers at its cloud tops would see the sun as merely the brightest star in a black sky.

At this distance, Neptune is invisible to the naked eye. Galileo actually saw the planet in his telescope in 1613, recording it as a star. If cloudy skies had not prevented him from observing its motion over several nights, he might have been its discoverer. As it was, Neptune's discovery became a triumph of Newtonian calculation; two mathematicians, Britain's John Adams and France's Urbain Le Verrier independently worked out its position based on variations in the orbital motion of Uranus. A third astronomer, German Johann Galle, saw the planet through a telescope for the first time in 1846 using the position provided him by Le Verrier (see p. 37).

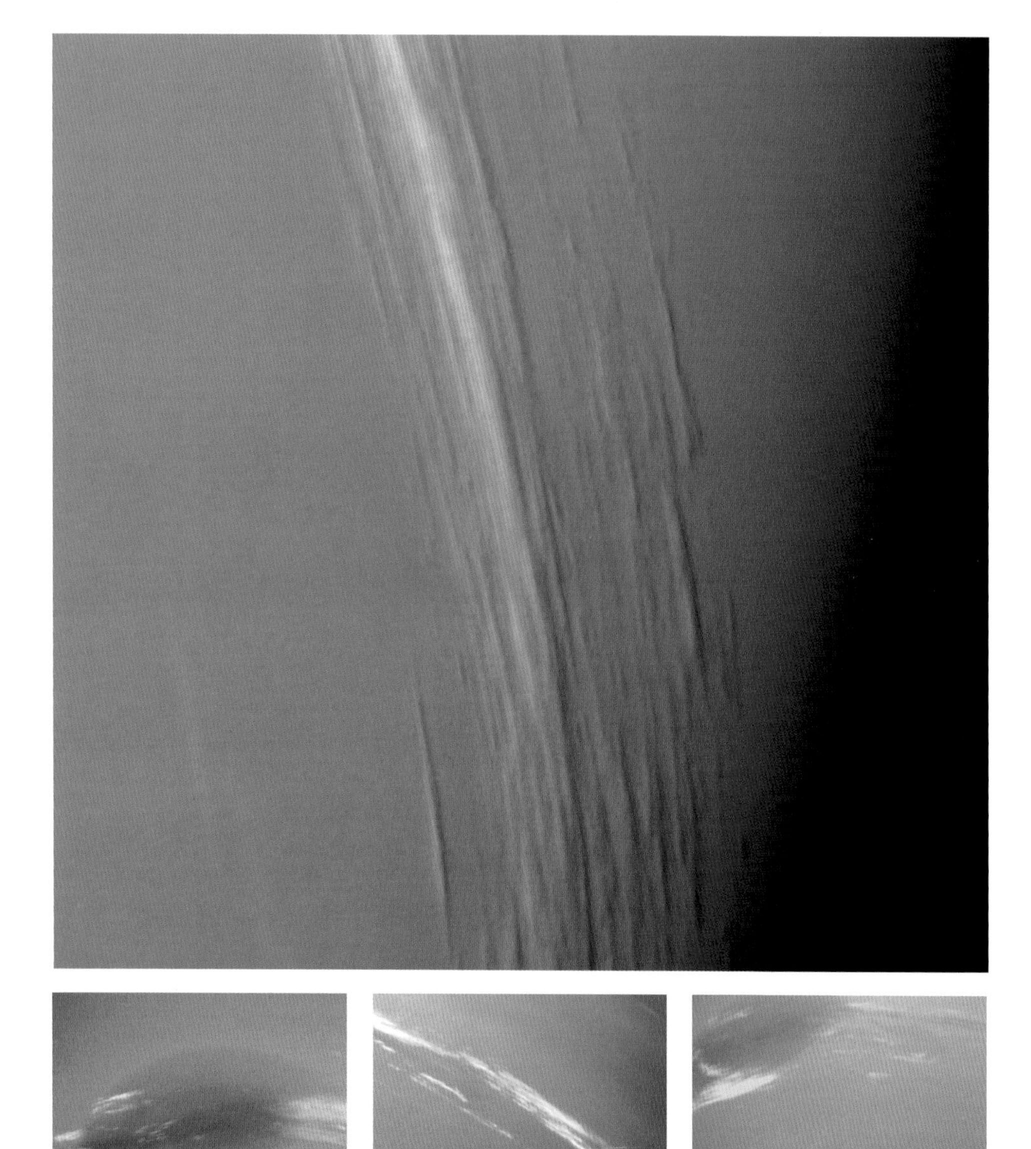

At top, Neptune's 50-km-high (31-mi) clouds; the Great Dark Spot (bottom left); clouds near the Spot (bottom middle); and the Great Dark Spot above Scooter and the Small Dark Spot (bottom right).

Only one spacecraft has visited Neptune, the indefatigable Voyager 2 in 1989. In recent years, the Hubble Space Telescope has contributed valuable images of the remote planet. No new missions have yet left the drawing board.

Neptune resembles its ice-giant sibling, Uranus, in many but not all ways. It's similar in size, about 17 Earth masses versus 14.5 for Uranus, and somewhat more dense. Like Uranus and the big gas giants, it has no solid surface but is enveloped by a deep gas atmosphere composed mainly of hydrogen and helium. Its indigo color comes from methane in its atmosphere (in rather higher concentrations than on Uranus). It rotates rapidly, every 16 hours, though its orbit is necessarily far longer than that of its sister planet: it takes almost 165 Earth years to complete one circuit around the sun. With its axis tilted at 29.6°, it has 41-year seasons.

When Voyager 2 reached Neptune in 1989, it found that the planet radiated more heat than it received from the sun. Though the temperature at its cloud tops is a numbing 59K (-214°C/-353°F), without an internal heat source it would be even colder at 46K (-227°C/-377°F). Scientists aren't sure where this extra heat is coming from. Perhaps warmth left over from its original formation is still leaking from its core and is insulated by the planet's methane.

STORMY WEATHER

This mysterious internal heat might explain another Neptunian puzzle: its intensely turbulent atmosphere. Despite receiving just a tiny bit of solar energy relative to a planet such as Earth, Neptune is a maelstrom. Winds blow around the planet at speeds of more than 2,000 kilometers an hour (1,200 mph), roughly twice the speed of sound on Earth. The entire atmosphere exhibits an odd differential rotation: Equatorial bands circle the world in 18 hours, but the planet underneath rotates every 17 hours, so the equatorial winds are effectively retrograde. At the poles, the winds flow the other way, circling every 12 hours, faster than the planet's rotation.

Enormous storms appear and vanish frequently. Around the time Voyager 2 visited, three big cyclones breached the top of its atmosphere. One, a counterclockwise oval the size of Earth, was dubbed the Great Dark Spot because of its resemblance to Jupiter's Great Red Spot. Observers named a second, fast-moving white storm near it Scooter, and an eyelike cyclone in the south the Small Dark Spot. Unlike the Great Red Spot, though, Neptune's storms did not last. By the time Hubble took a good look a few years after Voyager, the original storms had disappeared and others had taken their place. The rotating storms seem to be vortices that open wells into the darker lower atmosphere. Updrafts around them create white, high-level cirrus clouds of methane ice.

In 2007, a team of astronomers observing Neptune through the Very Large Telescope in Chile found that temperatures at the planet's south pole were 10°C (18°F) hotter than the rest of the planet. The added heat was enough to allow the pole's frozen methane to turn to gas and vent into space. The warmth probably built up over the course of Neptune's prolonged summer.

INSIDE NEPTUNE

From what little we know of Neptune, we assume it has a structure much like Uranus. Its frigid hydrogen, helium, and methane atmosphere changes with increasing depth to a hot, dense soup of water, ammonia, and methane. Water/ammonia clouds may layer the atmosphere. The planet's density suggests that, like Uranus, it has a rocky core about the size of Earth but ten times as massive.

Neptune's magnetic field is relatively strong, 27 times more powerful than Earth's. But like the field on Uranus, its axis is wonky, tilted at 47° relative to the planet's rotational axis and offset from the planet's center. The reason for this is unknown but presumably has to do with the way the field is generated inside Neptune, perhaps in its slushy, electrically conductive ices.

The field was quite useful in determining Neptune's rotation, given that the planet itself was hidden under a variety of clouds. Voyager used bursts of radio emissions generated by the planet's magnetic field to discover that its day was 16 hours, 7 minutes long.

HOT SPOTS

So far from any solar energy, Neptune should exist in a kind of frozen hibernation. And yet its roiling storms point to some source of internal heat, though scientists can only theorize what that might be. Infrared observations of the planet's deep atmosphere reveal irregular, dynamic weather systems that travel and change rapidly over the globe. The infrared images below, produced by the Center for Adaptive Optics, highlight the way in which Neptune's upper-level clouds changed over the course of a single month in 1999.

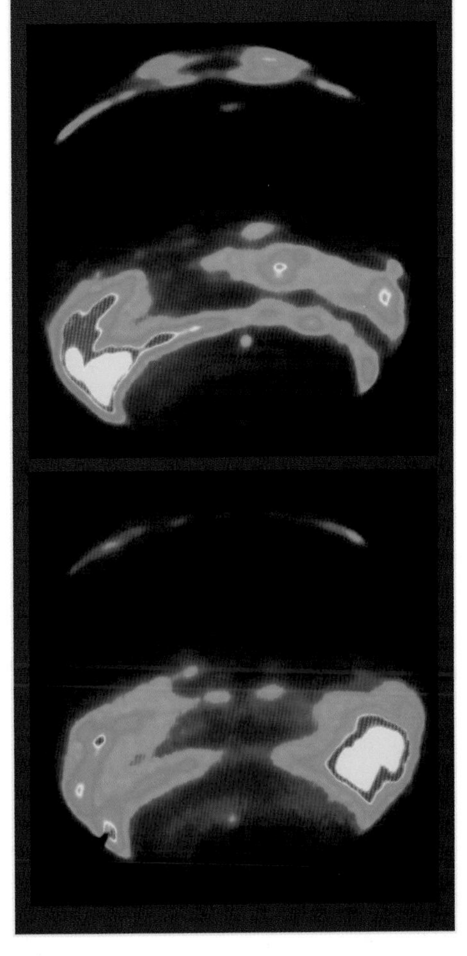

NEPTUNE'S RINGS

The surprising discovery of Uranus's rings in 1977 inspired astronomers in the 1980s to search for similar rings around Neptune using the same method of stellar occultation. But their observations yielded inconsistent results. About one-third of the time, something seemed to block the flickering starlight, but at other times nothing showed up. Perhaps, hypothesized scientists, Neptune had partial rings or arcs.

Voyager 2's visit in 1989 solved the mystery. Neptune indeed had full rings, but they were very thin and, in some cases, lumpy or twisted. The five rings are named in honor of Neptunian astronomers: From innermost to outermost, they are Galle, Le Verrier, Lassell, Arago, Adams. Voyager's closer look at the Adams ring explained the on-again, off-again nature of earlier observations. Unlike typical rings, which have a relatively even width all the way around, the Adams ring bulged distinctly in five locations around its circumference. Astronomers have even named the swollen arcs: The largest are Liberté, Égalité, and Fraternité, and the fourth, smaller arc is Courage. Ordinarily, such irregularities would disperse quickly into even rings, but astronomers believe they are confined by the gravity of Neptune's moon Galatea, orbiting just inside the Adams ring. Galatea may also be directly responsible for the unnamed, faint dust ring between Arago and Adams. Other small satellites orbit within the rings, possibly keeping them in shape as shepherd moons.

For the most part, the rings are thin as mist, with one million times less material than the rings of Saturn. The Galle and Lassell rings are thin but also very broad, like wide, powdery avenues. If the material for all the rings were combined into a single moon, it would be only a few kilometers wide.

BIRTH AND DEATH OF THE RINGS

Their origin is unknown, but Neptune's rings do appear to be much younger than the planet they encompass. One theory holds that they are ex-moons, the remains of colliding satellites or of a larger moon ripped apart by Neptune's tidal forces. In the years since Voyager 2 first spotted them, they have changed considerably. All of the arcs seem to have decayed, particularly the Liberté clump. Unless it is bolstered by new material, it will disappear within this century. The entire system, in fact, may eventually pound itself into a fine dust, spiral inward into Neptune's atmosphere, and vanish.

ICE GIANTS: DISTANT MOONS

The moons of Uranus and Neptune have both mystified and enlightened astronomers seeking to understand how the solar system was formed.

Uranus possesses 27 moons that we know of, and there are probably also smaller ones hidden within the planet's rings. They range in size from Titania, almost half the size of Earth's moon, to tiny Cordelia, just 26 kilometers (16 mi) wide. All orbit the planet on its perpendicular equatorial plane, so that at times they face the sun flat-on, like Uranus's rings. All are held in tidally locked orbits.

Uranus's discoverer, William Herschel, found the two biggest moons, Oberon and Titania, in 1787. Wealthy English brewer and amateur astronomer William Lassell discovered the next two, Ariel and Umbriel, in 1851. Almost a century passed before Gerard Kuiper spotted the fifth moon, Miranda, in 1948. Voyager 2, the Hubble Space Telescope, and other modern ground-based telescopes picked out most of the smaller ones. Despite protests from classicists, Herschel and his son John decided to name the satellites after characters in Shakespeare and in Alexander Pope's *The Rape of the Lock*. The tradition is carried on today in such recently named moons as Prospero, Mab, and Cupid.

Roughly half ice and half rock, the moons are surprisingly dark and dirty looking. They may be covered with some sort of interplanetary soot, or they may have undergone radiation darkening as high-energy particles interacted with their surfaces. Titania and Oberon, the biggest, are heavily cratered, with icy surfaces. Oberon sports a remarkably high mountain rising 6 kilometers (4 mi) above its surface, possibly the result of a collision. Ariel and Umbriel, similar in size, differ considerably in their surfaces. Ariel is bright and smooth, with long rift valleys. Some icy material, such as ammonia, may have welled up from within to resurface it. Umbriel, on the other hand, has a dark, old surface marked by a bright ringlike crater NASA calls the fluorescent Cheerio.

MIRANDA

Oddest of all is small Miranda, which looks like it was shaped and discarded by a frustrated toddler. Gouged, cratered, jumbled, its surface defies explanation. Canyons drop as deep as 20 kilometers (12 mi); ridges, valleys, and mountains rough up its terrain. Its peculiar composition may be the result of interrupted development. Perhaps in its hot early days it began the process of differentiation, with denser material sinking inward, but cooled and stopped with the process incomplete. Or it may have been the victim of repeated impacts, each one tossing the moon like a salad.

Cordelia and Ophelia, moons that shepherd Uranus's epsilon ring (see p. 155), float inside

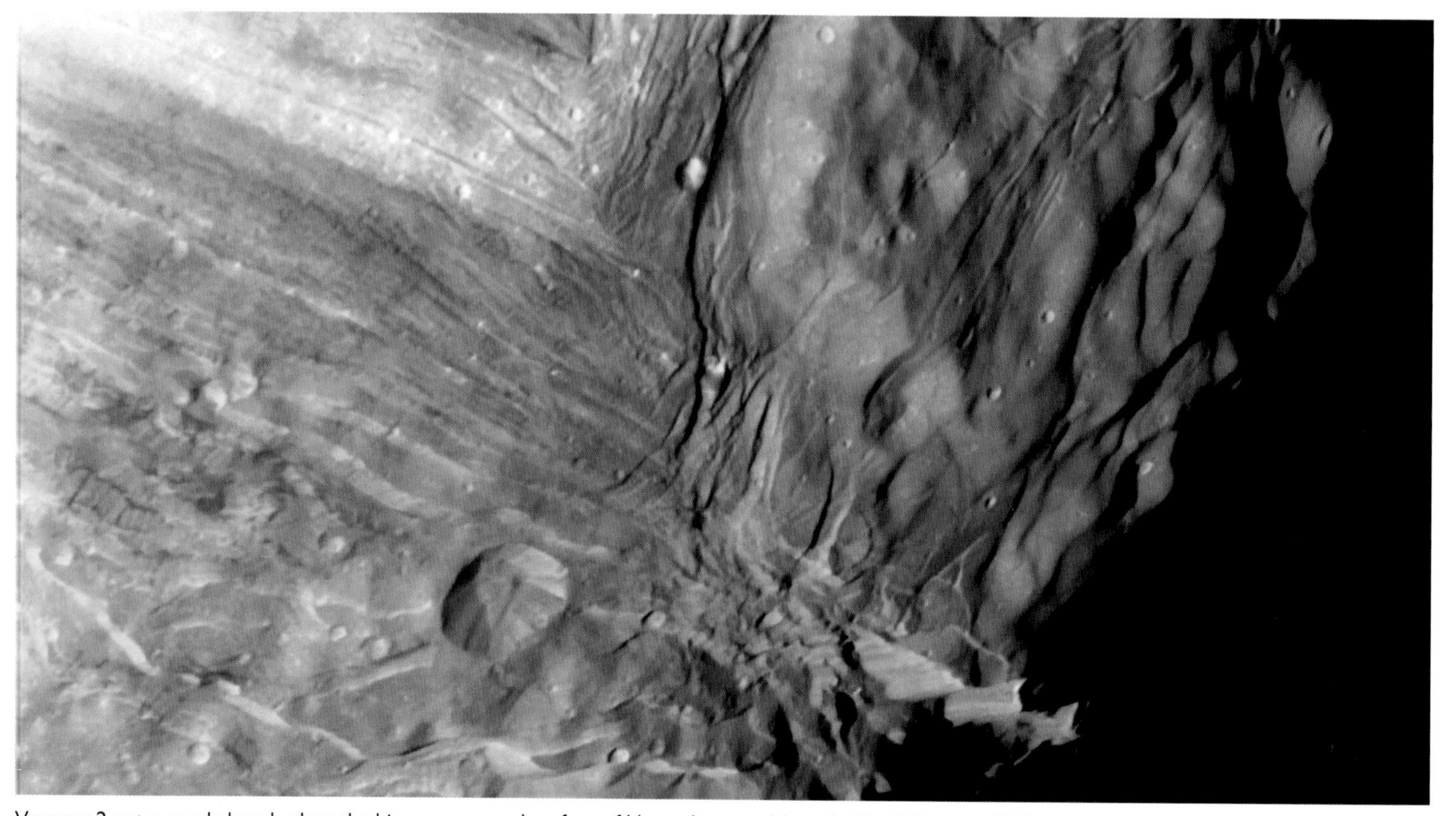

Voyager 2 got a good close look at the bizarre, grooved surface of Uranus's moon Miranda. The light gray cliff (bottom right) is 20 kilometers (12 mi) high; the crater is about 24 kilometers (15 mi) across. Disappearing into shadows at the upper right is more rugged, higher-elevation terrain.

Miranda's orbit. Other little moons cluster in the region as well, possibly exerting their own gravitational influences on the rings. The four outermost moons circle Uranus in retrograde, tilted orbits. They are probably straying asteroids pulled in by the planet's gravity.

NEPTUNE'S MOONS

Neptune's 13 moons, only one of them large, form a paltry satellite family by giant-planet standards. But what they lack in numbers they make up in strangeness.

In 1846, spurred by Neptune's discovery (by foreigners! Worse, a Frenchman!), William Lassell searched for and found its biggest satellite, Triton, just 17 days later. No other moon was discovered until Gerard Kuiper spotted Nereid in 1851. The rest are space age discoveries, including Proteus, the second largest Neptunian satellite.

ICE GIANT MOONS

To date, astronomers have found 27 satellites for Uranus and 13 for Neptune. Neptune's Triton and several of the smallest and most distant moons for both planets have retrograde orbits, which may mean they were captured by the planets' gravity.

Uranus	Trinculo
Cordelia	Sycorax
Ophelia	Margaret
Bianca	Prospero
Cressida	Setebos
Desdemona	Ferdinand
Juliet	
Portia	
Rosalind	**Neptune**
Cupid	Naiad
Belinda	Thalassa
Perdita	Despina
Puck	Galatea
Mab	Larissa
Miranda	Proteus
Ariel	Triton
Umbriel	Nereid
Titania	Halimede
Oberon	Sao
Francisco	Laomedeia
Caliban	Psamathe
Stephano	Neso

The six innermost moons are more typical satellites, orbiting in a prograde direction (in the same direction as the planet's rotation) not too far from the planet's cloud tops. Lumpy and irregular, they were not large enough to pull themselves into spherical shapes. Proteus, the largest of these, is so dark and sooty looking that it reflects only 6 percent of the sunlight that reaches it.

The outer six satellites are an odder bunch. Three orbit in a retrograde direction. Nereid, the outermost, has the most elongated orbit of any planetary moon. Its distance from Neptune varies by about eight million kilometers (five million mi) over the course of its year-long orbit and its path is tilted by 27° from Neptune's orbital plane. In time, it may collide with little Halimede. These oddball moons may be captured objects or debris left over from a collision between earlier moons and a passing body.

TRITON

Neptune's big moon, Triton, is one of the solar system's most remarkable satellites. It is the only large moon to orbit its planet in a retrograde direction, and its path is tilted at 157° to the orbital plane. About three-quarters the size of Earth's moon, it doesn't much resemble the other big outer moons in composition. Instead, it's relatively dense and rocky, like Pluto. Many astronomers theorize that it is, in a way, a purloined Pluto: a passing Kuiper belt object kidnapped by Neptune's gravity.

Triton is the coldest large body that we know of in the solar system. At about 38K (-200°C/-391°F), it exists in a deep freeze not far above absolute zero. And yet it shows signs of past and current activity. A scarred surface, with cracks, plains, a rindlike "cantaloupe" terrain, and a frosty methane polar cap, seems to be fairly young. Craterlike features may be icy lakes of water, cryovolcanic upwellings that froze as hard as steel.

Nitrogen frost colors the south polar region of Neptune's big moon Triton.

Triton even has a thin nitrogen atmosphere and deposits of nitrogen frost. As Voyager 2 passed the moon in 1989, Triton amazed observers by spraying great jets of nitrogen several kilometers into space from its polar cap. These frigid geysers may be vented through fissures in Triton's crust by some sort of pressures below the surface. Mixed with carbon particles, the plumes blow across the moon's surface and leave dark streaks below.

Triton may originally have had an elliptical orbit when it was first captured, but now it moves in a near circle. However, tidal interactions between Neptune and the retrograde Triton are causing the moon to gradually spiral into the big planet. In about 100 million years, Triton may reach the Roche limit and be torn apart. Since it contains more than 99 percent of the mass currently in orbit around Neptune, the rings that will result from its destruction may surpass Saturn's.

ICY DWARF PLANETS

ICY DWARFS

Tiny, distant, and profoundly cold, the icy world Pluto was always the runt of the solar system's litter—but like many another runt, it held a fond place in the hearts of Earth's inhabitants. Thus the public outcry when the International Astronomical Union (IAU) demoted Pluto from planetary status on August 24, 2006. Just 76 years after American astronomer Clyde Tombaugh added it to the planetary ranks, Pluto was removed and placed in a new class of solar system objects. At first called dwarf planets, the members of a new category of small worlds past the orbit of Neptune are now known as plutoids, in the ex-planet's honor.

At the time Pluto was found in 1930, no one knew that these distant bodies existed. Pluto's orbit seemed to mark the outer boundary of the solar system, an elliptical fence beyond which lay the empty reaches of interstellar space. But not all solar system bodies fit into this neat picture. Chief among them were comets, which had been puzzling astronomers for centuries.

In the 18th century, Edmund Halley had shown that comets orbited the sun, just as did the planets, though he could not say where the comets originated. The great French mathematician Pierre-Simon de Laplace, who pioneered theories of solar system formation, later noted correctly that some comets might have been deflected from their original, longer orbits into shorter ones by Jupiter's gravity. But these theories did not address the mystery of the comets' source, and no telescopes revealed any cometary nurseries within the known confines of the system.

Not until the 20th century did several astronomers independently come up with explanations for comets. In 1932, Estonian astronomer

1577
Tycho Brahe concludes comets are objects, not atmospheric phenomena.

1705
Halley believes a few historic comets are the same. Predicts a return in 1758.

1847
Astronomer Maria Mitchell makes first discovery of comet with telescope.

1932
Ernst J. Öpik theorizes existence of celestial reservoir beyond the planets.

1950
Jan Oort determines that a far reservoir of icy bodies must exist.

1976
Methane discovered on Pluto.

Comet McNaught decorates the night sky over Ashburton, New Zealand. Visits to Earth by comets such as this one have long inspired fear and wonder in equal parts. They also prompted astronomers to wonder where comets were born and what might lie beyond the visible planets.

Ernst Öpik suggested that comets start off in a vast reservoir of such objects beyond the orbit of newly discovered Pluto. In 1950, Dutch astronomer Jan Oort studied the orbits of 19 comets and postulated a similar theory in the *Bulletin of the Astronomical Institutes of the Netherlands.* "From a score of well-observed original orbits," he wrote, "it is shown that the 'new' long-period comets generally come from regions between about 50,000 and 150,000 AU distance. The sun must be surrounded by a general cloud of comets with a radius of this order, containing about 10^{11} comets of observable size. . . . Through the action of the stars fresh comets are continually being carried from this cloud into the vicinity of the sun." This distant shell of comets has come to be known as the Oort cloud, or sometimes the Öpik-Oort cloud.

Meanwhile, a complementary theory for the origins of short-period comets (those with orbits of less than 200 years) had been advanced by two other astronomers. In 1943, a little-known

1977
Charles Kowal discovers Chiron, the first known centaur object.

2000
Caltech team announces discovery of first truly large KBO, Quaoar.

2004
Caltech announces discovery of Sedna, most remote body in solar system.

2004
NASA's Stardust is the first spacecraft to collect comet particles.

2006
IAU creates new category of celestial bodies: dwarf planets.

2008
IAU terms trans-Neptunian dwarf planets similar to Pluto "plutoids."

Irish astronomer named Kenneth Edgeworth proposed the existence of a comet reservoir in a belt past the orbit of Neptune. Dutch-American astronomer Gerard Peter Kuiper, apparently unaware of Edgeworth's proposal, postulated a similar theory in 1951, suggesting that the belt might extend almost as far as the Oort cloud. Like the objects in the Oort cloud, these distant inert comets were too small and dark to be seen with the telescopes of the time. Both the Oort cloud and the Kuiper belt (or the Edgeworth-Kuiper belt, as it's often called) remained theories, though well-accepted ones.

The first proof of the existence of the Kuiper belt did not arrive until 1992. In that year, astronomers David Jewitt and Jane Luu at the University of Hawaii's Mauna Kea Observatory discovered a remote body orbiting past Pluto at roughly 44 AU. Unromantically dubbed 1992 QB1, it was small, perhaps 200 kilometers (125 mi) in diameter—but there it was, a sizable object orbiting just where a Kuiper belt object should be. Almost immediately, Jewitt, Luu, and other astronomers began to discover other trans-Neptunian objects. At first, these were whimsically called cubewanos after the first discovery, but soon they gained official IAU numbers or names. In 2002, astronomers Mike Brown and Chad Trujillo of the California Institute of Technology spotted a truly large object about half the size of Pluto, later named Quaoar (pronounced KWA-o-war), at 42 AU. Most of the large bodies found so far had fit into the predicted range of the Kuiper belt, but in 2004 Mike Brown's team announced the discovery of a far more distant world, now named Sedna. At 90 AU from the sun when discovered, Sedna will eventually reach 900 AU—not as far as the Oort cloud, but considerably farther than any other solar system object yet seen. The remote body may have been kicked into this extreme orbit long ago by the gravity of a passing star.

The discoveries highlighted an awkward fact: Astronomers never really had a good definition for planet. Planets were understood to be large, round bodies orbiting the sun, but what differentiated a small planet from a big asteroid? Did a planet have to be spherical? Did it have to orbit a star? Could huge satellites such as Titan or Ganymede qualify? Where did Pluto fit in?

Some astronomers had already removed Pluto from their own lists of planets. When the

GERARD PETER KUIPER
FATHER OF PLANETARY SCIENCE

Gerard Kuiper (1905–73), a Dutch-American astronomer, contributed much to planetary science, including discovering Neptune's moon Nereid, Uranus's moon Miranda, and Titan's atmosphere. He is best known for predicting the existence of a belt of icy bodies beyond Neptune. The hardworking Kuiper also served as the director of the Yerkes Observatory and Arizona's Lunar and Planetary Observatory.

Rose Center for Earth and Space opened in New York in 2000, for instance, the huge display of the solar system omitted Pluto, to considerable public disapproval.

In 2005, Mike Brown and his colleagues brought the issue to a head when they announced the dimensions of a new object that had been discovered more than twice as far from the Sun as Pluto. The icy body, temporarily called 2003 UB313 and now formally named Eris, was even bigger than our ninth planet. Would it become the tenth planet? And what about other Kuiper belt worlds, almost as large? Were they planets? Would we keep adding worlds to the solar system until no schoolchild could master them?

Hence the IAU's 2006 decision to redefine "planet" and exclude Pluto. The decision stirred up considerable controversy. Some astronomers felt the wording of the new definitions had been designed specifically to remove Pluto. Scientists involved with the ongoing New Horizons mission to Pluto were dismayed, seeing their target apparently downgraded to just one among many dwarf planets. Even more outraged were some members of the public, fond of the only American-discovered planet. (Yet others associated the planet with the familiar Disney character, though the names were unrelated.) Clyde Tombaugh's wife and son joined a group of demonstrators carrying signs reading "Size Doesn't Matter" in 2006. Legislators in the state of New Mexico, where Tombaugh had long made his home, introduced a resolution that Pluto again be "declared a planet."

So far, the IAU has not backed down with respect to its new categories, although it did add the "plutoid" class to the dwarf planet group in 2008 to further refine its new definitions. But the flurry of protest over Pluto may obscure a somewhat larger issue. With the discovery of 1992 QB1, an entirely new solar system realm has opened up to our view. If the terrestrial planets represent the inner solar system, and the giant planets the outer solar system, then Pluto and its icy cousins delineate a third zone, a vast and ancient frontier. Already our telescopic observations of these distant objects have surprised astronomers who did not expect their varied compositions or their unusual orbits. We have a whole new region to explore, one whose worlds may hold the key to the birth of solar systems.

KENNETH EDGEWORTH
ECONOMIST AND ASTRONOMER

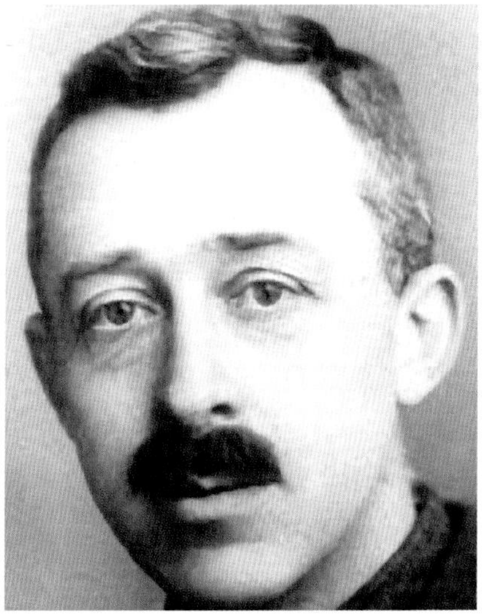

Kenneth Edgeworth (1880–1972), born in Ireland, had a varied career as a soldier, economist, and astronomer. After serving with distinction in Britain's Royal Engineers during World War I, he went on to publish on both international economics and astronomy. In 1943, his article suggesting that the solar system held a reservoir of comets beyond the planets anticipated Kuiper's prediction by several years.

PLUTO
FROZEN WORLD

With deep-freeze temperatures, three peculiar moons, and an atmosphere that turns to snow in winter, Pluto is one of the oddest members of the solar family.

Pluto, once a planet, is now officially a plutoid, the subcategory of dwarf planet, to which it has lent its name. It is thought to be the nearest large member of the orbiting collection of rocky debris and small worlds known as the Kuiper belt. Far smaller than the eight major planets, Pluto has a diameter of just 2,302 kilometers (1,430 mi), making it about one-fifth the size of Earth. In many ways it is a twin of Neptune's moon Triton, which itself may be a captured member of the Kuiper belt. Pluto has its own moons, three of them, and all four bodies orbit a single center of mass. • When the New Horizons mission reaches the dwarf planet in 2015, we'll see it clearly for the first time. Even the Hubble Space Telescope has been unable to discern many details on Pluto's surface, given that the planet orbits the sun at an average distance of 39 AU. Its orbit is highly eccentric, varying from 29.66 AU at perihelion (its closest point) to 49.3 AU at aphelion (its most distant point). For about 20 years out of every 248-year orbit, Pluto swings inside the orbit of Neptune: The last time this occurred was between 1979 and 1999. Pluto will remain outside Neptune's orbit until April 5, 2231.

Symbol: ♇
Discovered by: Clyde Tombaugh in 1930
Average distance from sun: 5,906,380,000 km (3,670,050,000 mi)
Rotation period: -6.39 Earth days (retrograde)
Orbital period: 247.92 Earth years

Equatorial diameter: 2,302 kilometers (1,430 mi)
Mass (Earth=1): 0.0022
Density: 2 g/cm³ (compared with Earth at 5.5)
Surface temperature: -233°C to -223°C (-387°F to -369°F)
Natural satellites: 3

SKYWATCH

• Pluto cannot be seen by the naked eye. To find it, consult planetary charts and use a ten-inch or larger telescope.

AMAZING FACT In 1.4 million years, the star Gliese 710 will pass through the outer Oort cloud.

1915
Percival Lowell predicts the existence of a planet beyond Neptune.

1930
Continuing Lowell's work, astronomer Clyde Tombaugh discovers Pluto.

1976
Methane discovered on Pluto.

1978
James Christy discovers Pluto's moon Charon.

2005
Hubble Space Telescope images lead to discovery of Pluto's two smaller moons.

2006
The IAU reclassfies Pluto as a dwarf planet.

An artist's conception of Pluto includes the reddish coloration of its surface (opposite). Art of Charon as seen from Pluto's surface shows the moon's nearness to the planet (above).

PLUTO: DOUBLE PLANET

Frigid little Pluto has had many identities in the short time astronomers have known it. Before it was even discovered, it existed in Percival Lowell's mind as "Planet X." Lowell, a wealthy Bostonian and talented amateur astronomer, founded the Lowell Observatory in Flagstaff, Arizona, in 1894. His studies of Neptune's orbit had convinced him that another planet, farther out in space, was perturbing the ice giant's path. For years he searched unsuccessfully for evidence of Planet X. In the year before his death he wrote "that X was not found was the sharpest disappointment of my life."

Another gifted amateur, Clyde Tombaugh, was hired at the observatory in 1929 and took up the search. Working with better equipment and patiently comparing sequential photographs of the night sky, Tombaugh found a traveling lintlike speck of light in images taken in January 1930. The official announcement that a ninth planet had been discovered came on March 13, 1930, Percival Lowell's birthday.

Lowell would have been pleased, but probably also chagrined to learn that his calculations (and those of other astronomers) were off. The orbits of Neptune and Uranus do not really show the irregularities he had worked out, and Pluto's mass is now known to be much too small to have affected them anyway. The planet's discovery was purely serendipitous.

Solar system cartographers work within the historic confines of the Lowell Observatory in Flagstaff, Arizona, where Pluto was first discovered.

ICY AND ECCENTRIC

Although observatories such as the Hubble Space Telescope have pulled in images of Pluto, the planet is so far away that details are lacking. Astronomers have cobbled together old observations, spectroscopic readings of its light, calculations based on its interactions with its big moon Charon, and occasional occultations of background stars to piece together what information they have. What we know so far gives us some intriguing clues about the origins and structure of the solar system.

Pluto's orbit, for instance, is a curious one. Its elongated, 248-year path is inclined by 17° to the plane of the solar system, so that Pluto swings from above to below the paths of all the other planets. The orbit is also locked into a 2:3 resonance with Neptune: Pluto orbits two times for every three Neptunian orbits. This means that although its orbital plane crosses that of Neptune, the planets will never collide. Their closest approach is 17 AU.

The plutoid's density implies that it is made of rock (perhaps 70 percent) and water ice, similar to the moons Io, Europa, and Triton. Pluto may consist of a large, cold, rocky interior overlaid by a mantle of water ice, though whether it is differentiated into internal layers like the terrestrial planets is unknown. The surface is quite bright and reflective and appears to be paved with methane, nitrogen, and carbon monoxide ices that shine a pale brownish orange. Variegated light and dark patches on its surface may include bright polar ice caps, but its darker features are still unresolved.

Pluto is, not surprisingly, very cold (though not as cold as Triton); its surface temperatures range from about 40K to 50K (-233°C to -223°C/-387°F to -369°F). In 1988, the dwarf planet passed in front of a star and revealed a very thin, extended atmosphere with a pressure only one one-millionth that of Earth's. Due to Pluto's low gravity—about 6 percent of Earth's—its atmosphere leaks steadily into space from its upper regions as fast-moving molecules escape the plutoid's weak hold. The atmosphere's three main gases—nitrogen, methane, and carbon monoxide—undergo phase transitions, changing between solid and gas

CLYDE TOMBAUGH

NOT IN KANSAS ANYMORE

Clyde William Tombaugh (1906–97) was the very ideal of an amateur astronomer. Living on a Kansas farm where he built his own telescopes, at the age of 22 Tombaugh (above, with a homemade reflector) was hired by Arizona's Lowell Observatory to photograph the sky in search of Percival Lowell's hypothetical Planet X. Painstakingly studying images of the night taken several days apart for months, in 1930 the young astronomer discovered the ninth planet. Tombaugh went on to find star clusters, galaxies, a comet, and many asteroids, as well as to found the astronomy program at New Mexico State University.

as the temperature warms or cools. When Pluto is closest to the sun, its atmosphere is gaseous; as it moves away from the sun in its long, eccentric orbit, those gases freeze and fall to the surface like snow in a very long, intensely frigid winter. This may begin to happen again as soon as 2010.

MOONS

In 1978, astronomers James Christy and Robert Harrington at the U.S. Naval Observatory in Flagstaff, Arizona, recognized that a bump sticking out of Pluto's side in earlier images was actually a moon. Named Charon, after the mythical ferryman of the dead, the moon is half the size of Pluto, which gives the two bodies the closest planet-to-moon size ratio in the solar system. (The runner-up, our moon, is roughly one-quarter the size of Earth.) Charon is only 19,600 kilometers (12,200 mi) from Pluto and is locked into a mutually synchronous orbit, unique among planetary satellites. Pluto and Charon rotate at exactly the same rate—once every 6.39 Earth days—and always present the same face to one another. An observer on either body would see the other one perpetually in the same place in the sky. Charon doesn't orbit Pluto—both circle a mutual center of mass, a barycenter. Many astronomers believe the two bodies should be considered a binary dwarf planet, rather than a planet and satellite. Charon seems to contain more ice than Pluto, and its surface is less colorful, implying that it is covered with water ice.

Almost 30 years after Charon's discovery, astronomers were in for another surprise. Scientists using the Hubble Space Telescope found two more satellites of Pluto in June and August of 2005. Named Nix, for the mother of Charon, and Hydra, for the many-headed serpent of the underworld, the two satellites orbit the center of mass in the Pluto-Charon system, but at greater distances than the big moon. Nix, about 48,700 kilometers (30,250 mi) out, may be slightly larger than Hydra, some 65,000 kilometers (40,400 mi) away from the barycenter. Both are small, with diameters in the range of 60 to 100 kilometers (40 to 60 mi). Little else is known about these satellites so far.

ORIGINS

Pluto's curious orbit, its composition, and its close relationship with its biggest moon have led to some informed speculation about its origins. Pluto's similarity to Neptune's moon Triton led to an early theory, espoused by astronomer R. A. Lyttleton in 1936, that Pluto was an escaped satellite of Neptune. However, Pluto's current orbit and the presence of Charon make this unlikely. A more plausible current scenario posits that Pluto was just one of hundreds of ice dwarf planets in the 30 AU range during the solar system's early formation. Charon could have been born from a collision between Pluto and another icy dwarf. Most planetesimals in that range may have been ejected into the Kuiper belt by the gravitational influence of Neptune and Uranus as the big planet formed and migrated outward, but the Pluto-Charon duo became trapped into the orbital resonance with Neptune that they show today.

Many questions about the former planet remain unanswered. Astronomers eagerly await findings from NASA's New Horizons spacecraft, launched in 2006 and due to fly past Pluto in 2015. New Horizons will look for rings around Pluto, observe its three moons, search for a magnetic field, map its surface, and more, before departing for further studies deeper in the Kuiper belt.

IAU RESOLUTIONS 5A & 6A

5A. The IAU therefore resolves that planets and other bodies in our Solar System, except satellites, be defined into three distinct categories in the following way:

(1) A "planet" is a celestial body that (a) is in orbit around the Sun, (b) has sufficient mass for its self-gravity to overcome rigid body forces so that it assumes a hydrostatic equilibrium (nearly round) shape, and (c) has cleared the neighbourhood around its orbit.

(2) A "dwarf planet" is a celestial body that (a) is in orbit around the Sun, (b) has sufficient mass for its self-gravity to overcome rigid body forces so that it assumes a hydrostatic equilibrium (nearly round) shape, (c) has not cleared the neighbourhood around its orbit, and (d) is not a satellite.

(3) All other objects, except satellites, orbiting the Sun shall be referred to collectively as "Small Solar-System Bodies."

6A. The IAU further resolves:

Pluto is a "dwarf planet" by the above definition and is recognized as the prototype of a new category of trans-Neptunian objects.

BEYOND NEPTUNE

TWILIGHT REALM

The frozen outer reaches of the solar system are only now coming into view. The distant neighborhood we're beginning to see contains a host of small and not-so-small bodies that have greatly expanded the solar system's family.

Perhaps one million icy bodies float beyond the orbit of Neptune in the solar system's shadowy outer zone. Most of these trans-Neptunian objects can be found within the wide, flattish orbital strip known as the Kuiper belt, which begins with Neptune at around 30 AU and extends past the orbit of Pluto to 50 AU. "Classical" Kuiper belt objects (KBOs) keep to fairly tidy orbits, most between about 38 and 48 AU. "Scattered" KBOs have wilder paths, ranging between 35 and 200 AU, some highly inclined to the orbital plane. The area they occupy is called the scattered disk. • Astronomers believe that at least 70,000 KBOs have diameters greater than 100 kilometers (60 mi); untold numbers are smaller and invisible, for now, to our telescopes. As numerous as they are, their total mass comes to no more than about 10 percent of Earth's. So far, more than a dozen of these icy bodies, which appear to be large and planetlike, have been discovered. Three of them, Eris, Makemake, and Haumea, have joined Pluto as officially designated plutoids, icy dwarf planets. Many more will undoubtedly be added to that category in years to come.

LARGE TRANS-NEPTUNIAN OBJECTS

Eris: ~2,400 km (1,500 mi) diameter
Pluto: 2,302 km (1,430 mi) diameter
Orcus: ~1,600 km (1,000 mi) diameter
Sedna: ~1,600 km (1,000 mi) diameter
Charon: 1,186 km (737 mi) diameter
Makemake: ~1,600 km (1,000 mi) diameter
Haumea: 1,320–1,550 km (820–960 mi) diameter
Quaoar: ~1,260 (780 mi) diameter
Ixion: ~1,064 km (660 mi) diameter

SKYWATCH

• No Kuiper belt object can be seen with the naked eye. You can track the New Horizons mission to Pluto at *http://pluto.jhuapl.edu/*.

AMAZING FACT Sedna's orbital period—its year—is at least 10,500 Earth years.

1943
Kenneth Edgeworth suggests there is a comet reservoir beyond Pluto.

1951
Gerard Kuiper proposes the existence of a distant belt of small icy bodies.

1992
David Jewitt and Jane Luu see first Kuiper belt object (KBO).

2005
Makemake is discovered by astronomer Mike Brown and team at Caltech.

2005
Eris, a KBO larger than Pluto, is discovered.

2008
Haumea, discovered in 2003, is the fifth body to be classified as a dwarf planet.

Artwork of dwarf planet Makemake brings out its reddish hue (opposite). From distant Sedna's surface (art, above), the sun and inner solar system would look like a dim disk.

BEYOND NEPTUNE: PLUTOIDS, CENTAURS, AND ICE

As Gerard Kuiper and Kenneth Edgeworth originally suggested, Kuiper belt objects (KBOs) seem to be primitive bodies left over from the solar system's formation. Astronomers are still debating just how they formed and whether they're still in their original orbits. One theory suggests that KBOs formed where we now see them, beyond the current orbit of Neptune. According to this scenario, even as the terrestrial and gas giant planets were building up closer to the sun, bodies of ice and rock were accreting into larger planetesimals around the distant edges of the system. In those remote regions, they would have formed slowly. Some would eventually achieve the size of the big objects that we see today, such as Eris or Quaoar. But their formation would at some point have been disrupted by the gravitational influence of the more rapidly growing Neptune. Most of the KBOs then turned on each other, colliding and smashing themselves into small fragments and dust. The solar wind could have blown most of the dusty bits farther out into space, leaving behind just a small percentage of the original mass of the outer solar nebula in the current Kuiper belt.

A second theory holds that the KBOs grew up next to the giant planets but were flung into the solar system's outskirts when those planets, particularly Neptune, moved outward to their current orbits. Or a combination of both theories may account for the present-day Kuiper belt: Some KBOs formed where they are now and others arrived on the Neptune gravity train later.

POPULATIONS

Astronomers have searched just a small fraction of the sky for KBOs so far, but they have found enough of them and tracked a sufficient number of orbits to gain a rough idea of their population and movements.

Most of the icy objects in the Kuiper belt orbit in relatively predictable, circular paths that don't interact with Neptune's orbit and aren't perturbed by the planet's gravitational influence. These are the classical Kuiper belt objects, clustered in a belt between 38 and 48 AU. Although they are a sedate group, many have highly tilted inclinations to the orbital plane, unlike all the planets except Pluto. The classical belt seems to have an oddly sharp edge around 50 AU, as if some outside force had sliced away any objects outside it. A passing star early on may have swept away these outer bodies.

Another population of KBOs at the inner edge of the belt is notable for being trapped into a 2:3 orbital resonance with Neptune. For every two orbits they complete around the sun, Neptune completes three. Pluto is the most prominent example of this kind of resonant body, and in its honor these KBOs are called plutinos. Yet other KBOs, farther out, orbit in a 1:2 resonance with

NASA's New Horizons spacecraft is equipped with instruments for imaging Kuiper belt objects in visible, ultraviolet, and infrared light; for mapping their surfaces; and for studying atmospheres and the solar wind. Its dish antenna allows it to communicate with Earth billions of kilometers away.

Neptune—for every single orbit they complete, Neptune completes two. These are informally known as twotinos. These resonant populations may have been trapped into Neptune-locked orbits when Neptune migrated out from the inner solar system and swept them up in its gravitational skirts.

Scattered disk objects make a rowdier group of KBOs. Their eccentric orbits take them to about 35 AU from the sun at perihelion (closest approach), which brings them within Neptune's influence, but their aphelia (the farthest points in their orbits) range out to 200 AU. Their orbits can be highly inclined as well, veering significantly above and below the orbital plane. It appears that their close encounters with Neptune speed them up and vary their trajectories, rendering them unpredictable.

CENTAURS

In 1977, American astronomer Charles Kowal spotted a sizable object, about 200 kilometers (125 mi) wide, between the orbits of Neptune and Saturn. Named Chiron and originally classified as an asteroid, it was later discovered to have a cloudlike coma and so was reclassified as a comet. In 1992, a second, similar object was discovered and named Pholus. Since then, several dozen such bodies have been discovered. Now they are called centaurs, because they seem to be half asteroid, half comet, just as a centaur is half human, half horse. Their composition and unstable orbits lead astronomers to think that centaurs are recently escaped Kuiper belt objects. In time, they will probably either fall into the sun or be ejected entirely from the solar system.

Centaurs come close enough to Earth to show up in our telescopes, but Kuiper belt objects are very difficult to see. Astronomers don't know much yet about their materials or surfaces. But the remote bodies are far from uniform, ranging in color from a neutral, bright gray to dark red. The redder surfaces may be somewhat older and darkened by interactions with cosmic rays. Brighter ones may be scarred by collisions or covered with young ices of carbon monoxide and methane.

PLUTOIDS AND OTHER POSSIBLE DWARF PLANETS

Among the Kuiper belt objects are some very sizable bodies, big enough to be considered dwarf planets. In addition to Pluto, three have been formally classified by the International Astronomical Union as plutoids: Eris, Makemake, and Haumea. But at least 50 more have been discovered with diameters greater than 400 kilometers (250 mi).

Eris, discovered by astronomers Mike Brown, Chad Trujillo, and David Rabinowitz in 2003, was the upstart planetoid that knocked Pluto off of the planetary lists when it was discovered to be larger than Pluto, with a diameter of about 2,400 kilometers (1,500 mi). Brown had temporarily named the body after the campy television character Xena, the warrior princess. When it came time for an official christening, because it caused so much trouble, the dwarf planet was named after the goddess of strife; its moon, Dysnomia, is named for the spirit of lawlessness. Eris has a highly eccentric orbit stretching as far as 97 AU (where it is now), making it the most distant solar system object ever discovered. Its white, bright surface may be covered with its frozen atmosphere.

Mike Brown and team also discovered Makemake, named for a Polynesian fertility god. The distinctly red planet may be covered with icy methane. Perhaps half the diameter of Pluto, it has an orbit that is not as remote as Eris's, with an average distance of about 46 AU.

Haumea (named for the Hawaiian goddess of childbirth) is bigger than Makemake, perhaps as wide as Pluto in its longest dimension, but shaped like an elongated egg. The odd plutoid rotates very rapidly end over end every four hours or so. Although it's not spherical, and thus would seem to be excluded from the definition of dwarf planet, its regular ovoid shape is deemed to be in hydrostatic equilibrium. Haumea appears to be made of rock glazed with ice, and has two moons, thought to have been created in the same collision that started it spinning.

OTHER POTENTIAL PLUTOIDS

The Kuiper belt and regions beyond contain many sizable bodies that may soon be raised to the rank of dwarf planet. Among them are Quaoar, one of the first big KBOs discovered (below, its discovery image). Quaoar (named for a Tongva Indian creation force) is about the size of Pluto's moon Charon and orbits 43 AU from the sun.

One of the most remarkable of the potential dwarf planets is Sedna, discovered in 2003. Its highly elongated orbit brings it to 76 AU from the sun at its nearest approach and a staggering 900 AU at its most distant point. The entire orbit lies far outside the Kuiper belt, but not as far as the unseen Oort cloud, home of comets.

LONG-HAIRED STARS COMETS

Comets have been regular visitors to our skies throughout history, recognized as harbingers of great deeds and catastrophes. Modern science knows them as icy messengers from the Kuiper belt and Oort cloud.

Comets are among the most beautiful of astronomical sights. The ancient Greeks referred to them as *aster kometes,* or "long-haired stars," from which comes our modern name. Roughly every ten years or so, the glowing coma and tail of a particularly spectacular long-haired star appears in our skies, inspiring awe—and sometimes fear—in watchers on the ground. Many more slingshot around the sun unseen. Thanks to a number of daring spacecraft missions to investigate these visitors, we are learning much more about just what comets are, and what they are not. • Comets belong to the large family of rocky, icy debris left over from the outer solar system's formation. They become recognizable as comets only when they are knocked out of their orbits and sent toward the sun in long, elliptical orbits. As comets approach our star, the volatile ices embedded in their bodies emerge as glowing clouds and tails of gas and dust. • Some comets have relatively small orbits, circling the sun within the bounds of the outer solar system in less than 200 years. These short-period comets usually originate in the Kuiper belt, just beyond the orbit of Neptune. But most comets come from much farther away in the Oort cloud comet reservoir at distances of 50,000 AU or more. The orbits of these long-period comets around the sun can take millions of years, making them essentially onetime visitors to the human race.

SELECTED SHORT-PERIOD COMETS

1P Halley: Orbital period 76.01 years
2P Encke: Orbital period 3.3 years
6P d'Arrest: Orbital period 6.51 years
9P Tempel 1: Orbital period 5.51 years
19P Borrelly: Orbital period 6.88 years
21P Giacobini-Zinner: Orbital period 6.61 years
26P Grigg-Sjkellerup: Orbital period 5.11 years
81P Wild 2: Orbital period 6.39
46P Wirtanen: Orbital period 5.46 years

SKYWATCH

• Sky-watching websites and magazines can alert you to the approach of bright comets. Binoculars and telescopes on a clear night can reveal their details.

AMAZING FACT Dust grains from comets can be found in our air, food, water, and hair.

1456
Pope Calixtus III excommunicates Halley's comet.

1680
Isaac Newton suggests that comets move along predictable orbits.

1758
Johann Georg Palitzsch observes return of Halley's comet as predicted.

1786
Encke's comet, with shortest known orbital period (3.3 years), is discovered.

1994
Comet Hale-Bopp's unusually large nucleus makes it easily visible.

2006
Comet Wild 2 particles captured by Stardust contain organic chemicals.

An artist's view show the dark nucleus of Comet Tempel 1 (opposite).
Comet Hale-Bopp sported a bright, wide dust tail and a blue gas tail (above).

LONG-HAIRED STARS: COMET PATHWAYS

Comets have trailed through our skies throughout history, but observers didn't begin to ascertain their true natures until the days of Tycho, Kepler, and Halley. To most ancient cultures they were harbingers of disaster and portents of great events. A comet appeared overhead the year Julius Caesar was assassinated, an event commemorated by Shakespeare: "When beggars die, there are not comets seen; / The heavens themselves blaze forth the death of princes." The comet later known as Halley blazed forth in 1066, during the Norman conquest of England. Even after comets were known to be just traveling solar system bodies, they continued to alarm spectators. In the early 20th century, the news that comet tails contain small amounts of cyanogen, a poison, panicked many members of the public into barricading themselves in their houses when comet Halley came around in 1910. (Those who stayed out to see the Great Comet suffered no ill effects and saw a great spectacle, to boot.) And a tragic incident in 1997 pointed out that such irrational fears live on. After an amateur astronomer mistakenly identified an ordinary star as a possible UFO following the bright comet Hale-Bopp, 39 members of the religious group Heaven's Gate committed suicide in San Diego, believing that Hale-Bopp was bringing the spaceship that would take them to another level of existence.

More scientifically oriented sky-watchers debated the nature of comets for centuries. Aristotle believed they were fiery atmospheric phenomena. The great astronomer Tycho Brahe, observing a bright comet in 1577, showed that it was too distant to exist in the atmosphere—but Galileo scoffed, suggesting in his *Assayer* that comets were refracted atmospheric phenomena. But as Johannes Kepler and then Isaac Newton worked out the laws of solar system gravitation, it became clear that comets followed those laws and must be orbiting bodies, like the planets.

British astronomer Edmund Halley followed up on these observations with the logical notion that if comets followed regular orbits, a historical survey of past comets might reveal a pattern to their appearances. He soon uncovered such a pattern: Bright comets sighted in 1456, 1531, 1607, and 1682 all had similar retrograde orbits. Halley concluded that these visitations represented the same comet, appearing at 76-year intervals. He further predicted that the comet would return at the end of 1758, though he would not live to see it. On Christmas Day, 1758, his prediction came true. Although Halley could not be said to be its discoverer, exactly, the comet was named after him and remains the most famous of the periodic comets.

EDMUND HALLEY

ASTRONOMER AND GEOGRAPHER

Edmund Halley (1656–1742), son of a wealthy English soapmaker, was a talented mathematician and astronomer by the time he reached his teens. Elected a fellow of London's Royal Society in 1678, he soon became one of the country's leading scientists. With Robert Hooke, Halley deduced much of the mathematics of planetary orbits. He then helped Isaac Newton to publish the *Principia*, which worked out the theory more completely. His 1705 book *A Synopsis of the Astronomy of Comets* showed that four historic comets were actually repeated visits of just one, which came to bear his name. Halley also made major contributions to navigation and meteorology.

ORIGINS AND ORBITS

Comet Halley's 76-year orbit is unusual among comets. Most fly in from a far-distant reservoir of icy bodies, the Oort cloud. This spherical shell of frigid detritus encases the solar system approximately 50,000 to 100,000 AU from the sun. Perhaps a trillion primordial chunks of rock and ice, invisible comets, make up the Oort cloud. Loosely bound by the sun's gravity, the cold, dark cometary bodies are easily disturbed by massive objects passing by, such as nearby stars or molecular clouds. These gravitational disturbances knock a few of the comets out of their orbits and on a path toward the sun. These are long-period comets. Their orbits can take millions of years and will approach the sun from every direction.

Most short-period comets start out much closer to the sun (relatively speaking), in the Kuiper belt (see pp. 172–73). Rousted from their orbital beds by the gravitational influence of the giant planets, a few Kuiper belt objects will fall in toward the sun and adopt new orbits, becoming regularly appearing comets. The short-period comet Encke circles the sun in as little as three years. Comet Halley, with its 76-year period, represents a particular subset of comets, now known as Halley-type comets, that seem to have been caught up in intermediate-length orbits by the giant planets' gravities. Jupiter-family comets

NAMING COMETS

With a few exceptions, comets now are named after their discoverers (last name only), such as comet Bennett. Two simultaneous discoveries may yield a hyphenated name (comet Ikeya-Seki, for instance), and a team can ask either that one person's name or a team acronym be given to the comet. Short-period comets are preceded by a *P* and a number giving the order of discovery in that category: thus 1P/Halley. Other comets are preceded by a *C*. Finally, each comet is labeled with the year of its discovery, followed by letters and numerals indicating the half month of the discovery announcement. So comet C/1997 B3 (SOHO) is a non-short-period comet discovered by the SOHO team, the third found in the second half of January 1997.

have periods under 20 years. Shoemaker-Levy 9, which broke apart and smashed into Jupiter in 1994, was one.

Some comets live dangerously and skim right past the sun or even plunge directly in to be destroyed. These are known as "sungrazers." Astronomers on the ground and instruments aboard the SOHO satellite have observed hundreds of these self-immolating travelers. Curiously, almost all seem to be fragments of a single big comet that began to break apart near the sun thousands of years ago. They're known as Kreutz sungrazers, after the German astronomer who first theorized that they had a common origin.

GREAT COMETS

Before the naming of comets was standardized, the brightest, most spectacular apparitions were typically known as Great Comets, such as the Great Comet of 1618 or the Great Comet of 1811. Astronomers have now listed certain criteria that a Great Comet should display. It should have a large nucleus and coma; a large active surface area; reach perihelion near the sun; pass close to Earth; and provide a long enough period of good viewing for observers on Earth. Halley is a Great Comet, although its relatively dim appearance in 1986 disappointed viewers because it reached perihelion, the closest point to the sun, on the opposite side from Earth. Great Comets Hyakutake (1996) and Hale-Bopp (1997) made up for Halley's underperformance. Hyakutake came within 0.1 AU of Earth. At its closest approach, it was easily visible to the naked eye, with a tail that stretched one-quarter of the way across the sky. Hale-Bopp was visible for months, even in the late afternoon. Unfortunately for those with fond memories of the two Great Comets, their long orbital periods ensure that they won't pay a return visit for 2,400 years, in the case of Hale-Bopp, or for 72,000 years for Hyakutake.

Other presumed Great Comets have turned out to be great duds. The most notorious recent example was Comet Kohoutek. Unusually bright when first spotted, the comet promised to put on a fine display when it came close to Earth in 1973–74. It was highly touted by astronomers but turned out to be dim and unremarkable; the very name Kohoutek became a synonym for an overhyped fizzle. Scientists later learned its initial brightness was due to a layer of frost that fluoresced and then evaporated as it approached the sun. Astronomers have since learned more about what might make a Great Comet. They've also become more cautious about predicting one.

New comets are discovered all the time, however. Most won't be spotted until they come close enough to the sun to develop their bright comae and tails, perhaps weeks before they make a close approach to Earth.

The extraordinarily long dust tail of Great Comet McNaught curves across the Chilean sky in January 2007. Visible to observers in the Southern Hemisphere even in daylight, McNaught was the brightest comet in decades, outshining even Sirius, the sky's brightest star.

LONG-HAIRED STARS: INSIDE A COMET

An illustration of the solar nebula depicts the dark, dusty outer edges where comets are forming. Far from the growing sun, these chunks of rocky ice have changed little since the solar system's infancy and represent a valuable record of primordial conditions.

Before they begin their journey toward the sun, comets are dark, inert objects, most of them invisible to our telescopes. A few kilometers wide, on average, they seem to be a mixture of ices and rocky dust. In 1950, pioneering astronomer Fred Whipple proposed the "dirty snowball" theory of comet composition, which suggested that a comet is made mainly of water ice laced with mineral dust. Certainly comets are lightweight and icy. Analysis of their gases shows not only water ice but also methane, carbon dioxide, ammonia, and other chemical ices common in the outer solar system. But the "dirty" part applies as well: Comets are sooty black because their chemicals react over the course of time to irradiation and accumulation of interstellar dust. Close-up, they look more like rough, lumpy boulders than Terran snowballs.

ANATOMY OF A COMET

After a comet has been nudged from its distant bed to fall toward the sun, it remains inactive until it gets close enough to the sun for its ices to begin to sublimate—in other words, to turn directly into gas without becoming liquid first. This starts to happen somewhere around the orbit of Pluto and continues as the comet warms, with different ices sublimating at different temperatures. Now the comet blooms. A solid, frozen nucleus remains, but around it grows a coma, a huge, misty cloud of gas and dust. As the comet approaches the sun, the coma brightens and expands for thousands of kilometers. Some comae are as wide as

1,000,000 kilometers (620,000 mi) across. Even larger than the coma is the comet's hydrogen envelope, which is a sphere of gas surrounding it for millions of kilometers.

Comets that don't approach the sun closely may remain roundish and bushy, never developing the trademark tail. But those that venture within about 1.5 AU usually form a long tail—in fact, they often form two tails, which can stretch as far as 1 AU. The dust tail, typically wide and ivory colored, consists of a thin mist of silicate and carbon particles blown off the nucleus by escaping gas. The dust particles follow their own orbits as they are released and so the dust tail often fans out and curves. The ion tail, straight and blue, is made of fluorescing gases ionized by solar radiation. This tail is not pulled by the sun's gravity, but is blown straight away from the sun by the solar wind and its magnetic field lines. As comets circle the sun and head back toward the outer reaches, their tails always point away from the sun (the dust tail curved and the ion tail straight). This means that the comet's tail does not trail behind it on the return trip but leads the way as the comet departs the inner solar system.

THE DEATH OF COMETS

The brightest comets, as seen by observers on Earth, are those that come closest to our planet, of course, and those with the largest and most active nuclei. Only a small percentage of a comet's icy surface will sublimate as it approaches the sun; younger comets with more exposed ice will burn away more gas than older ones. Comet Hale-Bopp, for instance, is relatively young and fairly active. But even these comets are fading, orbit by orbit, as about one-thousandth of their mass vanishes with each turn about the sun. Halley may dwindle to nothing within another 1,000 orbits. Some comets also flare up and disintegrate as they approach the sun or are pulled apart by planetary gravity, such as Shoemaker-Levy 9.

We can see cometary pieces in our skies throughout the year as meteors. The fragments spread out over the comet's orbit and remain there, some intersecting Earth's path. As the Earth plows through these swarms of debris, the particles enter the atmosphere and flare up as meteors, or "shooting stars." We encounter some of these cometary dust clouds at the same time every year. The densest are the source of spectacular annual meteor showers, named for the constellation in which they appear to originate. For instance, the Perseid meteor shower, which reaches its peak in August, is caused by fragments of comet Swift-Tuttle.

OTHER STARS' CASTOFF COMETS

Scientists believe that the solar system originally held many more comets than it does today. Disturbances from the outer planets would have flung some of them far into space, away from the sun's gravitational grasp. Presumably these orphaned ice balls would sail on through the millennia until captured by another body's gravity.

But do other stars' comets visit our solar system as well? Observations of some stars, such as Beta Pictoris, reveal dust clouds that may represent young solar systems with the capacity to form comets. And a recent cometary visitor, comet Machholz 1, revealed an odd carbon-poor chemical composition. One possible explanation: It formed around another star, was ejected, and has been captured by our sun.

MISSIONS TO COMETS

Beginning with Halley's visit in 1986, various space agencies launched a fleet of spacecraft with the purpose of intercepting that comet as well as others. Among them was the ESA's Giotto, which came as close as 605 kilometers (376 mi) to comet Halley. Although the spacecraft was shotblasted with grit as it approached the comet, it sent back the first close pictures of Halley's nucleus to earth-bound viewers. The famous comet was revealed as an irregular black object, some 15 by 10 kilometers (9 by 6 mi) around, surrounded by a mist of dust. Glowing jets of gas and dust sprayed from its sunlit side, propelling the comet into a 53-hour rotation.

The European Space Agency's Giotto spacecraft studied comet Halley at close range during the comet's 1986 swing past the sun.

NASA's Stardust mission flew within 236 kilometers (147 mi) of the youthful Comet Wild 2 in 2004 and snagged samples of its dust within an ingenious foamlike gel. The comet looked surprisingly cohesive and rocklike upon close examination. Even more surprising were the dust samples returned to Earth. Among other things, the samples confirmed that comets carried organic compounds, lending support to the notion that they could have brought these compounds to the infant Earth. And instead of holding materials found only in the outer solar system, they also contained silicates that probably formed under high heat in the inner solar system. The findings suggest that materials from the inner and outer solar system mixed it up at some point in the system's formation. Even more dramatic was NASA's Deep Impact mission. In 2005, the flyby spacecraft released an impactor which smashed into comet Tempel-1 at 37,000 kilometers an hour (23,000 mph). The impact kicked up powdery dust and indicated that the comet was a light, snowlike compilation of ice and rock. Future missions include the up-close-and-personal visit of the Rosetta spacecraft in 2014, when the craft will drop a lander on comet 67P/Churyumov-Gerasimenko.

FOR MORE ON COMETS, GO TO WWW.SKYANDTELESCOPE.COM/OBSERVING/OBJECTS/COMETS

CHAPTER 7

ARE WE ALONE?

ARE WE ALONE?

So many of the questions in astronomy circle back to this one: Are we alone in the universe? Does life on Earth represent a unique accident, or is it an inevitable consequence of natural law? These queries drive much of space exploration, from probes to the outer moons to searches for extrasolar planets.

Many scientists invoke the Copernican principle to support their belief that life will be found elsewhere. Named for the astronomer who proved that Earth does not hold a special place in the universe, it is sometimes known as the assumption of mediocrity: Our planet formed from common elements around a typical star, and there is no reason why many others like it should not exist. Opposing this view is the so-called Fermi paradox. Supposedly, the great physicist responded to a discussion about intelligent alien life by asking simply: So where is everybody? So far we haven't detected, nor have we been contacted by, other life-forms: Perhaps life on Earth is singular, arising from a series of rare coincidences of chemistry and environment.

In the absence of evidence one way or another, most scientists prefer to proceed as though life is possible elsewhere. It's logical to use terrestrial life as a template for extraterrestrial life until we find out otherwise.

The search for life is conducted with a series of questions and assumptions, the first of which is: Just what is life, anyway? Many definitions exist, and none is universally accepted. Most incorporate the following basics: life reproduces; life takes in and metabolizes nutrients; life evolves. On Earth, life requires three things of its environment: liquid water, energy, and nutrients. Liquid water is the universal solvent needed for biochemical reactions. Energy (such as sunlight or

1584
Monk Giordano Bruno suggests other planets, suns exist. Accused of heresy.

1953
Miller, Urey create organic molecules simulating conditions on early Earth.

1960
First search for extraterrestrial life, Project Ozma, seeks radio signals at stars.

1974
Arecibo Observatory sends broadcast into space detailing life as we understand it.

1961
The Drake equation (estimate of extraterrestrial civilizations) is proposed.

To find a truly green world, with life thriving on its surface (as in this illustration), we will have to depart the solar system. Planets and moons in our own system may turn out to hold life, but if so it will most likely be found in hidden stashes of subsurface water.

heat) propels the chemical processes of metabolism. And nutrients are the raw materials from which energy is extracted.

Just a few elements form the basis for almost all life on Earth, and of these the most important is carbon. Carbon atoms, with relatively few electrons, easily form compounds with other elements. Carbon is the backbone for amino acids, DNA and RNA, proteins, and nutrients such as carbohydrates, fats, and oils. Carbon-containing compounds are known as organic; those without carbon, inorganic.

It's possible, naturally, that life on other planets uses different building blocks. Silicon, for instance, is a common element with similarities to carbon; perhaps silicon creatures arose on other worlds. Ammonia, another prevalent compound, might work as a liquid solvent in place of water. But neither of these substances is as flexible and useful as carbon and water, so until we learn otherwise, carbon and water will make up our working model.

Another question basic to the search for life is: How did life arise on Earth? If we assume that it evolved from simple organic chemicals, where did those chemicals come from? Two theories approach this question from different

1977
SETI researches detect "Wow" signal. Now presumed interference.

1977
Scientists discover new forms of life near undersea hydrothermal vents.

1982
Steven Spielberg backs first privately funded search for extraterrestrials.

1992
NASA starts ten-year SETI program. Congress cuts funding one year later.

1993–2004
SETI Institute runs Project Phoenix, an ambitious and sensitive SETI search.

directions. The first suggests that conditions on the primordial Earth triggered chemical reactions that produced organic compounds from simpler, inorganic ones. In the classic Miller-Urey experiment of 1953, researchers at the University of Chicago energized a primordial soup and atmosphere of water, methane, carbon dioxide, and ammonia with an electrical charge resembling lightning. Within a few days, the liquid was enriched by amino acids, the building blocks of proteins. Some of their assumptions about the early atmosphere later proved to be wrong, but the experiment has been successfully repeated with updated elements. Although there is a big step between amino acids and living organisms, the experiment was proof that organic molecules could have formed on the early Earth.

Another theory gaining more acceptance recently, known as panspermia, suggests that the chemicals for life reached Earth from outer space. Scientists have found a surprising number of complex organic molecules in space; spectral analysis of interstellar molecular clouds reveals organic chemicals, such as sugars, and organic compounds have also been recovered from comets and meteorites. It's not unlikely—maybe even probable—that comets and other icy space debris smacking into Earth in its early years brought ready-made organics to our surface.

No matter where it began, life spread in Earth's oceans and onto land, almost all of it then and now in the shape of microbes. We now know that life is capable of thriving in a remarkably wide variety of terrestrial habitats. The discovery of microbes living in intense heat and cold, in darkness, in acidic, salty, and radioactive environments, has greatly encouraged astrobiologists who can now expand their search parameters for life on other worlds.

Biologists call these life-forms extremophiles, and the most heralded are those that live near deep-ocean hydrothermal vents. In 1977 biologists first discovered colonies of planets and animals thriving in the deep sea where cracks in the ocean crust allow superheated, mineral-laden water to emerge. Among the life-forms discovered happily basking in 315°C (600°F) water are microbes that produce energy through chemical processes, or chemosynthesis, rather than using sunlight as in photosynthesis. They represent a third domain of life, now called archaea, that may be older in their lineage than bacteria, plants, and animals.

STANLEY MILLER

CREATING LIFE IN THE LAB

Stanley Miller (1930–2007) was the American chemist who made life in a bottle—or at least, organic molecules in a bottle. As a graduate student at the University of Chicago in 1953, Miller (above) studied under physicist Harold C. Urey. When Urey suggested in a lecture that someone should try to re-create the formation of organic chemicals in the conditions of primordial Earth, Miller volunteered. The successful experiment was instantly famous. Urey went on to win the Nobel Prize for his discovery of deuterium. Miller spent most of his career at the University of California, San Diego.

At the other end of the temperature spectrum, bacteria have been recovered from deep within the ice just above Lake Vostok, a huge and buried Antarctic lake, as well as within permafrost soil. Microbes can exist in a dormant state in deserts for decades, springing back to life with the first touch of water. In Yellowstone National Park, microbes live within pools with the pH of battery acid; in African soda lakes, different creatures thrive in extreme alkalinity. Microorganisms have even been found three kilometers (two mi) down in the depths of a South African gold mine, metabolizing the chemical by-products of the radioactive breakdown of water.

The fact that life can thrive at these extremes makes the sub-ice oceans of the moon Europa or the possible groundwater reserves on Mars much more promising targets for life-search missions. It also expands our parameters for habitable extrasolar planets.

Finding life in any form on another world would be an enormously important event. Finding intelligent life would profoundly change our future. The odds of discovering other civilizations in our Milky Way are almost impossible to calculate, because so many factors are unknown, but in 1961 astronomer Frank Drake drew up an equation that attempted to do just that. It looks like this:

$$N=R^* \times f_p \times n_e \times f_l \times f_i \times f_e \times L.$$

Spelled out, it means: The number of civilizations in the galaxy = the number of stars forming each year x the fraction of stars with planetary systems x the average number of habitable planets in the system x the fraction of habitable planets on which life arises x the fraction of lifeforms that develop intelligence x the fraction of intelligent lifeforms that develop civilization x the average lifetime of such a civilization. In general, the longer that we estimate a civilization will survive, the greater the number that will coexist in the galaxy, and the more time they will have to get messages from one world to another.

In hopes of detecting just such messages, organizations like the SETI (Search for Extraterrestrial Intelligence) Institute monitor the heavens for radio signals using radio telescopes such as the Allen Telescope Array. SETI's public program, SETI@home, links home computer users into a network that allows all participants to analyze signals in search of significant messages. Someday a user at home, in the tradition of amateur astronomers, may become the first human to hear from an alien intelligence.

FRANK DRAKE
HUNTING FOR OTHER INTELLIGENCES

Frank Drake (1930–) is an American radio astronomer and pioneer in the search for extraterrestrial intelligence. Born in Chicago, Drake went to graduate school at Harvard and developed an interest in radio astronomy. In 1960, he carried out Project Ozma, the first systematic search for signals from other intelligences. He and astronomer Carl Sagan also collaborated in an attempt to detect such signals using the Arecibo radio telescope. Drake became dean of natural sciences at the University of California, Santa Cruz, in 1984 and was the first chairman of the board of trustees of the SETI Institute.

THE SEARCH FOR WATER
PLANETS & MOONS

Once considered a haven for advanced civilizations, then dismissed as a lifeless desert, Mars is again a favorite in the search for life due to clear signs of ancient floods. But joining it in the life hunt are the distant moons Titan, Europa, and Enceladus.

The search for life on other planets has traditionally been the search for liquid water—in particular, liquid water on a planet's surface. The best place to look for this in any solar system is traditionally within the habitable zone. This is the distance from the sun at which temperatures allow surface water to exist, and it's a relatively narrow range. For our solar system, the habitable zone is typically held to span a region from just outside Venus's orbit to just inside that of Mars, with Earth securely in the middle. A broader interpretation of the zone might include Mars. Although the other terrestrial planets are worth consideration, Mars remains our best nearby bet for finding evidence of past or current life. • But if it's water you want, the surprising moons of Saturn and Jupiter might provide more than is contained in all Earth's oceans. Europa and Enceladus show clear signs of sub-ice seas, kept liquid by heat from gravitational stresses. Meanwhile, Titan's orange atmosphere is rich with organic chemicals and its surface is dotted with lakes of liquid methane. Though not in the habitable zone, the outer solar system may prove to have habitable environments after all.

COMMON ELEMENTS, BY MASS, IN BACTERIA:
- **Oxygen:** 68%
- **Carbon:** 15%
- **Hydrogen:** 10%
- **Nitrogen:** 4%
- **Phosphorus:** 0.8%
- **Potassium:** 0.45%
- **Sodium:** 0.4%
- **Sulfur:** 0.3%
- **Calcium:** 0.25%

SKYWATCH

* You can see Saturn's satellite Titan, the only moon with an atmosphere, with good binoculars or a small telescope, circling Saturn every 16 days.

AMAZING FACT The bacterium *Deinococcus radiodurans* can survive radiation at more than 1,000 times the lethal dose for humans.

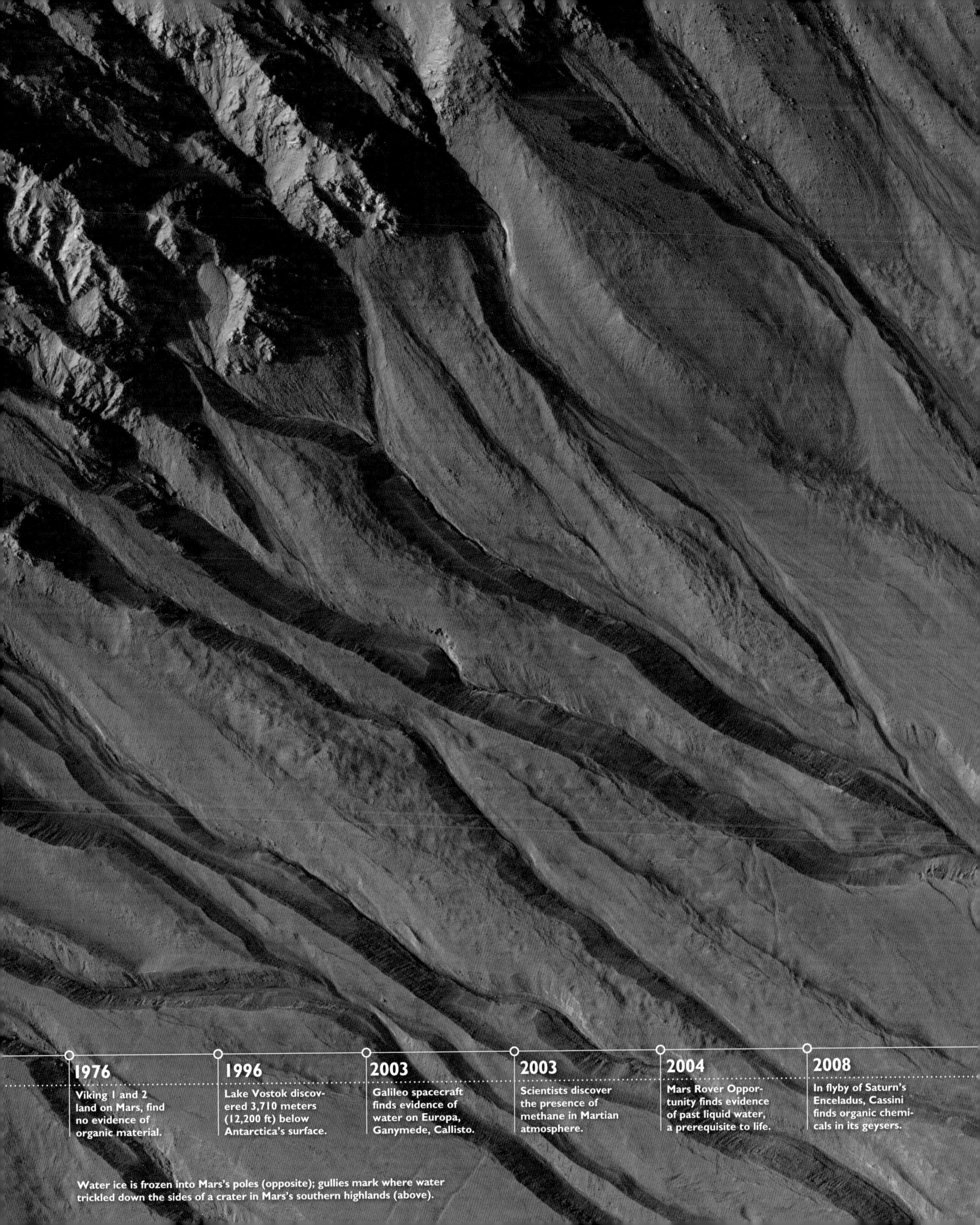

1976
Viking 1 and 2 land on Mars, find no evidence of organic material.

1996
Lake Vostok discovered 3,710 meters (12,200 ft) below Antarctica's surface.

2003
Galileo spacecraft finds evidence of water on Europa, Ganymede, Callisto.

2003
Scientists discover the presence of methane in Martian atmosphere.

2004
Mars Rover Opportunity finds evidence of past liquid water, a prerequisite to life.

2008
In flyby of Saturn's Enceladus, Cassini finds organic chemicals in its geysers.

Water ice is frozen into Mars's poles (opposite); gullies mark where water trickled down the sides of a crater in Mars's southern highlands (above).

PLANETS AND MOONS: NEAR NEIGHBORS

Looking for life on the solar system's terrestrial planets (and moons) means looking for Earthlike environments in general and liquid water in particular. "Earthlike" in this context means any habitat that falls within the extreme limits for life. It should exist within the temperature range that keeps water liquid. It should offer energy to power growth and metabolism. It should provide nutrients to feed life-forms. It should be free of sterilizing radiation or deadly toxins.

MOON, MERCURY, AND VENUS

We can almost certainly rule out Mercury and our moon as suitable habitats for life. Airless, they show no evidence of liquid water now or in the past; temperatures on their surfaces are either too hot or too cold to maintain water now, although they may harbor ice in shadowy regions on their poles. Ultraviolet radiation, cosmic rays, and the solar wind bombard their unprotected surfaces.

Venus is a little more intriguing. At present it's an inferno. Temperatures over 450°C (840°F), grinding atmospheric pressures, an arid surface and sulfuric acid clouds would instantly and brutally kill any Terran creature who found itself on the planet's surface. However, it is possible that Venus had liquid oceans perhaps four billion years ago, when the runaway greenhouse effect had not boiled away its water with blistering temperatures. Conceivably, microbial life might have developed in those oceans then.

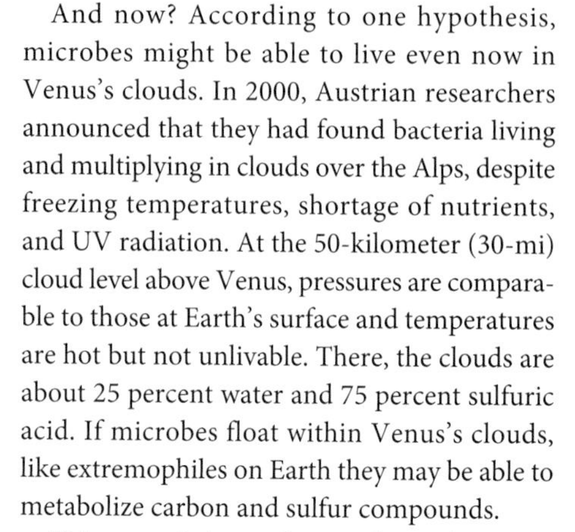

And now? According to one hypothesis, microbes might be able to live even now in Venus's clouds. In 2000, Austrian researchers announced that they had found bacteria living and multiplying in clouds over the Alps, despite freezing temperatures, shortage of nutrients, and UV radiation. At the 50-kilometer (30-mi) cloud level above Venus, pressures are comparable to those at Earth's surface and temperatures are hot but not unlivable. There, the clouds are about 25 percent water and 75 percent sulfuric acid. If microbes float within Venus's clouds, like extremophiles on Earth they may be able to metabolize carbon and sulfur compounds.

This scenario is purely speculative, however. No evidence yet exists for Venusian cloud life, and the odds for finding any life on Venus still remain dauntingly low.

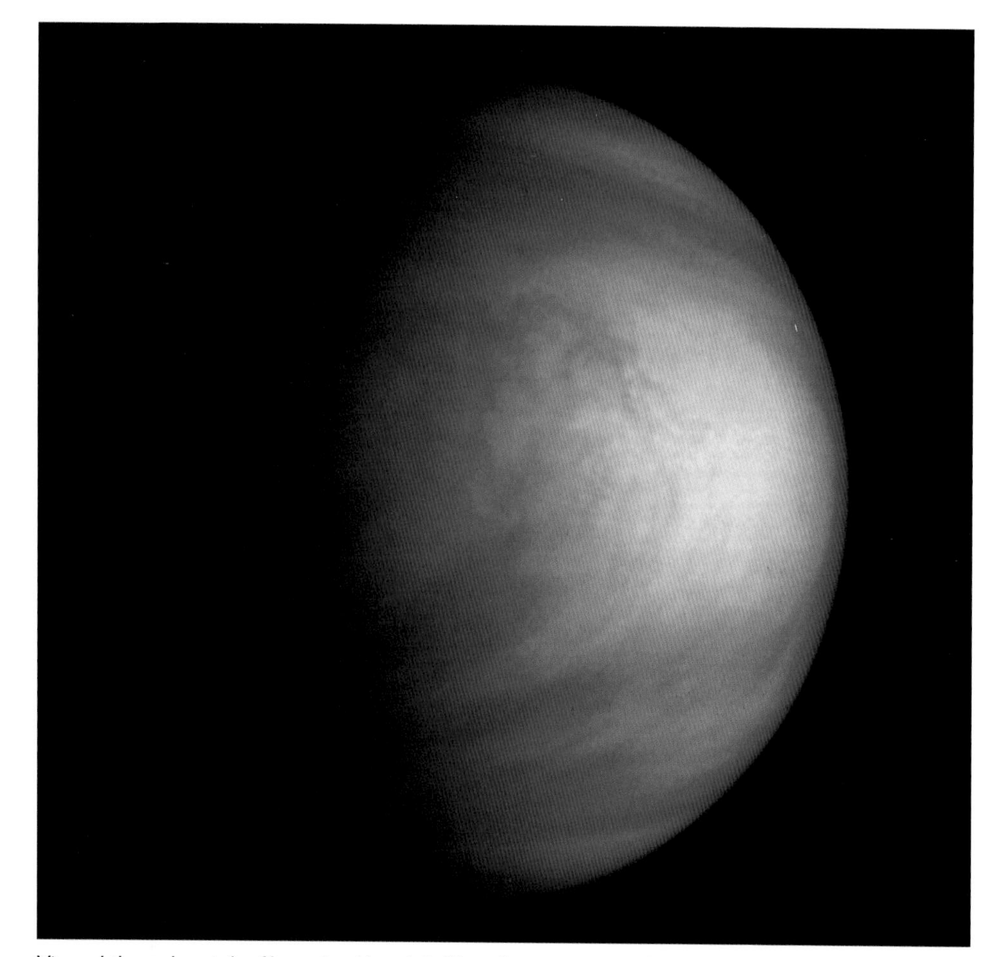

Viewed through a violet filter, cloud bands in Venus's upper atmosphere stand out. Although the clouds are mostly sulfuric acid, some also contain water vapor and may provide a habitat for bacterial life.

MARS

Mars is another matter. Of all the planets (aside from Earth), Mars is the most likely candidate for past or present life. Orbiting at the edge of the solar system's habitable zone, at first glance Mars seems a hostile environment. Its atmosphere is thin and dominated by carbon dioxide. Radiation bombards its unprotected surface. Temperatures average well below zero on any scale.

But Mars scores high on life criterion number one: the presence of water. Observations from orbit and soil excavation by Mars rovers have confirmed that large quantities of water ice exist at the poles and just under the surface of the soil, at least in some locations. Snow falls in Mars's upper atmosphere. Furthermore, it is almost certain that Mars contained liquid rivers and oceans in the past, when its climate was warmer and its atmosphere was thicker. Features found in various regions of the planet include dry gullies, floodplains, and riverbeds. Outflow channels in Elysium Plains may have funneled floodwater with 100 times the volume of the Mississippi. Teardrop-shaped erosion features are probably islands carved out by flowing waters. Some of these channels may date to catastrophic flooding during Mars's early, warmer Noachian epoch, which ended about 3.5 billion years ago. However, others appear to have been carved out as recently as 20 million years ago.

Certain kinds of minerals also tell us that the surface has known liquid water. In 2004, the Mars rover Opportunity, patrolling Meridiani Planum, sent back pictures of marble-size spheres stuck in rock outcroppings. Nicknamed blueberries by scientists on Earth, the little balls contained the iron-bearing mineral hematite, which forms in a watery environment. The Mars Express spacecraft detected clay minerals in other regions, also associated with watery terrains. And images from the orbiting Mars Odyssey spacecraft reveal chloride salt deposits—like table salt—in low-lying regions of the southern hemisphere. The salty patches may be the remains of saltwater lakes.

These findings make it increasingly likely that early Mars, at least, had an environment favorable for the development of life. This is a far step from saying that life actually arose there. If anything grew on young Mars, and if it followed the evolutionary process we know on Earth, it would likely have evolved to no more than microbial status before the climate cooled, the atmosphere thinned, and the water dried up.

What about current life? It would be difficult, but not impossible, for life as we know it to exist somewhere on Mars, most probably beneath the surface. Microbes, possibly dormant, might exist in frozen Martian soil. On Earth, bacteria have been recovered from polar permafrost millions of years old. Microbes have also been extracted from Lake Vostok, a frigid lake trapped beneath the ice of Antarctica. Liquid water might even be found in pockets beneath the Martian surface. Warmed by subsurface volcanic activity, the water might host small creatures like the extremophiles found near Earth's hydrothermal vents.

Proof of ancient Martian life could come in the form of fossils. To date, only one highly controversial claim has been made for the existence of Martian fossils. A 2-kilogram (4.75-lb) meteorite discovered in Antarctica in 1984 proved to be an arrival from Mars and a rare example of the early Martian surface from about four billion years ago. Some scientists believe that tiny tubular structures in the ALH 84001 meteorite are fossils of ancient Martian bacteria. Others are sure that the little formations are inorganic. Arguments continue, although the bulk of scientific opinion weighs against an organic explanation.

If life currently exists on Mars but lurks unseen beneath the surface, we might still be able to detect the chemical signatures of a living metabolism. That's what NASA's Viking landers attempted to do in their famous biology experiments of 1976. Three experiments tested samples of the Martian soil for the presence of carbon dioxide take-up (as in photosynthesis), gases released by digestion, or consumption of organic chemicals. The first results were positive, but disappointed scientists eventually realized that other, nonorganic reactions could just as easily have brought about the same results. The consensus now is that the Viking experiments failed to show the presence of life.

But hope does not die easily. In 2009, researchers confirmed that plumes of methane had been detected in the Martian atmosphere several years earlier. Geologic processes involving volcanic heat can produce the gas; it may have been trapped in pockets below the surface and released in bursts. Methane is also a by-product of animal metabolism: Terran cows are notorious emitters of the pungent gas. Conceivably, microbes living in warm, watery pockets deep in the Martian soil could have metabolized carbon dioxide to produce the gas. We have no way of knowing the methane's source without the aid of future Mars missions.

The Martian northern polar ice cap contains water ice overlaid by carbon dioxide ice. The areas around the poles are also believed to hold large amounts of subsurface ice.

CROSS CONTAMINATION

In 1969, the crew of Apollo 12 retrieved a camera from a Surveyor 3 spacecraft that had been stranded on the moon for three years. When the camera was examined back on Earth, it proved to have little hitchhikers: streptococcus bacteria that had traveled with the camera from Earth, survived in the airless, frigid, radiation-blasted environment for years, and returned home unharmed. The hardy bacteria were a living example of the hazards of cross contamination. The possibility of accidentally carrying Terran organisms to another world (forward contamination), or of bringing alien organisms back home (back contamination), is a very real one. Space agencies have procedures for sterilizing and quarantining spacecraft and samples in hopes of avoiding this. And in fact the Outer Space Treaty of 1967 requires such procedures: Article 9 of the treaty requires signatories to "conduct exploration of [celestial bodies] so as to avoid their harmful contamination and also adverse changes in the environment of the Earth resulting from the introduction of extraterrestrial matter."

FUTURE MISSIONS TO MARS

The turn of the 21st century brought a flotilla of spacecraft to Mars and moved the search for water into high gear. Martian life—more specifically, Martian habitability—is the goal of NASA's next mission, the Mars Science Laboratory, scheduled for launch in the fall of 2011. That rover will parachute to the Martian surface in 2012 with a package of scientific instruments designed to test the atmosphere and soils for the chemicals of life. NASA hopes to follow in the following decade with a Mars Sample Return Mission. Robotic devices in that endeavor would not only collect samples of the soil, rocks, and atmosphere, but would blast off the Martian surface with them and return to Earth—an ambitious task that has been accomplished only by manned moon missions.

PLANETS AND MOONS: ICY OCEANS

The mysterious moon Titan, with its hydrocarbon haze, has long been cited as the best possibility for hosting extraterrestrial life aside from Mars. Until recently, the rest of the outer solar system was dismissed as too cold and, quite simply, just too alien to support Earthlike life. But with the turn of the century came discoveries that changed that mind-set. In 2000, the Galileo spacecraft, measuring the magnetic field of Jupiter's moon Europa, brought in evidence that beneath its icy surface lay an enormous liquid ocean. And in 2005, Saturn's moon Enceladus sprayed the Cassini orbiter with towering geysers from its south pole. Combined with the knowledge that living organisms exist in sub-ice lakes and deep-ocean vents on Earth, these discoveries brought the icy moons to the top of the list for future exploration.

The outer solar system, although deeply cold, is not actually as inhospitable as it might seem. For one thing, the distant reaches of the system are much richer in carbon, necessary for organic molecules, than the inner planets and moons. Even comets have been shown to carry organic compounds. Although little energy reaches the worlds far flung from the sun, other forces may instead heat those worlds, including gravitational compression and the tidal flexing exerted by the influence of a giant planet on its moon. Liquid water is the other necessary ingredient, and with the discovery of subsurface oceans that requirement seems to be met.

What about the big planets themselves? So far, they look less welcoming than the moons. Lacking solid surfaces, they don't provide sheltered regions in—or on—which life could develop. They do possess some water vapor in their clouds, but in general it exists at very high pressures and in dim, windy environments. In 1976, astronomers Carl Sagan and Edwin Salpeter published a paper in which they speculated about the kinds of life-forms that might evolve in Jupiter's atmosphere. Like Earth's oceans, they suggested, the Jovian atmosphere might support a vast menagerie of floating multicellular life. Gas-filled balloonlike creatures could possibly fill different niches in a cloud ecosystem as "sinkers," "floaters," and "hunters." Our instruments have not peered deeply in Jupiter's clouds, but for now this charming and imaginative scenario is not supported by any evidence and seems unlikely. Other, colder gas giants appear to be even more inhospitable.

TITAN

Much more intriguing is Saturn's big moon Titan. For more than a century, astronomers have known that Titan has an atmosphere. The dense, smoggy, orange haze that surrounds the moon resembles ancient Earth's primordial air, a mixture of nitrogen and methane, with other carbon compounds mixed in. Energy from the sun's ultraviolet light and energetic particles spun off from Saturn split the molecules, which recombine into heavy hydrocarbons. Some hang about to thicken the atmospheric smog, while others fall to Titan's surface as liquid rain, filling seasonal lakes with ethane and methane.

Without any form of liquid water on its surface, life on Titan might have developed in liquid methane, subsisting on hydrocarbons for nourishment. Ice composed of water and frozen hard as rock does form part of its crust, so conceivably giant impacts might occasionally melt that ice into life-producing pools. Some evidence also points to a subsurface ocean, possibly of water and ammonia. Sandwiched between two layers of ice, the ocean would be frigid and dark, a difficult but not impossible home for life.

Some scientists have speculated that floating life might exist in the windy bands and belts of Jupiter's atmosphere. The narrow-angle camera aboard NASA's Cassini spacecraft pulled in this close-up view of one cloudy strip, including the Great Red Spot and stormy ovals and vortices.

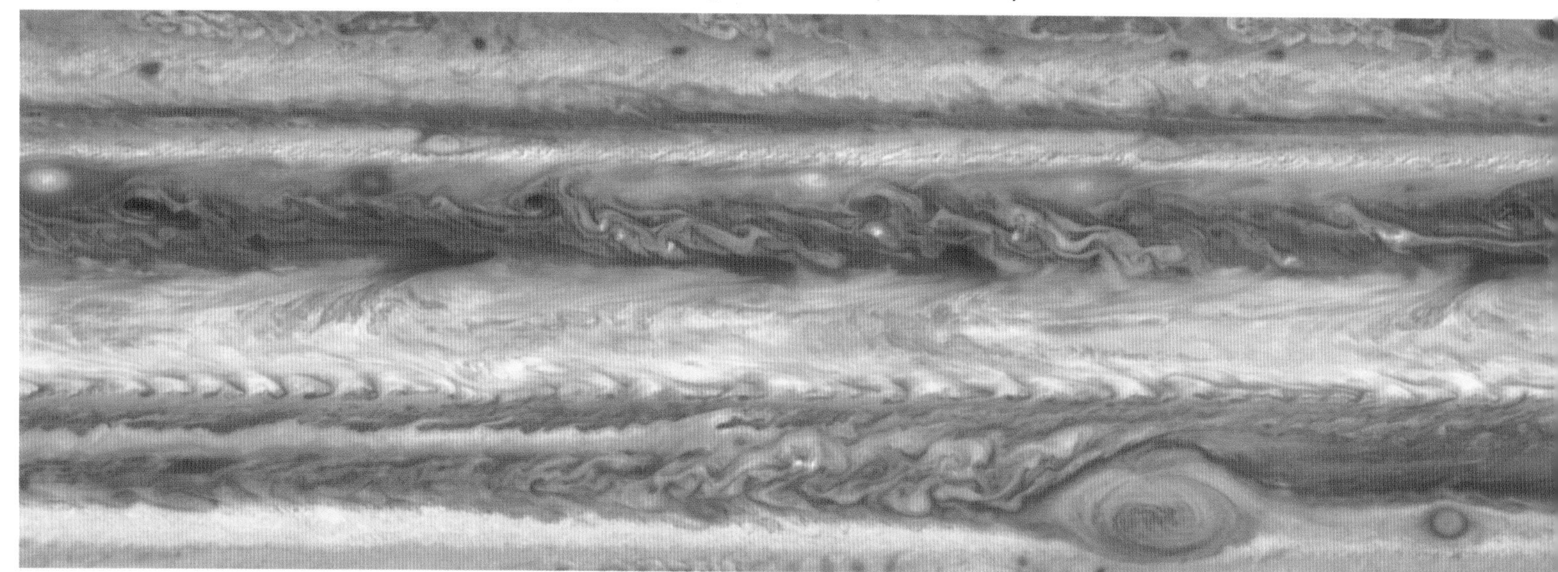

EUROPA

At first glance, Jupiter's moon Europa is a much less promising candidate for a life search than atmospheric Titan. Surface temperatures can drop as low as -200°C (-328°F). The airless satellite, slightly smaller than Earth's moon, is completely encased in water ice. Dark fractures crisscross the icy surface. Because Europa's surface seems fairly young, with few craters, and because its cracks resemble those that form in glass or ice on Earth, scientists believe that the moon's ice is an easily fractured veneer, perhaps a few miles thick, over a vast ocean. Darker terrain may represent areas where water has welled up from underneath.

If Europa does have liquid water, it could be 160 kilometers (100 mi) deep, giving it a greater volume than Earth's ocean in a moon-encompassing layer between the icy crust and a solid rocky mantle. The satellite's magnetic field suggests that the liquid conducts electricity and is therefore salty. What likely keeps it liquid is the gravitational tugging and flexing the moon undergoes as its moves toward and away from Jupiter's great mass in the course of its orbit. If Europa is tilted on its axis—unknown right now—the tidal forces might propel waves within the locked-in seas, and this motion would release heat as well. If the ice is not thick, its cracks might also allow in solar energy and radiation to warm the waters. Possibly, photosynthetic life may exist on the dim sunlight filtering through the cracks into Europa's ocean. Or hydrothermal vents, like those in Earth's oceans, might sustain underwater colonies living off hydrogen and methane.

The strong likelihood of oceans on Europa has moved the satellite up in line for future missions to the outer planets. NASA and the ESA are currently planning a joint mission to Jupiter's four big satellites, launching in 2020. The NASA orbiter would settle in to study Europa, while the ESA probe would home in on Ganymede.

ENCELADUS

Like Europa, Saturn's moon Enceladus is an icy globe squeezed in the gravitational fist of its parent planet. The small satellite is even smoother than Europa, a chilly, reflective ball decorated with four big, bluish tiger-stripe fissures at its south pole. Remarkably, ice particles and vapor surround the little moon. As the Cassini orbiter flew past Enceladus in 2005, it found out why: Geysers of water, carbon dioxide, and organic molecules spray periodically out of the tiger stripes as far as 160 kilometers (100 mi). Scientists guess that, like Europa, Enceladus is sufficiently heated by tidal forces to maintain a liquid ocean beneath its ice. A second fly-through of the geysers in 2008 revealed even richer complex organic chemicals. Moreover, the plumes in space registered temperatures of -93°C (-135°F); too cold for skinny-dipping, but surprisingly warm for such a distant object. Given liquid water, an energy source, and organic nutrients, such an environment might sustain chemosynthetic life, such as that found in Earth's oceans.

SUBMARINES IN SPACE

To explore the waters of Jupiter's moon Europa, we may need to send a robotic submarine probe beneath the ice (below). Toward this end, several groups on Earth are already developing prototype subs. The Woods Hole Oceanographic Institution is testing Sentry, a man-size autonomous submarine whose chemical sensors can detect underwater methane. NASA's submersible probe ENDURANCE has successfully explored the terrain under icy Terran lakes. But both of these subs are too massive for a spacecraft to carry to Europa, and they don't have the tools to drill through its thick ice. A Swedish technology institute is tackling the size issue with a prototype mini-submersible the dimensions of a bottle rocket, equipped with miniaturized sensors. Next up: a way to punch through Europa's icy surface and survive the pressures, temperatures, and possible acidity of its water.

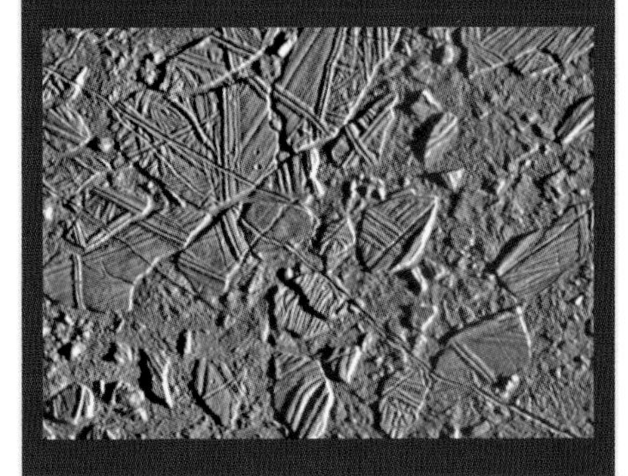

OTHER STARS, OTHER WORLDS
EXOPLANETS

Only recently did we have proof that our planetary system was not unique. Astronomers have found hundreds of extrasolar planets—planets orbiting other stars—in our galactic neighborhood. The hunt is on for one showing the hallmarks of life.

Finding planets around other stars has long been one of astronomy's main goals. As early as the 19th century astronomers claimed that orbital wobbles in the binary star system 70 Ophiuchi meant that planets were tugging on the stars. Similar claims were made for Barnard's star in the 1950s. Unfortunately for planet seekers, neither assertion held up under closer examination. • When the first true extrasolar planet was discovered, it was stranger than anyone might have imagined. In 1991, radio astronomer Alexander Wolszczan and his colleagues used differences in the timing of radio signals from the star PSR B1257 to show that it must be orbited by a planetary companion; eventually three small planets were detected. What made this discovery so odd was that PSR B1257 is a pulsar: a tiny, dense, rapidly rotating supernova remnant. Though the planets are Earthlike in their masses, the constant high-energy radiation that bombards them would keep them sterile. • The bizarre system set the stage for exoplanet discoveries in the next decades. Each new planet seemed odder than the last, overturning our expectations of what a solar system should look like.

SELECTED EXOPLANETS
Pulsar planet: PSR 1257 b
Gas giant: GQ Lup b
Hot Jupiter: 51 Pegasi b
Hot Neptune: HD 219828 b
Terrestrial: Gliese 581 d
Stars with multiple planets: 55 Cancri
Closest system: Epsilon Eridani (10.5 ly)
Around giant star: HD 47536 c
Around binary stars: gamma Cephei b

SKYWATCH

* Nearby star system Epsilon Eridani can be seen by the naked eye in the constellation Eridanus, just to the west of the bright star Rigel.

AMAZING FACT The "lava world" COROT-Exo-7b completes one orbit around its star in 20 hours.

1959	1994	1995	2004	2006	2009
Physicists Giuseppe Cocconi, Philip Morrison first search for extraterrestrial life.	Alexander Wolszczan discovers first extrasolar planets, orbiting a pulsar.	Michel Mayor, Didier Queloz discover first exoplanet to orbit a sun similar to ours.	NASA discovers new type of exoplanets, much smaller than hot Jupiters.	French COROT spacecraft launched; able to detect terrestrial exoplanets.	COROT spacecraft discovers smallest exoplanet yet found.

Artists' conceptions of two notable exoplanets depict the terrestrial Gliese 876 d (opposite) and "hot Jupiter" 51 Pegasi b (above).

EXOPLANETS: SEARCH FOR ANOTHER EARTH

The detection of pulsar planets opened the door to the discovery of hundreds of new extrasolar planets, although for a long time none of them were seen, in the sense of appearing in a direct visual image. This isn't surprising when you consider that a star is a billion times brighter than a planet. Looking for extrasolar worlds is like looking for a firefly in the sun; the star's glare completely overpowers the planet's dim reflected light.

HOW TO FIND A PLANET

Astronomers have devised a number of ingenious ways to get around this barrier. In 1995, Swiss scientists Michel Mayor and Didier Queloz used the most successful of the techniques to date to find planets orbiting the normal (nonpulsar) star 51 Pegasi. Their method, radial velocity, measures changes in the velocity of a star toward or away from us—its wobble, in other words. We can detect this using the Doppler effect. Light from a star moving toward us will be compressed and shifted into the blue end of the spectrum. When a star moves away, wavelengths from its light are expanded and shift toward the red. Gravity from a sufficiently massive planet would pull on its parent star, making it wobble about the system's center of gravity: Jupiter does this to our own sun. The bigger the planet and the closer it orbits to its star, the bigger the shift. The 51 Pegasi planet was typical of many that would come to be detected this way. At least half the size of Jupiter, it orbits perilously close to its sun at only 0.05 AU, so it must suffer temperatures over 1000°K (727°C/1340°F). This was the first of many "hot Jupiters" discovered, planets that confounded astronomers' expectations of what a solar system should look like.

A technique similar to radial velocity has also found exoplanets. Astrometric measurement picks up stars that wobble back and forth against the background sky; this is the method that detected the first pulsar planets. The measurements involved are painstakingly fine, so the best tools are space-based telescopes, like Hubble, or others that collate observations from multiple locations.

The transit method uses a technique that would have been familiar to Kepler and Galileo. If a planet passes directly between its star and an observer, it will reduce that light just a little. In our own solar system, we can see the disks of Venus and Mercury as they cross the sun, but in distant stars we have to measure the slight dimming of the star's light. The biggest planets will block the most light, so this method favors the discovery of more hot Jupiters. The planet spotted crossing the star OGLE-TR-56, for instance, has a mass almost as great as Jupiter's and whirls around its sun every 1.2 days at a blistering range of 0.02 AU.

Planet Earth is the ideal in our search for livable planets.

The opposite effect is seen when astronomers use the gravitational microlensing method. Drawing on Einstein's observation that a massive object will bend the light that passes near it, observers look for a brief brightening in stellar light that tells them that a planet is passing in front of the star and acting as a lens. In 2004, two teams of astronomers announced the discovery of the first exoplanet found using that technique, a Jupiter-mass world orbiting a red dwarf star about 10,000 light-years distant.

But for sheer satisfaction, nothing beats being able to see a planet with our own eyes (or at least in an image captured by vastly more sensitive telescopic eyes). At stellar distances, planets appear to be buried within their blindingly bright parent star. Among the earliest and most encouraging extrasolar images, seen in 1984, were those of the star Beta Pictoris. When the central sun was covered up, two warm disks of dusts were revealed surrounding it, complete with infalling comets—probably the planetary nebulae around the young star. And now the world's best telescopes are beginning to see actual exoplanets for the first time. In 2008, astronomers published the first visual images of an exoplanet, a small but distinct speck of light orbiting the star Fomalhaut, 25 light-years away. For years, observers had suspected that such a planet existed, because Fomalhaut was known to have a disk of dust around it whose inside edge was cut off sharply, implying the influence of a nearby planet. The Hubble Space Telescope finally captured the image of the Jupiter-size planet, which orbits its sun every 872 years.

A PLANETARY MENAGERIE

Thanks to this useful toolbox of detection methods, astronomers have found more than 340 extrasolar planets and are still going strong. Most of these extrasolar systems are a peculiar bunch, surprisingly different from our own. To some extent, this is due to the techniques we're using, which strongly favor finding massive,

The COROT space telescope was launched by the European Space Agency in 2006. It searches for gas giant and terrestrial-type exoplanets by measuring changes in starlight caused by transiting planets.

close-in planets. Earthlike planets are far more difficult to detect. However, the strange dimensions of these other systems are making us wonder if our own planetary family is so ordinary after all.

Most exoplanets found so far are massive, on the scale of Saturn, Jupiter, or larger. Many of those orbit far closer to the parent star than our gas giants do to our sun, earning them the name hot Jupiters. Infrared telescopes have measured temperatures on at least one at an incendiary 2038°C (3700°F). Now astronomers are beginning to find "hot Neptunes": less massive stars on the order of Neptune, such as Gliese 436 b, whose eccentric orbit practically skims its star. Far more weighty planets are also coming to light. Planet COROT-exo-3b, for instance, is 20 times as massive as Jupiter and may be a failed star.

Dust clouds and asteroid belts have also come to light. Even before the first exoplanet was discovered, astronomers had found a dust belt around the star Beta Pictoris; the star Epsilon Eridani, 10.5 light-years from us, has not just one, but two asteroid belts and an icy outer ring of debris. Astronomers have also found multi-planet systems, planets circling giant stars, and planets circling white dwarf stars.

TERRA NOVA

As exciting as all these new discoveries are, observers will not be satisfied until they find the holy grail of exoplanets: another Earth. Increasingly sensitive search methods have found increasingly smaller and more Earth-like planets in recent years, although none has the hallmarks of a terrestrial world. The planet would have to be located in that star's habitable zone, where temperatures would allow for liquid water; it should provide a hard, rocky surface; and preferably it would possess an atmosphere containing water vapor, carbon dioxide, and ozone (a form of oxygen). The Hubble Space Telescope has already found one planet, HD 189733 b, with both methane and carbon dioxide in its atmosphere, though the Jupiter-size world is too hot for life.

The COROT space telescope has found a planet (COROT-exo-7b) that is Earthlike in size, perhaps twice the mass of our planet. Unfortunately it is so close to its star that even if it is a rocky planet, its surface would be molten lava. Astronomers at the European Southern Observatory in Chile spotted three interesting planets around the star HD 40307, some 42 light-years away. Between four and ten times the mass of Earth, they've been dubbed super-Earths. However, these are scorchingly hot as well.

The odds are good that many planetary systems exist with Earthlike worlds—we just haven't had the instruments that could find them. This is starting to change. France's COROT spacecraft, launched in 2006, has already begun to detect transiting exoplanets. NASA launched its Kepler mission in 2009. Designed to survey and monitor 100,000 sun-like stars, it is expected to detect smaller planets using the transit method.

NASA's planned Terrestrial Planet Finder and the ESA's Darwin will use space-based telescopes to scrutinize the spectra of distant planets for the chemicals of life. If and when we do find them, we won't know at first if they're signs of extraterrestrial algae or bug-eyed aliens or life-forms profoundly different from our own. But we will know that we are not alone.

WATERWORLDS

If planets can migrate from orbit to orbit during the early days of a planetary system—and studies of our own solar system suggest that they do—then our search for extrasolar planets may turn up "waterworlds." Ice giant planets, such as Uranus and Neptune, should typically form in the outer regions of a solar nebula, where temperatures allow ice to remain solid and the early sun's radiation doesn't boil water away. If ice planets then moved in toward the sun, solar heat would melt that ice and the ice giant would become an ocean planet. Such a world would be an excellent candidate for a search for life.

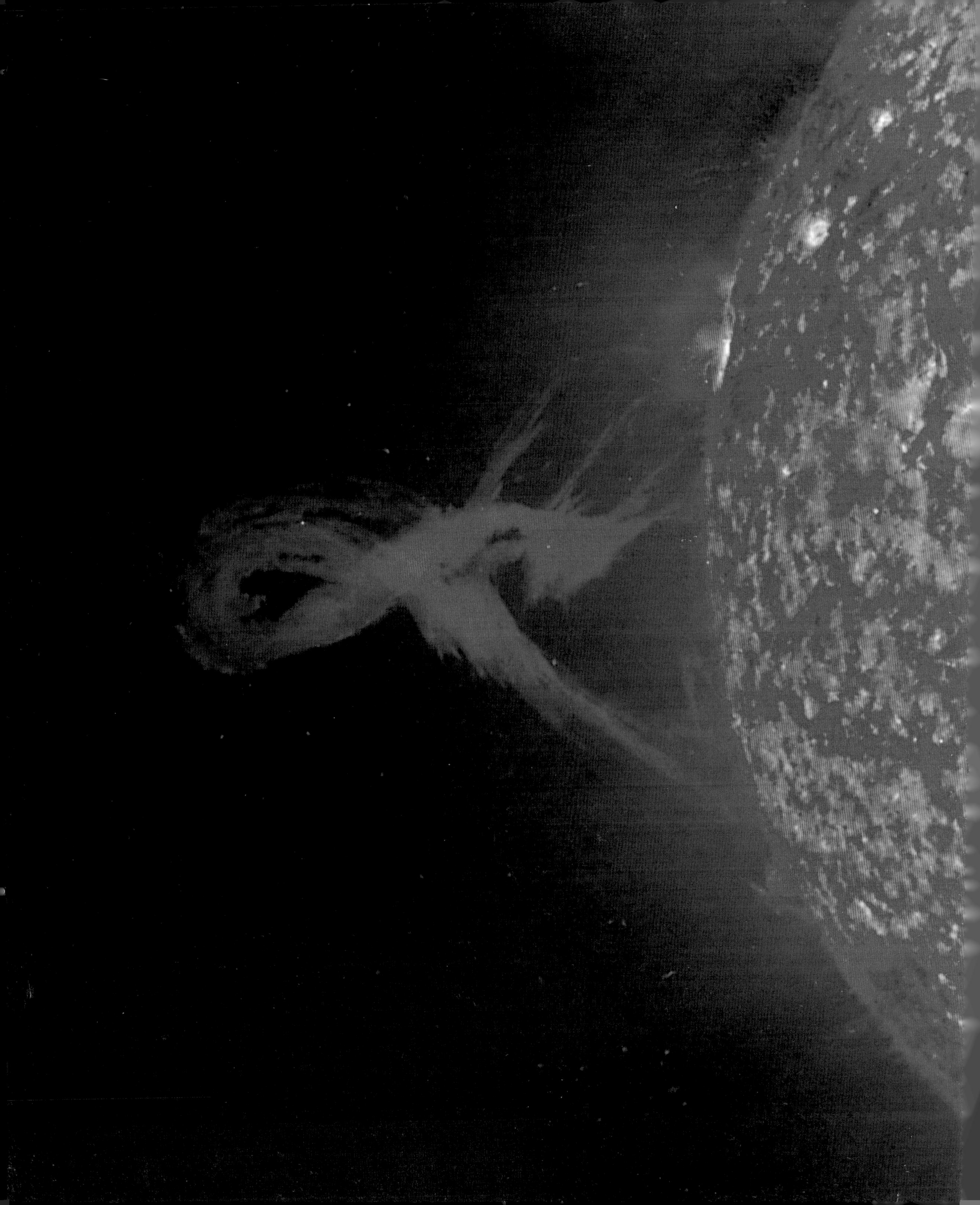

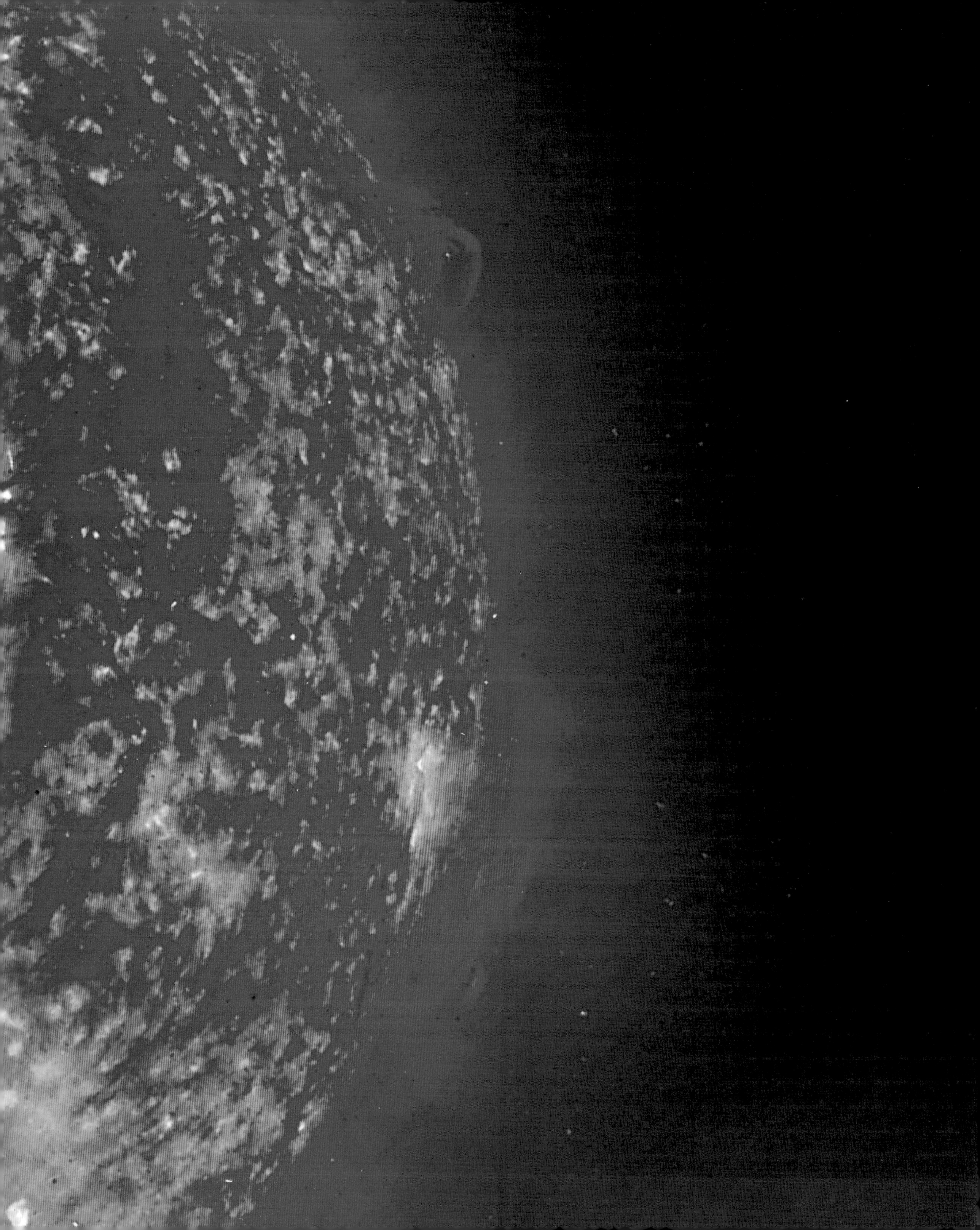

ACKNOWLEDGMENTS

MANY THANKS TO THE DEDICATED FOLKS at National Geographic Books who helped to make this book a reality. In particular, I'd like to thank Barbara Brownell Grogan, whose encouragement and organizational skills launched this project; editor Susan Straight, who shepherded the many lagging pieces of the process with unfailing patience and cheer; photo editor Kevin Eans, who found so many worthy images; art director Sam Serebin and designer Cameron Zotter, who took craters and clouds and made them more than just beautiful.

A big thank-you as well to my consultant, astronomy guru, and longtime friend Robert Burnham, whose guidance from outline to final copy was invaluable. Any errors herein are entirely my own.

And last, but never least, my love and gratitude to my husband, Jim Tybout, for his support and uncomplaining sufferance of my pre-dawn writing sessions. My love to my sons Andy and Sam as well, whose talents are always an inspiration to me.

ABOUT THE AUTHORS

PATRICIA DANIELS is a writer and editor specializing in science and history. Among her books are *The National Geographic Encyclopedia of Space; Pocket Guide to the Constellations; Body: the Complete Human;* and *The National Geographic Almanac of World History.* She was also an editor for the *Voyage Through the Universe* astronomy series. Daniels has written for magazines and newspapers about sky-watching and space propulsion and has helped to create science exhibits at Philadelphia's Franklin Institute. She currently lives in State College, Pennsylvania, with her husband, a college professor, and two sons.

ROBERT BURNHAM is a science writer for the Mars Space Flight Facility in Arizona State University's School of Earth and Space Exploration. In addition to news releases and media relations activities, he writes feature articles on Mars geology and geoscience for the Mars Space Flight Facility's THEMIS (Thermal Emission Imaging System) instrument website (themis.asu.edu). He is also program manager for ASU's Mars Education Program *(marsed.asu.edu).*

Before joining ASU in 2005, he was an editor at *Astronomy* magazine, holding various positions starting in 1978; he was editor in chief of the magazine from 1992 to 1996. His books include *Exploring the Starry Sky* (2004), *Reader's Digest Children's Atlas of the Universe* (2000), and *Great Comets* (2000).

GLOSSARY

Total solar eclipse

ALBEDO: the measure of an object's shininess; the ratio of reflected light to light falling on the surface.

ALLEN TELESCOPE ARRAY (ATA): a radio interferometer dedicated to astronomical research and the search for extraterrestrial life, located at the Hat Creek Radio Observatory, north of San Francisco.

ANNULAR ECLIPSE: an eclipse, during which the angular size of the moon is too small to completely block out the entire sun, leaving a ring of light around the dark disk of the moon (see also *eclipse*).

ANNULUS: the ring of visible light seen around the moon during an annular eclipse.

APHELION: the point in an object's orbit when it is farthest from the sun.

APOLLO ASTEROID: one of a group of asteroids whose orbits cross that of Earth.

ARCHAEA: primitive organisms, composed of one or more prokaryotic cells, typically found in extreme environments.

ARECIBO RADIO TELESCOPE: the world's most sensitive radio telescope, possessing a 305-meter-long (1,000-ft) reflecting surface or radio mirror. The telescope is run by the National Astronomy and Ionosphere Center in Puerto Rico and enables the study of astronomy, the planets, and space and atmospheric sciences.

ASTEROID: small, rocky body revolving around the sun. The majority of asteroids orbit between Mars and Jupiter in the asteroid belt.

ASTEROID BELT: the region between Mars and Jupiter in which most asteroids orbit.

ASTHENOSPHERE: the layer of mantle located below the lithosphere in rocky planets and in which low resistance allows for convective flow that drives plate tectonics.

ASTROBIOLOGY: the study of life in the universe and the search for extraterrestrial life.

ASTRONOMICAL UNIT (AU): the average distance from Earth to the sun, 149,597,870 kilometers (92,955,730 mi). Employed as a standard unit of astronomical measurement.

ASTRONOMY: from the Greek *astronomos*, meaning "star-arranging"; the study of the chemical and physical properties of objects outside of Earth's atmosphere, including celestial objects, space, and the universe as a whole.

ATMOSPHERE: the gaseous layer surrounding a planet or star that is retained due to the object's gravitational field.

AURORA: The light displays seen over the south and north magnetic poles, created by incoming charged particles channeled through Earth's magnetic field.

AXIAL TILT: the angle at which the axis of a planet is tilted relative to the plane on which it orbits the sun.

BAND: striated zones in a giant planet's upper atmosphere due to underlying latitudinal cloud currents; broken down into belts and zones.

Barringer (Meteor) Crater

BARYCENTER: the center of mass for a system of orbiting objects.

BASALT: hard, black volcanic rock.

BIG BANG MODEL: the widely accepted theory of the origin and evolution of the universe that states that the universe began in an infinitely compact state and is now expanding.

BINARY STARS: two stars in orbit around a common center of gravity. Often made up of one massive star with a smaller partner; also called a double star system.

BROWN DWARF: an object intermediate in size beteen a planet and a star and lacking sufficient mass to ignite fusion in its core.

CARBONACEOUS METEOROID: a rare type of meteoroid with large amounts of carbon.

CASSINI DIVISION: the large, visible gap between the A and B sections of Saturn's rings, discovered by French-Italian astronomer Giovanni Cassini.

CELESTIAL SPHERE: the imaginary surface surrounding Earth upon which all celestial objects appear and which is used to describe the positions of objects in the sky.

CENTAUR: small objects in orbit between Jupiter and Neptune that have characteristics of both asteroids and comets.

CHARGE-COUPLED DEVICE (CCD): an electronic device for detecting photons electronically using a silicon chip covered with light-sensitive material.

CHARGED PARTICLE: an ion or a subatomic particle, such as a proton or electron, with a positive or negative electric charge.

CHEMOSYNTHESIS: the process by which certain microbes produce energy, using chemical reactions rather than photosynthesis.

CHROMOSPHERE: meaning "color sphere"; the region of a star's atmosphere between its corona and photosphere, named for the red light visible during a total eclipse.

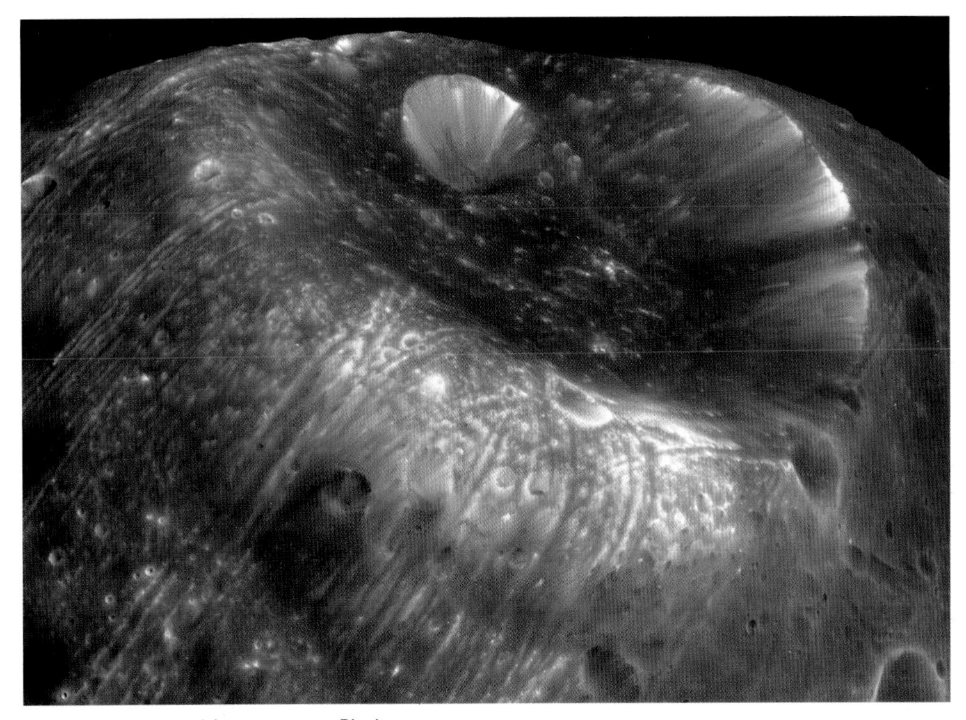

Stickney crater on Martian moon Phobos

COMA: the thin cloud of gas and dust that forms around the nucleus of a comet when the comet's ice sublimates near the sun.

COMET: A small, icy body orbiting the sun; short-period comets originate in the Kuiper belt, while long-period comets originate in the Oort cloud.

CONSTELLATION: one of 88 shapes assigned to an arbitrary arrangement of stars that serves as a tool to identify the stars.

CONVECTION: the circular movement of a liquid caused by the transfer of heat and changes in density.

CONVECTION ZONE: a layer in the sun between the radiative zone and the photosphere where convection currents transport energy outward.

CORE: the dense, innermost part of a planet, large moon, asteroid, or star.

CORONA: the sun's outermost atmosphere; the visible light from the sun, seen around the disk of the moon as it covers the sun during a total eclipse.

CORONAL MASS EJECTIONS: bubble shaped bursts of ionized gas from the solar corona; they are often associated with large flares.

COSMOLOGY: the study of the origins and structure of the universe.

CRUST: the solid, outermost layer of a planet or moon.

DAMOCLOID: a type of asteroid with an extremely elliptical, inclined orbit resembling a comet, but without a comet's coma or tail.

DENSITY: the ratio of the mass of an object to its volume, commonly measured in grams per cubic centimeter (g/cm^3) or grams per liter (g/L).

DEUTERIUM: a heavy form of hydrogen whose nucleus contains a proton and a neutron.

DIFFERENTIAL ROTATION: the rotation of an object whose different parts spin at different rates.

DOPPLER SHIFT: the observed change in the wavelength of light or sound caused by motion in the observer, the object, or both.

DRAKE EQUATION: the equation proposed by astrophysicist Frank Drake in 1961 that attempts to determine the number of technically advanced civilizations in the Milky Way galaxy.

DWARF PLANETS: bodies orbiting the sun that have achieved hydrostatic equilibrium but have not cleared their orbital lanes.

DWARF STARS: all nongiant stars, including the sun; the category includes white dwarfs and red dwarfs.

ECLIPSE: the partial or total covering of one celestial body by another, such as the sun by the moon.

ECLIPTIC: the apparent path of the sun on the celestial sphere; also, the plane of the Earth's orbit around the sun.

EJECTA: material that is ejected, such as lava from a volcano.

ELLIPSE: an elongated circle; closed orbits, such as those of the planets, are elliptical.

EMISSION LINE: a bright line in the spectrum caused by the emission of light at a specific wavelength from a luminous object.

EPICYCLE: in the geocentric solar model, the circle inside which a planet orbits, whose center follows another, larger circle.

EXOPLANET: extrasolar planet; a planet orbiting a star other than the sun.

EXTREMOPHILE: an organism that lives in extreme environmental conditions.

FALSE COLORS: colors in an image that do not exist in the actual object, but are used to enhance, contrast, or distinguish features or details in the image

FLYBY: a space mission in which a spacecraft passes close to its intended target, but does not land on or enter into orbit around it.

FUSION: a nuclear reaction that occurs light nuclei join to produce a heavier nucleus; nuclear fusion.

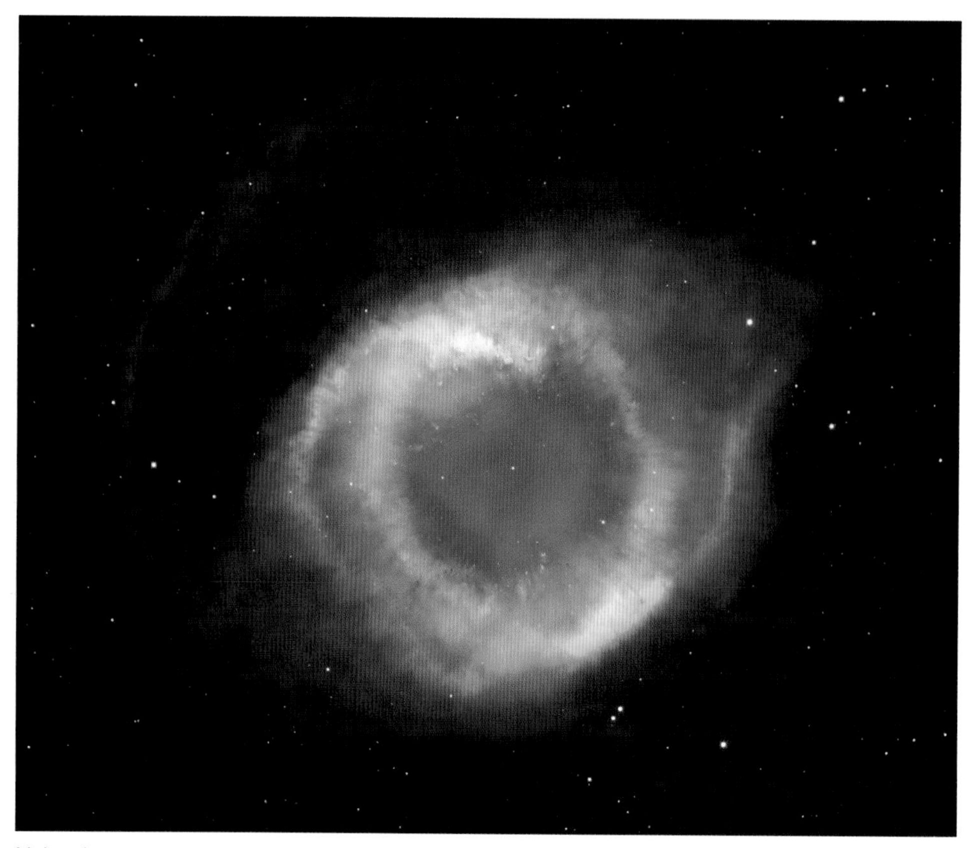

Helix planetary nebula

GALAXY: a large collection of stars bound by mutual gravitational attraction.

GALILEAN MOONS: Jupiter's four largest moons, Io, Europa, Ganymede, and Callisto, named after the Italian astronomer Galileo Galilei, who is credited with their discovery; also known as the Galilean satellites.

GAMMA RAY: the shortest wavelength and the most energetic type of electromagnetic radiation.

GAS GIANT: a large, low-density planet composed primarily of hydrogen and helium; in our solar system Jupiter, Saturn, Uranus, and Neptune are considered gas giants, with Uranus and Neptune also called ice giants.

GEOCENTRIC: measured from or referring to Earth as the center; related to Earth's center.

GRAVITY: the attractive force that every massive object in the universe has on every other massive object. Also called gravitation.

GRAVITY TRACTOR: a hypothetical spacecraft that can use gravitational attraction to change the velocity of another object.

GREAT DARK SPOT: a massive storm system in Neptune's atmosphere that was similar in latitude and shape to the Great Red Spot on Jupiter.

GREAT RED SPOT: a large storm system in Jupiter's atmosphere that is at least 300 years old and characterized by a large, red, swirling pattern.

GREENHOUSE EFFECT: the trapping of solar radiation by gasses in a planetary atmosphere.

HABITABLE ZONE: the orbital distance within which liquid water exists on a planet's surface; Earth exists inside the habitable zone of our solar system.

HELIOCENTRIC: relating to sun as the center; measured from or referring to the center of the sun.

HELIOPAUSE: the border to interstellar space; the boundary between interstellar winds and solar winds, where pressure is equal.

HELIOSEISMOLOGY: the study of the inner structure of the sun through the analysis of internal waves.

HELIOSHEATH: in the heliosphere, a region of transition before the heliopause where solar winds slow to subsonic speeds.

HELIOSPHERE: the region of space directly influenced by the sun, its gravitation, and solar wind.

HERTZSPRUNG-RUSSELL (H-R) DIAGRAM: the graph on which the luminosity and temperatures of stars are plotted.

HOT JUPITER: a Jupiter-size extrasolar planet orbiting within 0.05 AU of its parent star.

HYDROCARBON: an organic compound containing only carbon and hydrogen.

HYDROSTATIC EQUILIBRIUM: a state in which the outward pressure of hot gas or liquid is in balance with the inward pressure of gravity.

IMPACT CRATER: the depression formed as a result of a high-speed solid hitting a rigid surface, such as the circular craters on the surface of the moon.

Methane ice worm

INFRARED: having a wavelength longer than the red end of the visible light spectrum and shorter than that of microwaves.

INNER PLANETS: the four small, rocky planets nearest to the sun; also called terrestrial planets.

INORGANIC: belonging to the class of compounds that do not contain carbon in their chemical makeup.

INTERFEROMETER: an instrument that uses the interference of waves collected from two or more different vantage points to determine precise measurements.

INTERSTELLAR SPACE: the region between stars; also called interstellar medium.

INTERSTELLAR WIND: the dust and gas between stars, which pushes gently against the heliopause.

INVERSE-SQUARE LAW: the law stating that physical quantities, such as light, decrease in proportion to the square of the distance from their source.

ION: an electrically charged atom; one that has gained or lost an electron.

IONIZE: to convert into an ion or ions.

KIRCHHOFF'S LAWS: the rules governing the formation of spectra by solids, liquids, or gases under pressure.

KIRKWOOD GAPS: gaps in the asteroid belt caused by the gravitational pull of the large planets, particularly Jupiter.

KUIPER BELT: (pronounced KI-per) a disk-shaped region located 30 to 50 AU from the sun that is filled with icy bodies; the source of most short-period comets.

KUIPER BELT OBJECT: (KBO) a small icy object orbiting in the Kuiper belt.

LAGRANGIAN POINT: a gravitational point in a two-body system at which a smaller third body can remain in equilibrium.

LIGHT-YEAR: the distance light travels in a vacuum in one year; equivalent to about 9.5 trillion kilometers (6 trillion mi).

LITHOSPHERE: Earth's outer, solid part, made up of the crust and the uppermost mantle.

Barred spiral galaxy NGC 1300

LIQUID METALLIC HYDROGEN: an intensely compressed, electrically conductive phase of hydrogen, found inside Jupiter and Saturn.

LONG-PERIOD COMET: a comet that takes more than 200 years—and perhaps as many as several million years—to complete its orbit and that is believed to originate in the Oort cloud.

LUNAR: from Latin *luna;* belonging to or related to the moon.

MAGNETIC FIELD: the region in which magnetic and electric forces act.

MAGNETOSPHERE: the region around a celestial body that is dominated by that object's magnetic field and associated charged particles.

MAGNITUDE: the measurement of the brightness of a celestial object; lower numbers

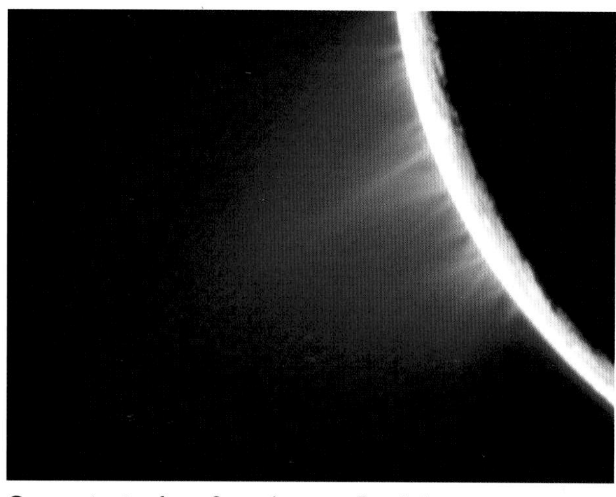
Geysers jetting from Saturn's moon Enceladus

indicate greater brightness; apparent magnitude is the object's brightness seen from Earth, absolute magnitude, the object's intrinsic brightness.

MANTLE: the layer between the crust and the core of a planet or moon.

MARE: from Latin meaning "sea"; any of the dark, flat, low-elevation regions on the surface of the moon.

MASCON: a contraction of "mass concentration"; a region of higher gravitational attraction detected on Earth's moon.

MAUNDER MINIMUM: the time of solar inactivity between 1646 and 1715 associated with an extended cold spell.

METEOR: the streak of light produced by an object entering Earth's atmosphere and traveling so fast that friction causes it to ignite.

METEORITE: a meteoroid that has reached the ground.

METEOROID: a small fragment of dust or rock in interplanetary space.

METONIC CYCLE: period of 19 years, after which the moon returns to its original position relative to the sun.

MICROLENSING: a method to find extrasolar planets using massive objects as gravitational lenses to bend the light from more distant stars.

MILKY WAY: the large spiral galaxy that contains over 200 billion stars and is home to the sun.

MOON: a body in orbit around a planet, dwarf planet, or asteroid; a natural satellite.

NEBULA: a cloud of gas and dust in interstellar space.

NEAR EARTH OBJECT (NEO): an asteroid or short-period comet whose orbit brings it into the Earth's orbital neighborhood where it could collide with our planet.

NEUTRINO: chargeless elementary particles with a very small mass that travel at the speed of light but interact very weakly with matter.

NUCLEUS: in a comet, the solid body of ice and dust that forms the center of its head.

OBSERVATORY: a facility that houses telescopes and other such research instruments.

OORT CLOUD: an enormous spherical cloud of icy objects surrounding our solar system at distances up to 100,000 AU; thought to be the region where comets originate.

OPPOSITION: the position occupied by a planet or comet in the opposite part of the sky from the sun.

ORBIT: the path that is taken by a body moving in a gravitation field, as in the path of a planet around the sun or a moon around a planet.

ORBITER: a spacecraft in orbit around a planet or moon.

OUTGASSING: the expulsion of gas from the crust of a planet or moon, thought to be the means by which secondary atmospheres are formed.

PARTIAL ECLIPSE: an eclipse in which the more distant of the bodies is only partially blocked.

PENUMBRA: the shadow cast over an area by Earth or the moon during a partial eclipse; the lighter, outer edge of a sunspot, surrounding the umbra or dark center.

Arecibo radio telescope

PERIHELION: the closest point to the sun in a celestial body's orbit.

PHOTOSPHERE: the visible surface of the sun or other stars; the outer layer of the sun.

PERIOD: the time it takes for a celestial body to make a complete rotation around its axis or to make a complete orbit.

PHASE TRANSITION: the change of matter from one state to another, such as from liquid to gas.

PHOTON: an individual packet of electromagnetic radiation; a particle of light.

PHOTOSPHERE: the part of the sun that emits light.

PLANET: a body in orbit around the sun, possessing sufficient mass for its self-gravity to bring it to a nearly round shape, and gravitationally dominant, meaning it will have cleared its path of other bodies of comparable size, other than its satellites.

PLANETARY NEBULA: a shell of ejected material moving away from an extremely hot, dying star.

PLANETARY RINGS: billions of pieces of rock and ice organized into thin, flat rings orbiting planets.

PLANETESIMAL: small primordial bodies formed by accretion in the early solar system.

PLASMA: a gas made up of charged particles.

PLASMA TORUS: a doughnut-shaped region of charged particles around a planet or moon.

PLATE TECTONICS: a geological model that describes the movements of rigid planetary plates in relation to each other.

PLUTINO: one of a group of Kuiper belt objects that share Pluto's 3:2 orbital resonance with Neptune.

PLUTOID: dwarf planets in orbit around the sun outside the orbit of Neptune.

Giant tube worms near hydrothermal vent

POSITRON: a positively charged subatomic particle with the same mass and spin as an electron; the antiparticle of an electron.

PRECESSION: the gradual change in the angle of the axis of a rotating object.

PROBE: an unmanned, exploratory device that is used to collect information about celestial bodies.

PROTON-PROTON CHAIN: a series of fusion reactions in which four protons combine to form helium; an important source of energy in the core of the sun.

PROTOPLANET: the early stages of an accreting object in the solar nebula before it grows into a planet.

PROTOSUN: a central concentration of gas in the early stages of the formation of the sun.

PTOLEMAIC MODEL: a theory of the universe developed by Greek astronomer Ptolemy that describes Earth as the motionless center of the universe.

PULSAR: a rapidly rotating neutron star that emits regular pulses of radio waves.

P-WAVE: a pressure wave that travels through gases, solids, and liquids.

QUASAR: a contraction of quasi-stellar object; a distant, very luminous, active galactic nucleus emitting radio waves.

RADIAL VELOCITY: a star or other body's speed of as it moves toward or away from an observer.

RADIATIVE ZONE: the zone in a star between the core and the convective zone.

RADIATION: energy emitted in the form of waves or particles.

RADIO WAVE: the longest wavelengths in the electromagnetic spectrum.

RADIOACTIVE DECAY: a method of dating geological samples by measuring the rate of decay of nitrogen in carbon 14 .

RED GIANT: a star that has used up its hydrogen core, and expanded to more than 100 times its original size.

REFLECTING TELESCOPE: a telescope that uses mirrors to collect and focus light from a distant object.

REFRACTING TELESCOPE: a telescope that uses lenses to gather visible light; also known as a refractor.

REFRACTION: the bending of light from its original path as it passes from one medium through another; the change in apparent position of a celestial body due to the deflection of light rays as they pass through the atmosphere.

REGOLITH: loose, fragmented soil-like material on the surface of the moon or other airless body.

RESONANCE: the increase in the strength of a vibration that occurs when the frequency of an applied force becomes equal with the natural frequency of an object.

RETROGRADE: moving in a direction opposite to the general motion of similar objects; retrograde orbits or rotations in the solar system appear to move clockwise as seen from north of the ecliptic.

REVOLUTION: a single complete course of one celestial body around another.

ROCHE LIMIT: the minimum distance at which a satellite can approach the center of a planet without getting torn apart by tidal forces.

Comet Hale-Bopp over Monument Valley, Arizona

ROTATION: the turn of a body around an axis.

SCATTERED DISK: a collection of icy bodies past Neptune in eccentric orbits ranging from 50 to 100 AU.

SEARCH FOR EXTRATERRESTRIAL INTELLIGENCE (SETI): an institute, partially funded by the U.S. government and partially privately funded, based in California with the mission to explore, understand, and explain life in the universe.

SHEPHERD SATELLITE: a moon with gravitational forces strong enough to constrain planetary rings; also called a shepherd moon.

SHORT-PERIOD COMET: a comet with an orbital period less than 200 years, subdivided into Jupiter-type comets, with periods under 20 years, and Halley-type comets, with periods of 20 to 200 years.

SILICATE MINERAL: a rock-forming mineral containing oxygen and silicon.

SOLAR CYCLE: a 22-year cycle of sunspot activity, during which such activity rises and falls and the polarities of sunspots reverse and then return to their original polarities. Solar cycle #24 began in 2008.

SOLAR CONSTANT: the total radiant energy, per unit area, Earth's outer atmosphere receives from the sun.

SOLAR FLARES: sudden and intense outbursts of the lower layers of the sun's atmosphere, usually near a sunspot group.

SOLAR MASS: the mass of the sun, which is used to measure other celestial objects; one solar mass = $(1.989 \pm 0.004) \times 10^{30}$ kg

SOLAR NEBULA: the spinning gaseous cloud that condensed to form the solar system.

SOLAR SYSTEM: the sun and all the celestial bodies that are influenced by it, including planets, asteroids, and comets.

SOLAR WIND: an outflow of charged particles from the sun.

SOLSTICE: either of two points along the ecliptic, and midway between the equinoxes, during which time the sun reaches its northernmost declination (June 21) and southernmost declination (December 22).

SPECTROMETER: an instrument used to measure light spectra by separating them according to their frequencies; a spectrometer can be connected to a telescope to analyze the composition of stellar and planetary gases.

SPECTROSCOPY: the study of emission and absorption of light and other radiant matter.

SPECTRUM: the arrangement of light according to its wavelengths.

SPHERICAL ABERRATION: the loss of definition in an image produced by a telescope due to an irregularity in the reflecting surfaces.

STELLAR NURSERY: dense, gaseous clouds in which stars are forming.

STELLAR WINDS: charged particles, mostly protons and electrons, that flow out from a star into space.

STROMATOLITES: dome-shaped, layered, limestone deposits formed by fossilized blue-

Grand Prismatic Spring, Yellowstone National Park

green algae, some dating back more than three billion years.

SUNSPOT: a large, dark, relatively cool spot on the surface of the sun, associated with strong magnetic activity.

SUPERGIANT STAR: a member of a class of the largest and most luminous stars known, with radii between 100 and 1,000 times that of the sun.

SUPERNOVA: an enormous stellar explosion; the death of a massive star.

SYNCHRONOUS ROTATION: the motion that results when the time it takes a satellite to complete its orbit equals the time it takes for it to revolve on its axis; a moon in synchronous rotation around a planet will always present the same face to the planet.

T TAURI STAR: any of the very young stars (named for the prototype in the constellation Taurus) that represent early stages in stellar evolution and are characterized by erratic changes in brightness.

TACHOCLINE: the thin layer between the radiative zone and the convective zone of a star; thought to be where a star's magnetic field is generated.

TERRAE: from the Latin *terra,* meaning "earth." A highland region of the moon, characterized by brighter, higher terrain.

TERRESTRIAL: from the Latin *terrestris*, meaning "earthly"; of or relating to Earth.

TERRESTRIAL PLANET: the four small, inner planets: Mercury, Venus, Earth, and Mars.

TIDAL BULGE: the deformation in a body created by the gravitational pull of a nearby object; on Earth, the rising of the sea due to the gravity of Earth and the sun.

TIDAL FORCE: the varying gravitational force one massive object exerts on another.

TORINO IMPACT HAZARD SCALE: a scale of one through ten used to assess the hazards of potential asteroid and comet impacts.

TOTAL ECLIPSE: an eclipse during which one celestial body is completely obscured from view by either the body or the shadow of another.

TRAJECTORY: the path of a body in motion.

TRANS-NEPTUNIAN OBJECT: (TNO) an object that orbits the sun at an average distance greater than the orbit of Neptune.

TRANSIT: the passage of one celestial body in front of another, as when Venus passes in front of the sun.

TRANSITION ZONE: the region between the sun's chromosphere and its corona.

TROJAN ASTEROIDS: asteroids located at the Lagrangian points, or gravitationally stable places, of Jupiter's orbit

ULTRAVIOLET RADIATION: radiation just beyond the blue end of the visible range of the electromagnetic spectrum.

UMBRA: the dark, central region of a celestial body's shadow.

VAN ALLEN BELTS: zones of high-energy particles, mostly protons and electrons, held in doughnut-shaped regions in the high altitudes of Earth's magnetic field; also known as Van Allen radiation belts.

VISIBLE LIGHT: electromagnetic waves that can be detected by the human eye.

WAVELENGTH: the length of a wave as measured from crest to crest or trough to trough.

WHITE DWARF: a star that has burned up all of its nuclear fuel and has collapsed to a fraction of its former size while still retaining a significant mass.

ZEEMAN EFFECT: the splitting or widening of spectral lines into different frequencies when the light source is placed in a magnetic field.

ZODIAC: an imaginary band around the sky, centered on the ecliptic and in which the planets, the moon, and the sun move.

ZONE: bright, high-pressure bands in the upper atmosphere of a gas giant planet.

SOLAR SYSTEM STATISTICS

SUN

Average distance from sun
--

Perihelion
--

Aphelion
--

Mass (Earth=1)
332,900

Density
1.409 g/cm3

Equatorial radius
695,500 km (432,200 mi)

Equatorial circumference
4,379,000 km (2,715,000 mi)

Orbital period
--

Rotation period
25.38 Earth days (at 16° lat)

Axial tilt
--

Min./max. surface temperature
5,500 °C (10,000 °F)

Volume (Earth = 1)
1,300,000

Equatorial Surface Gravity (Earth = 1)
28

Spectral Type
G2 V

Luminosity
3.83 x 10^{33} ergs/sec.

Atmosphere
hydrogen, helium

Natural satellites
none

MERCURY

Average distance from sun
57,909,175 km (35,983,095 mi)

Perihelion
46,000,000 km (28,580,000 mi)

Aphelion
9,820,000 km (43,380,000 mi)

Mass (Earth=1)
0.055

Density
5.427 g/cm3

Equatorial radius
2,439.7 km (1,516.0 mi)

Equatorial circumference
5,329.1 km (9,525.1 mi)

Orbital period
87.97 Earth days

Rotation period
58.646 Earth days

Axial tilt
0 degrees

Min./max. surface temperature
173/427 °C (-279/801 °F)

Volume (Earth = 1)
0.054

Equatorial Surface Gravity (Earth = 1)
0.38

Orbital Eccentricity
0.20563069

Orbital Inclination to Ecliptic
7 degrees

Atmosphere
trace

Natural satellites
none

VENUS

Average distance from sun
108,208,930 km (67,237,910 mi)

Perihelion
107,476,000 km (66,782,000 mi)

Aphelion
108,942,000 km (67,693,000 mi)

Mass (Earth=1)
0.815

Density
5.24 g/cm3

Equatorial radius
6,051.8 km (3,760.4 mi)

Equatorial circumference
38,025 km (23,627 mi)

Orbital period
224.7 Earth days

Rotation period
-243 Earth days (retrograde)

Axial tilt
177.3 degrees

Min./max. surface temperature
462 °C (864 °F)

Volume (Earth = 1)
0.88

Equatorial Surface Gravity (Earth = 1)
0.91

Orbital Eccentricity
0.0068

Orbital Inclination to Ecliptic
3.39 degrees

Atmosphere
carbon dioxide, nitrogen

Natural satellites
none

EARTH

Average distance from sun
149,597,890 km (92,955,820 mi)

Perihelion
147,100,000 km (91,400,000 mi)

Aphelion
152,100,000 km (94,500,000 mi)

Mass (Earth=1)
1

Density
5.515 g/cm3

Equatorial radius
6,378.14 km (3,963.19 mi)

Equatorial circumference
40,075 km (24,901 mi)

Orbital period
365.24 Earth days

Rotation period
23.934 hours

Axial tilt
23.45 degrees

Min./max. surface temperature
-88/58 °C (-126/136 °F)

Volume
1.0832 x 10^{12} km^3

Equatorial Surface Gravity
9.766 m/s^2

Orbital Eccentricity
0.0167

Orbital Inclination to Ecliptic
0.00 degrees

Atmosphere
nitrogen, oxygen

Natural satellites
1

MARS

Average distance from sun
227,936,640 km (141,633,260 mi)

Perihelion
206,600,000 km (128,400,000 mi)

Aphelion
249,200,000 km (154,900,000 mi)

Mass (Earth=1)
0.10744

Density
3.94 g/cm3

Equatorial radius
3,397 km (2,111 mi)

Equatorial circumference
21,344 km (13,263 mi)

Orbital period
686.93 Earth days

Rotation period
24.62 hours

Axial tilt
25.19

Min./max. surface temperature
-87 to -5 °C (-125 to 23 °F)

Volume (Earth = 1)
0.150

Equatorial Surface Gravity (Earth = 1)
0.38

Orbital Eccentricity
0.0934

Orbital Inclination to Ecliptic
1.8 degrees

Atmosphere
carbon dioxide, nitrogen, argon

Natural satellites
2

JUPITER

Average distance from sun
778,412,020 km (483,682,810 mi)

Perihelion
740,742,600 km (460,276,100 mi)

Aphelion
816,081,400 km (507,089,500 mi)

Mass (Earth=1)
317.82

Density
1.33 g/cm3

Equatorial radius
71,492 km (44,423 mile)

Equatorial circumference
449,197 km (279,118 mi)

Orbital period
11.8565 Earth years

Rotation period
9.925 hours

Axial tilt
3.12 degrees

Min./max. surface temperature
-148 °C (-234 °F)

Volume (Earth = 1)
1316

Equatorial Surface Gravity (Earth = 1)
2.14

Orbital Eccentricity
0.04839

Orbital Inclination to Ecliptic
1.305 degrees

Atmosphere
hydrogen, helium

Natural satellites
63

SATURN

Average distance from sun
1,426,725,400 km (885,904,700 mi)

Perihelion
1,349,467,000 km (838,519,000 mi)

Aphelion
1,503,983,000 km (934,530,000 mi)

Mass (Earth=1)
95.16

Density
0.70 g/cm3

Equatorial radius
60,268 km (37,449 mi)

Equatorial circumference
378,675 km (235,298 mi)

Orbital period
29.4 Earth years

Rotation period
10.656 hours

Axial tilt
26.73 degrees

Min./max. surface temperature
-178 °C (-288 °F)

Volume (Earth = 1)
763.6

Equatorial Surface Gravity (Earth = 1)
0.91

Orbital Eccentricity
.0541506

Orbital Inclination to Ecliptic
2.484 degrees

Atmosphere
hydrogen, helium

Natural satellites
61

URANUS

Average distance from sun
2,870,972,200 km (1,783,939,400 mi)

Perihelion
2,735,560,000 km (1,699,800,000 mi)

Aphelion
3,006,390,000 km (1,868,080,000 mi)

Mass (Earth=1)
14.371

Density
1.30 g/cm3

Equatorial radius
25,559 km (15,882 mi)

Equatorial circumference
160,592 km(99,787 mi)

Orbital period
84.02 Earth years

Rotation period
-17.24 hours (retrograde)

Axial tilt
97.86 degrees

Min./max. surface temperature
-216 °C (-357 °F)

Volume (Eath = 1)
63.1

Equatorial Surface Gravity (Earth = 1)
0.86

Orbital Eccentricity
0.047168

Orbital Inclination to Ecliptic
0.770 degrees

Atmosphere
hydrogen, helium, methane

Natural satellites
27

SOLAR SYSTEM STATISTICS (CONT.)

	NEPTUNE	CERES	PLUTO	HAUMEA
Average distance from sun	4,498,252,900 km (2,795,084,800 mi)	2.767 AU	5,906,380,000 km (3,670,050,000 mi)	43.34 AU
Perihelion	4,459,630,000 km (2,771,087,000 mi)	381,419,582 km 237,003,140	4,436,820,000 km (2,756,902,000 mi)	--
Aphelion	4,536,870,000 km (2,819,080,000 mi)	447,838,164 km 278,273,734	7,375,930,000 km (4,583,190,000 mi)	--
Mass (Earth=1)	17.147	0.00016	0.0022	0.00070
Density	1.76 g/cm3	2.1 g/cm3	2 g/cm3	--
Equatorial radius	24,764 km (15,388 mi)	474 km (295 mi)	1,151 km (715 mi)	660–775 km (410–480 mi)
Equatorial circumference	155,597 km (96,683 mi)	~2900 km	7,232 km (4,494 mi)	~7900 km
Orbital period	164.79 Earth years	4.60 yrs	247.92 Earth years	285.4 years
Rotation period	16.11 hours	9.075 hrs	-6.387 Earth days (retrograde)	3.9 hours
Axial tilt	29.58 degrees	3 degrees	119.61 degrees	--
Min./max. surface temperature	-214 °C (-353 °F)	~167 K	-233/-223 °C (-387/-369 °F)	~-241°C (-402°F)
Volume (Earth = 1)	57.7	Volume: --	0.0059	Volume: --
Equatorial Surface Gravity (Earth = 1)	1.10	0.03	0.08	0.05
Orbital Eccentricity	0.00859	0.0789	0.249	0.195
Orbital Inclination to Ecliptic	1.769 degrees	10.58 degrees	17.14 degrees	Orbital Orbital Inclination to Ecliptic: 28.22 degrees
Atmosphere	hydrogen, helium, methane	possible trace	trace nitrogen, carbon monoxide, methane	--
Natural satellites	13	none	3	2

	MAKEMAKE	ERIS
Average distance from sun	45.8 AU	67.67 AU
Perihelion	--	--
Aphelion	--	--
Mass (Earth=1)	0.00067	0.0028
Density	2g/cm3	2g/cm3
Equatorial radius	~800 km (500 mi)	1,200 km (745 mi)
Equatorial circumference	~4700 km	~8200 km
Orbital period	309.88 Earth years	557 Earth years
Rotation period	--	--
Axial tilt	--	--
Min./max. surface temperature	-240°C (-400°F)	-230°C (-382°F)
Volume	--	--
Equatorial Surface Gravity (Earth = 1)	0.05	0.08
Orbital Eccentricity	0.159	0.44
Orbital Orbital Inclination to Ecliptic / Orbital Inclination to Ecliptic	28.96 degrees	44.19 degrees
Atmosphere	--	--
Natural satellites	none	1

PLANETARY SATELLITES

NAME	DATE DISCOVERED	DISTANCE FROM PLANET (KM)	RADIUS (KM)
EARTH			
Moon	--	384400	1737
MARS			
Phobos	1877	9376	11.1 ± 0.15
Deimos	1877	23458	6.2 ± 0.18
JUPITER			
Io	1610	421800	1821.6 ± 0.5
Europa	1610	671100	1560.8 ± 0.5
Ganymede	1610	1070400	2631.2 ± 1.7
Callisto	1610	1882700	2410.3 ± 1.5
Amalthea	1892	181400	83.45 ± 2.4
Himalia	1904	11461000	85
Elara	1905	11741000	43
Pasiphae	1908	23624000	30
Sinope	1914	23939000	19
Lysithea	1938	11717000	18
Carme	1938	23404000	23
Ananke	1951	21276000	14
Leda	1974	11165000	10
Thebe	1980	221900	49.3 ± 2.0
Adrastea	1979	129000	8.2 ± 2.
Metis	1980	128000	21.5 ± 2.0
Callirrhoe	1999	24103000	4.3
Themisto	1975/2000	7284000	4.0
Megaclite	2000	23493000	2.7
Taygete	2000	23280000	2.5
Chaldene	2000	23100000	1.9
Harpalyke	2000	20858000	2.2
Kalyke	2000	23483000	2.6
Iocaste	2000	21060000	2.6
Erinome	2000	23196000	1.6
Isonoe	2000	23155000	1.9
Praxidike	2000	20908000	3.4
Autonoe	2001	24046000	2.0
Thyone	2001	20939000	2.0
Hermippe	2001	21131000	2.0
Aitne	2001	23229000	1.5
Eurydome	2001	22865000	1.5
Euanthe	2001	20797000	1.5
Euporie	2001	19304000	1.0
Orthosie	2001	20720000	1.0
Sponde	2001	23487000	1.0
Kale	2001	23217000	1.0
Pasithee	2001	23004000	1.0
Hegemone	2003	23577000	1.5

NAME	DATE DISCOVERED	DISTANCE FROM PLANET (KM)	RADIUS (KM)
Mneme	2003	21035000	1.0
Aoede	2003	23980000	2.0
Thelxinoe	2003	21164000	1.0
Arche	2003	23355000	1.0
Kallichore	2003	23288000	2.0
Helike	2003	21069000	2.0
Carpo	2003	17058000	1.5
Eukelade	2003	23328000	2.0
Cyllene	2003	23809000	1.0
S/2003 J2	2003	28455000	1.0
S/2003 J3	2003	20224000	1.0
S/2003 J4	2003	23933000	1.0
S/2003 J5	2003	23498000	2.0
S/2003 J9	2003	23388000	0.5
S/2003 J10	2003	23044000	1.0
S/2003 J12	2003	17833000	0.5
S/2003 J14	2003	24543000	1.0
S/2003 J15	2003	22630000	1.0
S/2003 J16	2003	20956000	1.0
S/2003 J17	2003	22983000	1.0
S/2003 J18	2003	20426000	1.0
S/2003 J19	2003	23535000	1.0
S/2003 J23	2003	23566000	1.0
SATURN			
Mimas	1789	185540	198.20 ± 0.2
Enceladus	1789	238040	252.10 ± 0.10
Tethys	1684	294670	533.00 ± 0.70
Dione	1684	377420	561.70 ± 0.45
Rhea	1672	527070	764.30 ± 1.10
Titan	1655	1221870	2575.50 ± 2.00
Hyperion	1848	1500880	135.00 ± 4.00
Iapetus	1671	3560840	735.60 ± 1.50
Phoebe	1898	12947780	106.60 ± 1.00
Janus	1966	151460	89.4 ± 3.0
Epimetheus	1980	151410	56.7 ± 3.1
Helene	1980	377420	16. ± 4.
Telesto	1980	294710	11.8 ± 1.0
Calypso	1980	294710	10.7 ± 1.0
Atlas	1980	137670	15.3 ± 1.2
Prometheus	1980	139380	43.1 ± 2.0
Pandora	1980	141720	40.3 ± 2.2
Pan	1990	133580	14.8 ± 2.0
Ymir	2000	23040000	9
Paaliaq	2000	15200000	11.0
Tarvos	2000	17983000	7.5
Ijiraq	2000	11124000	6.0

NAME	DATE DISCOVERED	DISTANCE FROM PLANET (KM)	RADIUS (KM)
Suttungr	2000	19459000	3.5
Kiviuq	2000	11110000	8.0
Mundilfari	2000	18628000	3.5
Albiorix	2000	16182000	16.0
Skathi	2000	15540000	4.0
Erriapo	2000	17343000	5.0
Siarnaq	2000	17531000	20
Thrymr	2000	20314000	3.5
Narvi	2003	19007000	3.5
Methone	2004	194440	1.5
Pallene	2004	212280	2.0
Polydeuces	2004	377200	2.0
Daphnis	2005	136500	3.5
Aegir	2005	20751000	3.0
Bebhionn	2005	17119000	3.0
Bergelmir	2005	19336000	3.0
Bestla	2005	20192000	3.5
Farbauti	2005	20377000	2.5
Fenrir	2005	22454000	2.0
Fornjot	2005	25146000	3.0
Hati	2005	19846000	3.0
Hyrrokkin	2006	18437000	4.0
Kari	2006	22089000	3.5
Loge	2006	23058000	3.0
Skoll	2006	17665000	3.0
Surtur	2006	22704000	3.0
Anthe	2007	197,700	0.5
Jarnsaxa	2006	18811000	3.0
Greip	2006	18206000	3.0
Tarqeq	2007	18009000	3.5
S/2004 S7	2005	20999000	3.0
S/2004 S12	2005	19878000	2.5
S/2004 S13	2005	18404000	3.0
S/2004 S17	2005	19447000	3.5
S/2006 S1	2006	18009000	3.0
S/2006 S3	2006	16725000	3.0
S/2007 S2	2007	16725000	3.0
S/2007 S3	2007	18975000	2.5
S/2008 S1	2008	167500	0.25
URANUS			
Ariel	1851	190900	578.9 ± 0.6
Umbriel	1851	266000	584.7 ± 2.8
Titania	1787	436300	788.9 ± 1.8
Oberon	1787	583500	761.4 ± 2.6
Miranda	1948	129900	235.8 ± 0.7
Cordelia	1986	49800	20.1 ± 3

NAME	DATE DISCOVERED	DISTANCE FROM PLANET (KM)	RADIUS (KM)
Ophelia	1986	53800	21.4 ± 4.
Bianca	1986	59200	25.7 ± 2
Cressida	1986	61800	39.8 ± 2
Desdemona	1986	62700	32.0 ± 4
Juliet	1986	64400	46.8 ± 4
Portia	1986	66100	67.6 ± 4
Rosalind	1986	69900	36. ± 6
Belinda	1986	75300	40.3 ± 8
Puck	1985	86000	81. ± 2
Caliban	1997	7231000	36.0
Sycorax	1997	12179000	75.0
Prospero	1999	16256000	25.0
Setebos	1999	17418000	240
Stephano	1999	8004000	160
Trinculo	2001	8504000	9.0
Francisco	2001	4276000	11.0
Margaret	2003	14.0345000	10
Ferdinand	2001	20901000	10.0
Perdita	1999	76417	13.0
Mab	2003	97736	6
Cupid	2003	74392	6
NEPTUNE			
Triton	1846	354759	1352.6 ± 2.4
Nereid	1949	5513787	170. ± 25
Naiad	1989	48227	33. ± 3
Thalassa	1989	50074	41. ± 3
Despina	1989	52526	75. ± 3
Galatea	1989	61953	88. ± 4
Larissa	1989	73548	97. ± 3
Proteus	1989	117647	210. ± 7
Halimede	2002	16611000	31.0
Psamathe	2003	48096000	20.0
Sao	2002	22228000	22.0
Laomedeia	2002	23567000	21.0
Neso	2002	49285000	30.0
PLUTO			
Charon	1978	19600	593
Nix	2005	48,680	44.0 ± 5.0
Hydra	2005	64,780	36.0 ± 5.0
ERIS			
Dysnomia	2005	33,000	200
HAUMEA			
Hi'iaka	2005	49500	~310
Namaka	2005	39000	~170

SELECTED BIBLIOGRAPHY

BOOKS AND PERIODICALS:

Arnett, David. *Supernovae and Nucleosynthesis:An Investigation of the History of Matter, From the Big Bang to the Present.* Princeton, N.J.: Princeton University Press, 1996.

Boorstin, Daniel J. *The Discoverers: A History of Man's Search to Know His World and Himself.* New York: Random House, 1983.

Burnham, Robert. *Great Comets.* Cambridge: Cambridge University Press, 2000.

Chaisson, Eric, and Steve McMillan. *Astronomy Today, Vol. 1: The Solar System,* 6th ed. San Francisco: Pearson Addison-Wesley, 2008.

Comins, Neil, and William J. Kaufmann. *Discovering the Universe,* 5th ed. New York: W. H. Freeman and Company, 2000.

Corfield, Richard. *Lives of the Planets: A Natural History of the Solar System.* New York: Basic Books, 2007.

Darling, David. *The Universal Book of Astronomy.* Hoboken, N.J.: John Wiley and Sons. 2004.

DeVorkin, David and Robert Smith. *Hubble: Imaging Space and Time.* Washington, D.C.: National Geographic, 2008.

Dickinson, Terence. *NightWatch: A Practical Guide to Viewing the Universe.* Buffalo, N.Y.: Firefly Books, 1998.

Editors of Time-Life Books. *The Far Planets.* Alexandria, VA: Time-Life Books, 1988.

———. *Life Search.* Alexandria, Va.: Time-Life Books, 1988.

———. *The Near Planets.* Alexandria, Va.: Time-Life Books, 1992.

———. *The Sun.* Alexandria, Va.: Time-Life Books, 1990.

Glover, Linda, Patricia S. Daniels, Andrea Gianopoulos, and Jonathan T. Malay. *National Geographic Encyclopedia of Space.* Washington, D.C.: National Geographic, 2005.

Hoskin, Michael, ed. *The Cambridge Concise History of Astronomy.* Cambridge: Cambridge University Press, 1999.

Lang, Kenneth R. *The Cambridge Guide to the Solar System.* Cambridge: Cambridge University Press, 2003.

McFadden, Lucy-Ann, Paul R. Weissman, and Torrence V. Johnson, eds. *Encyclopedia of the Solar System.* 2nd ed. San Diego: Academic Press, 2007.

Mitton, Jacqueline. *Cambridge Dictionary of Astronomy.* Cambridge: Cambridge University Press, 2001.

The Solar System. Astronomy magazine special issue, 2008.

Verger, Fernand, Isabelle Sourbes-Verger, and Raymond Ghirardi. *The Cambridge Encyclopedia of Space: Missions, Applications and Exploration.* Cambridge: Cambridge University Press, 2003.

WEBSITES:

Astrobiology magazine. http://www.astrobio.net/

Astronomy magazine. http://www.astronomy.com/

European Space Agency. http://sci.esa.int/

The Extrasolar Planets Encyclopaedia. http://exoplanet.eu/

The Galileo Project. http://galileo.rice.edu/

The Giant Planet Satellite and Moon Page. http://www.dtm.ciw.edu/sheppard/satellites/

IAU Minor Planet Center. http://www.cfa.harvard.edu/iau/mpc.html

International Astronomical Union. http://www.iau.org/

The Internet Encyclopedia of Science. http://www.daviddarling.info/encyclopedia/ETEmain.html

JPL Solar System Dynamics. http://ssd.jpl.nasa.gov/?bodies

Kuiper belt pages. http://www.ifa.hawaii.edu/faculty/jewitt/kb.html

Marshall Space Flight Center: Solar Physics. http://solarscience.msfc.nasa.gov/

Mike Brown Dwarf Planets. http://www.gps.caltech.edu/~mbrown/dwarfplanets/

NASA Astrobiology Institute. http://astrobiology.nasa.gov/nai/

NASA Heliophysics. http://sec.gsfc.nasa.gov/

NASA Phoenix Mars Mission. http://phoenix.lpl.arizona.edu/mission.php

NASA Solar System Exploration. http://solarsystem.nasa.gov/planets/

NASA Sun-Earth Day. http://sunearthday.nasa.gov/

NASA Human Space Flight History. http://spaceflight.nasa.gov/history/

NASA/Jet Propulsion Laboratory. http://www.jpl.nasa.gov/

NASA/JPL Near Earth Object Program. http://neo.jpl.nasa.gov/

NASA/JPL PlanetQuest. http://planetquest.jpl.nasa.gov/

National Science Foundation. X-treme Microbes. http://www.nsf.gov/news/special_reports/microbes/index.jsp

New Scientist space. http://space.newscientist.com/channel/solar-system

Nine Planets. http://www.nineplanets.org/

NOAA Space Weather Prediction Center. http://www.swpc.noaa.gov/

The Planetary Society. http://www.planetary.org/explore/

Scientific American. http://www.sciam.com/

SETI institute. http://www.seti.org/

Solar and Heliospheric Observatory (SOHO) homepage. http://sohowww.nascom.nasa.gov/

SPACE.com. http://www.space.com/solarsystem/

Stanford Solar Center. http://solar-center.stanford.edu/about/

USGS Gazetteer of Planetary Nomenclature. http://planetarynames.wr.usgs.gov/

Views of the Solar System. http://www.solarviews.com/

ILLUSTRATIONS CREDITS

Cover, NASA/JPL/Space Science Institute; 1, James P. Blair; 2-3, NASA/JPL/Space Science Institute; 4-5, NASA/JPL/Space Science Institute; 7, NASA/JPL/Space Science Institute; 8-9, David Aguilar; 10-11, NASA/JPL/USGS; 12-13, NASA/JPL/Space Science Institute; 14-15, John R. Foster/Photo Researchers, Inc.; 17, Robert W. Madden; 21, Kenneth Geiger; 22, Richard T. Nowitz; 23, Stapleton Collection/CORBIS; 24, The Art Archive/CORBIS; 25, Photo Researchers, Inc.; 26, Stefano Bianchetti/CORBIS; 27, K.M. Westermann/CORBIS; 28, NASA/JPL/MSSS; 29, Roger Ressmeyer/CORBIS; 30 (UP LE), Hulton Archive/Getty Images; 30 (UP RT), Hulton Archive/Getty Images; 30 (LO), Mary Evans Picture Library/Photo Researchers, Inc.; 31, Jean-Leon Huens; 32, CORBIS; 33, NASA/JPL/Space Science Institute; 34, Jean-Leon Huens; 35 (RT), Jim Sugar/CORBIS; 36, Royal Astronomical Society/Photo Researchers, Inc.; 37, NASA/JPL; 38, Mark Thiessen; 39, NASA; 40, NASA; 41, NASA; 42-43, NASA/JPL-Caltech/Cornell; 43 (UP), NASA/Walt Feimer; 44, David Aguilar; 45, NASA; 46-47, Nicolas Reynard; 50, Chrystal Henkaline/iStockphoto.com; 52, NASA, P. Challis, R. Kirshner (Harvard-Smithsonian Center for Astrophysics) and B. Sugerman (STScI); 53, NASA; 54, Davis Meltzer; 55, NASA; 56, ESA/NASA/SOHO; 57, NASA; 58, Thaddeus Bowling; 59, Tim Laman; 60, Bettmann/CORBIS; 61 (LE), Fabrizio Zanier/iStockphoto.com; 61 (RT), ESA/NASA/SOHO; 62, NASA; 63 (A), NASA; 63 (B), NASA; 63 (C), NASA; 63 (D), NASA; 64, David Aguilar; 65, NASA; 66, ESA/NASA; 67, Alex Lutkus/NASA; 68, ESA/NASA/SOHO; 69, ESA/NASA/SOHO; 70, NASA; 71, ESA/NASA/SOHO; 72, NASA; 73, David Aguilar; 74, ESA/NASA; 75, NANASA/JPL-Caltech/NRL/GSFC; 76-77, Photo Researchers, Inc.; 79 (A), NASA/Johns Hopkins University Applied Physics Laboratory/Carnegie Institution of Washington; 79 (B), NASA/JPL; 79 (C), NASA; 79 (D), NASA/JPL/USGS; 79 (E), NASA/JPL/MSSS; 80, Michael Nicholson/CORBIS; 81 (A), NASA/JPL/MSSS; 81 (B), NASA; 81 (C), NASA/JPL; 81 (D), NASA/Johns Hopkins University Applied Physics Laboratory/Carnegie Institution of Washington; 82, NASA/Johns Hopkins University Applied Physics Laboratory/Carnegie Institution of Washington; 83, NASA/Johns Hopkins University Applied Physics Laboratory/Arizona State University/Carnegie Institution of Washington. Image reproduced courtesy of Science/AAAS.; 84 (A), NASA/Johns Hopkins University Applied Physics Laboratory/Carnegie Institution of Washington; 84 (B), NASA/Johns Hopkins University Applied Physics Laboratory/Carnegie Institution of Washington; 84 (C), NASA/Johns Hopkins University Applied Physics Laboratory/Carnegie Institution of Washington; 84 (D), NASA/Johns Hopkins University Applied Physics Laboratory/Carnegie Institution of Washington; 85 (A), NASA; 85 (B), NASA/Johns Hopkins University Applied Physics Laboratory/Carnegie Institution of Washington; 86, Mark Garlick/Photo Researchers, Inc.; 87, NASA/Johns Hopkins University Applied Physics Laboratory/Arizona State University/Carnegie Institution of Washington. Image reproduced courtesy of Science/AAAS.; 88, NASA/JPL; 89, ESA; 90, ESA/VIRTIS/INAF-IASF/Obs. de Paris-LESIA/Univ. of Oxford; 91, Gary Braasch/CORBIS; 92, NASA/JPL/USGS; 93 (A), NASA/JPLa; 93 (B), NASA/JPLa; 93 (C), NASA/JPLa; 93 (D), NASA/JPLa; 93 (RT), Richard T. Nowitz; 94, NASA Goddard Space Flight Center Image by Reto Stöckli (land surface, shallow water, clouds). Enhancements by Robert Simmon (ocean color, compositing, 3D globes, animation). Data and technical support: MODIS Land Group; MODIS Science Data Support Team; MODIS Atmosphere Group; MODIS Ocean Group Additional data: USGS EROS Data Center (topography); USGS Terrestrial Remote Sensing Flagstaff Field Center (Antarctica); Defense Meteorological Satellite Program (city lights).; 95, Alaska Stock Images/National Geographic Stock; 96, James L. Stanfield; 97, J. Baylor Roberts; 98, Paul Nicklen; 99 (A), CORBIS; 99 (B), James Forte/National Geographic Stock; 100, David Marchal/iStockphoto.com; 101, Stephen Low Productions, Inc.; 102, NASA/JPL/USGS; 103, NASA/JPL/USGS; 104, NASA; 105, NASA/JPL/Northwestern University; 106 (UP), NASA/Getty Images; 106-107, NASA/JPL/MSSS; 108, NASA/JPL/MSSS; 109, NASA/JPL/MSSS; 110, NASA; 111, NASA/JPL/GSFC; 112 (A), Ira Block; 112 (B), Richard T. Nowitz; 113, NASA/JPL/MSSS; 115 (LE), David Aguilar; 115 (RT), NASA/JPL-Caltech/University of Arizona/ Texas A&M Univeristy; 116-117, NASA/JPL-Caltech/T. Pyle (SSC); 119, NASA/JPL-Caltech/T. Pyle (SSC); 121, Roger Ressmeyer/CORBIS; 122, Image/Animation courtesy of Gareth Williams, Minor Planet Center; 123, Roger Harris/Photo Researchers, Inc.; 124, Julian Baum/New Scientist/Photo Researchers, Inc.; 125, NASA; 126, Victor Habbick Visions/Photo Researchers, Inc.; 127, Sanford/Agliolo/CORBIS; 129, NASA, Dan Durda (FIAAA, B612 Foundation); 130, Thomas J. Abercrombie; 131 (LE), Jay Leviton/Time & Life Pictures/Getty Images; 131 (RT), Stephan Hoerold/iStockphoto.com; 132-133, NASA/JPL; 135, Photo Researchers, Inc.; 136, NASA/JPL/University of Arizona; 139, NASA/JPL; 140, NASA, M. Wong and I. de Pater (University of California, Berkeley; 141, NASA/JPL; 142, NASA/JPL/University of Arizona; 143 (A), NASA/JPL/DLR; 143 (B), NASA/JPL/DLR; 143 (C), NASA/JPL/DLR; 143 (D), NASA/JPL/DLR; 144, NASA/JPL/Space Science Institute; 145, NASA/JPL/Space Science Institute; 146, NASA/JPL/Space Science Institute; 147 (UP), NASA, ESA, J. Clarke (Boston University), and Z. Levay (STScI); 147 (LO), NASA/JPL; 148, NASA/JPL/Space Science Institute; 149, NASA/JPL/Space Science Institute; 150, NASA/JPL/Space Science Institute; 151, NASA/JPL; 152, NASA/JPL; 153, NASA/JPL/USGS; 154, Popperfoto/Getty Images; 155 (LE), NASA/JPL; 155 (RT), NASA/JPL/STScI; 156 (A), NASA/JPL; 156 (B), NASA/JPL; 156 (C), NASA/JPL; 156 (D), NASA/JPL; 157, NASA; 158, NASA; 159, NASA/JPL; 160-161, David Aguilar; 163, Euan G. Mason; 166, Friedrich Saurer/Photo Researchers, Inc.; 167, John R. Foster/Photo Researchers, Inc.; 168, Robert F. Sisson; 169 (LE), Bettmann/CORBIS; 169 (RT), NASA/ESA, Alan Stern (Southwest Research Institute), Marc Buie (Lowell Observatory),; 170, NASA/JPL; 171, NASA/ESA, Adolf Schaller; 172-173, Johns Hopkins University Applied Physics Laboratory/Southwest Research Institute; 173, NASA, Chad Trujillo & Michael Brown (Caltech); 175, Bill & Sally Fletcher; 176, Bettmann/CORBIS; 177, David Lillo/AFP/Getty Images; 178, NASA/JPL-Caltech; 179, NASA; 180-181, Norbert Wu/Minden Pictures; 183, James Thew/iStockphoto.com; 184, Roger Ressmeyer/CORBIS; 185, Roger Ressmeyer/CORBIS; 186, NASA; 187, NASA; 188, NASA/JPL; 189, NASA/JPL/USGS; 190-191, NASA/JPL/Space Science Institute; 191 (UP), NASA/JPL/ASU; 192, NASA/NSF; 193, Dr. Seth Shostak/Photo Researchers, Inc.; 194, NASA images by Reto Stöckli, based on data from NASA and NOAA; 195, CNES/D.Ducros; 196-197, ESA/NASA/SOHO; 198, SOHO (ESA & NASA); 200 (UP), Simon Podgorsek/iStockphoto.com; 200 (LO), Stephan Hoerold/iStockphoto.com; 201, NASA/JPL/University of Arizona; 202, NASA/ESA, C.R. O'Dell (Vanderbilt University), M. Meixner and P. McCullough (STScI); 203 (LE), NASA; 203 (RT), NASA, P. Challis, R. Kirshner (Harvard-Smithsonian Center for Astrophysics) and B. Sugerman (STScI); 204 (UP), NASA/JPL/Space Science Institute; 204 (LO), Bruce Dale; 205, Emory Kristof; 206, Kenneth C. Zirkel/iStockphoto.com; 207, Raymond Gehman.

INDEX

Page numbers in **bold** indicate illustrations.

D

E

F

G

K

L

M

N

O

P

Q

R

S

T

U

V

W

X

Y

Z

THE NEW SOLAR SYSTEM

PATRICIA DANIELS

The National Geographic Society is one of the world's largest nonprofit scientific and educational organizations. Founded in 1888 to "increase and diffuse geographic knowledge," the Society works to inspire people to care about the planet. It reaches more than 325 million people worldwide each month through its official journal, *National Geographic,* and other magazines; National Geographic Channel; television documentaries; music; radio; films; books; DVDs; maps; exhibitions; school publishing programs; interactive media; and merchandise. National Geographic has funded more than 9,000 scientific research, conservation and exploration projects and supports an education program combating geographic illiteracy. For more information, visit nationalgeographic.com.

For more information, please call 1-800-NGS LINE (647-5463) or write to the following address:

National Geographic Society
1145 17th Street N.W.
Washington, D.C. 20036-4688 U.S.A.

Visit us online at www.nationalgeographic.com

For information about special discounts for bulk purchases, please contact National Geographic Books Special Sales: ngspecsales@ngs.org

For rights or permissions inquiries, please contact National Geographic Books Subsidiary Rights: ngbookrights@ngs.org

ISBN: 978-1-4262-0462-3

Printed in China

09/RRDS/1